T0324979

METHODS IN MOLECULAR BIOLOGY

Series Editor
John M. Walker
School of Life and Medical Sciences
University of Hertfordshire
Hatfield, Hertfordshire, AL10 9AB, UK

For further volumes:
http://www.springer.com/series/7651

Dendritic Cell Protocols

Third Edition

Edited by

Elodie Segura

Pavillon Pasteur, Institut Curie INSERM U932, Paris, France

Nobuyuki Onai

Department of Biodefense, Tokyo Medical and Dental University, Tokyo, Japan

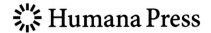

Editors
Elodie Segura
Pavillon Pasteur
Institut Curie INSERM U932
Paris, France

Nobuyuki Onai
Department of Biodefense
Tokyo Medical and Dental University
Tokyo, Japan

ISSN 1064-3745   ISSN 1940-6029   (electronic)
Methods in Molecular Biology
ISBN 978-1-4939-3604-5   ISBN 978-1-4939-3606-9   (eBook)
DOI 10.1007/978-1-4939-3606-9

Library of Congress Control Number: 2016939414

Printed on acid-free paper

This Humana Press imprint is published by Springer Nature
The registered company is Springer Science+Business Media LLC New York

Preface

Dendritic cells are fascinating cells but can be challenging to work with. In this new edition of *Dendritic Cell Protocols*, we aimed at complementing the previous edition in order to provide both beginners and more experienced researchers a choice of methods to isolate and analyze dendritic cells. An introductory review provides an overview of recent advances in the characterization of DC subsets in mouse and human.

Dendritic cells are rare, fragile, and their isolation is often a tedious procedure. For those who wish to generate dendritic cells in vitro, two chapters provide methods to culture human and mouse dendritic cells. Several chapters are devoted to protocols for the isolation of dendritic cells from various organs and tissues (lymphoid organs, intestine, skin, lung, liver), both in mouse and human. In addition, a chapter describes the isolation of dendritic cell progenitors from mouse, and another chapter the purification of dendritic cells from human blood.

Dendritic cells are often described as conductors of the immune response and, as such, perform a wide range of functions. We have compiled cutting-edge methods for the analysis of dendritic cell properties ex vivo. Some of these methods can be applied to dendritic cells from any species. We also included a "bioinformatics tutorial" chapter for the analysis of dendritic cell transcriptome by nonexperts.

In vivo mouse studies have significantly contributed to our knowledge of dendritic cells homeostasis, interactions with other immune cells, and division of labor between dendritic cell subsets. The last section contains several protocols for the in vivo analysis of dendritic cells through cell ablation, adoptive transfer, infection models, or in vivo imaging. Finally, we provide a protocol for the generation of humanized mice for analyzing human dendritic cells in a physiological setting.

The editors wish to thank all authors for their excellent contributions. We hope that this Protocols book will be a valuable tool for a better understanding of dendritic cell biology.

Paris, France   *Elodie Segura*
Tokyo, Japan   *Nobuyuki Onai*

Contents

Contributors

HANS ACHA-ORBEA • *Department of Biochemistry CIIL, University of Lausanne, Epalinges, Switzerland*

SOLANA ALCULUMBRE • *INSERM U932 Immunity and Cancer, Paris, France; Institut Curie, Centre de recherche, Paris, France*

SEBASTIAN AMIGORENA • *INSERM U932, Institut Curie, Centre de recherche, Paris, France*

KEIICHI ARIMURA • *Division of Immunology, Department of Infectious Diseases, Faculty of Medicine, University of Miyazaki, Miyazaki, Japan*

DEVIKA ASHOK • *Department of Biochemistry CIIL, University of Lausanne, Epalinges, Switzerland*

TEGEST AYCHEK • *Department of Immunology, The Weizmann Institute of Science, Rehovot, Israel*

SREEKUMAR BALAN • *Centre d'Immunologie de Marseille-Luminy (CIML), Marseille, France; INSERM U1104, Marseille, France; CNRS UMR7280, Marseille, France*

BIANA BERNSHTEIN • *Department of Immunology, The Weizmann Institute of Science, Rehovot, Israel*

CATERINA CURATO • *Department of Immunology, The Weizmann Institute of Science, Rehovot, Israel*

MARC DALOD • *Centre d'Immunologie de Marseille-Luminy (CIML)UNIV UM2, Aix Marseille Université, Marseille, France; INSERM U1104, Marseille, France; CNRS UMR7280, Marseille, France*

CLAIRE DUMONT • *Department of Biochemistry and Molecular Biology, Bio21 Molecular Science and Biotechnology Institute, The University of Melbourne, Parkville, VIC, Australia*

MÉLANIE DURAND • *INSERM U932, Paris, France; Institut Curie, Centre de recherche, Paris, France*

GYOHEI EGAWA • *Department of Dermatology, Kyoto University Graduate School of Medicine, Kyoto, Japan*

JANA M. ELLEGAST • *Division of Hematology, University Hospital Zurich, Zurich, Switzerland*

TOMOHIRO FUKAYA • *Division of Immunology, Department of Infectious Diseases, Faculty of Medicine, University of Miyazaki, Miyazaki, Japan*

FLORENT GINHOUX • *Singapore Immunology Network (SIgN), Agency for Science, Technology and Research (A*STAR), Singapore, Singapore*

MARTIN GUILLIAMS • *Laboratory of Immunoregulation, VIB Inflammation Research Center, Ghent University, Ghent, Belgium; Department of Biomedical Molecular, Biology, Ghent University, Ghent, Belgium*

MERRY GUNAWAN • *Human DC Lab, Institute of Cellular Medicine, Newcastle University, Newcastle upon Tyne, UK*

MUZLIFAH HANIFFA • *Human DC Lab, Institute of Cellular Medicine, Newcastle University, Newcastle upon Tyne, UK*

HIROAKI HEMMI • *Laboratory for Immune Regulation, WPI Immunology Frontier Research Center, Osaka University, Osaka, Japan; Laboratory of Inflammatory Regulation, RIKEN Center for Integrative Medical Sciences, Kanagawa, Japan; Department of Immunology, Institute of Advanced Medicine Wakayama Medical University, Wakayama, Japan*

SANDRINE HENRI • *Centre d'Immunologie de Marseille-Luminy (CIML), UM2 Aix-Marseille Université, Marseille Cedex 9, France; INSERM U1104, Marseille Cedex 9, France; CNRS UMR7280, Marseille Cedex 9, France*

KATSUAKI HOSHINO • *Laboratory for Immune Regulation, WPI Immunology Frontier Research Center, Osaka University, Osaka, Japan; Laboratory of Inflammatory Regulation, RIKEN Center for Integrative Medical Sciences, Kanagawa, Japan; Department of Immunology, Faculty of Medicine, Kagawa University, Kagawa, Japan*

BAPTISTE JANELA • *Singapore Immunology Network (SIgN), Agency for Science, Technology and Research (A*STAR), Singapore, Singapore*

LAURA JARDINE • *Human DC Lab, Institute of Cellular Medicine, Newcastle University, Newcastle upon Tyne, UK*

ANGUS P.R. JOHNSTON • *Drug Delivery, Disposition and Dynamics, Monash Institute of Pharmaceutical Sciences, Monash University, Parkville, VIC, Australia; ARC Centre of Excellence in Convergent Bio-Nano Science and Technology, Monash University, Parkville, VIC, Australia*

STEFFEN JUNG • *Department of Immunology, The Weizmann Institute of Science, Rehovot, Israel*

KENJI KABASHIMA • *Department of Dermatology, Kyoto University Graduate School of Medicine, Kyoto, Japan; Precursory Research for Embryonic Science and Technology, Japan science and Technology Agency, Saitama, Japan; Singapore Immunology Network (SIgN) and Institute of Medical Biology, Agency for Science, Technology and Research (A*STAR), Biopolis, Singapore*

TSUNEYASU KAISHO • *Laboratory for Immune Regulation, WPI Immunology Frontier Research Center, Osaka University, Osaka, Japan; Laboratory of Inflammatory Regulation, RIKEN Center for Integrative Medical Sciences, Kanagawa, Japan; Department of Immunology, Institute of Advanced Medicine, Wakayama Medical University, Wakayama, Japan*

HAIYIN LIU • *Department of Biochemistry and Molecular Biology, Bio21 Molecular Science and Biotechnology Institute, The University of Melbourne, Parkville, VIC, Australia; Drug Delivery, Disposition and Dynamics, Monash Institute of Pharmaceutical Sciences, Monash University, Parkville, VIC, Australia*

JOAO G. MAGALHAES • *INSERM U932, Institut Curie, Centre de recherche, Paris, France*

CAMILLE MALOSSE • *Centre d'Immunologie de Marseille-Luminy (CIML), UM2 Aix-Marseille Université, Marseille Cedex 9, France; INSERM U1104, Marseille Cedex 9, France; CNRS UMR7280, Marseille Cedex 9, France*

MARKUS G. MANZ • *Division of Hematology, University Hospital Zurich, Zurich, Switzerland*

NAOMI MCGOVERN • *Singapore Immunology Network (SIgN), Agency for Science, Technology and Research (A*STAR), Singapore, Singapore*

SIMON W.F. MILLING • *Centre for Immunobiology, Institute of Infection, Immunity and Inflammation, College of Veterinary, Medical and Life Sciences, University of Glasgow, Scotland, UK*

JUSTINE D. MINTERN • *Department of Biochemistry and Molecular Biology, Bio21 Molecular Science and Biotechnology Institute, The University of Melbourne, Parkville, VIC, Australia*

ALLAN McI MOWAT • *Centre for Immunobiology, Institute of Infection, Immunity and Inflammation, College of Veterinary, Medical and Life Sciences, University of Glasgow, Scotland, UK*

TOSHIAKI OHTEKI • *Department of Biodefense, Medical Research Institute, Tokyo Medical and Dental University, Tokyo, Japan*

NOBUYUKI ONAI • *Department of Biodefense, Medical Research Institute, Tokyo Medical and Dental University, Tokyo, Japan*

LUCIA PATTARINI • *INSERM U932 Immunity and Cancer, Paris, France; Institut Curie, Centre de recherche, Paris, France*

MATTEO PIGNI • *Department of Biochemistry CIIL, University of Lausanne, Epalinges, Switzerland*

CHRISTIANE RUEDL • *School of Biological Sciences, Nanyang Technological University, Singapore, Singapore*

YASUYUKI SAITO • *Division of Hematology, University Hospital Zurich, Zurich, Switzerland; Division of Molecular and Cellular Signaling, Department of Biochemistry and Molecular Biology, Kobe University Graduate School of Medicine, Kobe, Japan*

KATSUAKI SATO • *Division of Immunology, Department of Infectious Diseases, Faculty of Medicine, University of Miyazaki, Miyazaki, Japan*

ANDREAS SCHLITZER • *Singapore Immunology Network (SIgN), Agency for Science, Technology and Research (A*STAR), Singapore, Singapore*

CHARLOTTE L. SCOTT • *Centre for Immunobiology, Institute of Infection, Immunity and Inflammation, College of Veterinary, Medical and Life Sciences, University of Glasgow, Scotland, UK; Laboratory of Immunoregulation, Inflammation Research Centre (IRC), VIB Ghent University, Ghent, Belgium; Department of Respiratory Medicine, Ghent University Hospital, Ghent, Belgium*

ELODIE SEGURA • *INSERM U932, Paris, France; Institut Curie, Centre de recherche, Paris, France*

HIDEAKI TAKAGI • *Division of Immunology, Department of Infectious Diseases, Faculty of Medicine, University of Miyazaki, Miyazaki, Japan*

SIMON TAVERNIER • *Laboratory of Immunoregulation, VIB Inflammation Research Center, Ghent University, Ghent, Belgium; Department of Respiratory Medicine, Ghent University, Ghent, Belgium*

PIOTR TETLAK • *School of Biological Sciences, Nanyang Technological University, Singapore, Singapore*

TOMOFUMI UTO • *Division of Immunology, Department of Infectious Diseases, Faculty of Medicine, University of Miyazaki, Miyazaki, Japan*

LIANNE VAN DE LAAR • *Laboratory of Immunoregulation, VIB Inflammation Research Center, Ghent University, Ghent, Belgium; Department of Internal Medicine, Ghent University, Ghent, Belgium*

OMAR I. VIVAR • *INSERM U932, Institut Curie, Centre de recherche, Paris, France*

DAVID VREMEC • *The Walter and Eliza Hall Institute of Medical Research, Parkville, VIC, Australia*

THIEN-PHONG VU MANH • *Centre d'Immunologie de Marseille-Luminy (CIML) UNIV UM2, Aix Marseille Université, Marseille, France; INSERM U1104, Marseille, France; CNRS UMR7280, Marseille, France*

PAMELA B. WRIGHT • *Centre for Immunobiology, Institute of Infection, Immunity and Inflammation, College of Veterinary, Medical and Life Sciences, University of Glasgow, Scotland, UK*

Part I

Introduction

Chapter 1

Review of Mouse and Human Dendritic Cell Subsets

Elodie Segura

Abstract

Dendritic cells are specialized antigen-presenting cells that initiate and orient immune responses. Numerous studies in mice and humans have shown that dendritic cells are heterogeneous and comprise several subsets that can be distinguished by their surface phenotype, ontogeny, and molecular signature. This review gives an overview of mouse and human dendritic cell subsets and their defining features and summarizes the current knowledge of dendritic cell subsets' functional specialization in terms of antigen presentation.

Key words Dendritic cells, Subsets, Human, Mouse, Antigen presentation

1 Introduction

Dendritic cells (DCs) have long been known to be the best antigen-presenting cells [1]. DCs can be found in secondary lymphoid organs and in most peripheral tissues and non-lymphoid organs, such as the skin, muscle, lung, kidney, intestine, or liver. DCs are equipped with the machinery to capture and process antigens, present them to T lymphocytes, and provide co-stimulatory signals (membrane molecules and cytokines) that orient immune responses. DCs also express various receptors allowing them to sense tissue damage and recognize pathogens. After activation, DCs undergo a program of phenotypic and molecular changes termed "maturation."

Numerous studies, in mice and humans, have shown that DCs are a heterogeneous population composed of several distinct subsets. DC subsets have been initially separated according to their surface phenotype, but recent work has demonstrated that DC subsets can be further distinguished by their ontogeny and transcriptomic signature [2]. A large body of data has shown that mouse DC subsets display functional specializations in terms of antigen presentation or pathogen sensing, but the specific functions of human DC subsets are only beginning to be unraveled.

Elodie Segura and Nobuyuki Onai (eds.), *Dendritic Cell Protocols*, Methods in Molecular Biology, vol. 1423,
DOI 10.1007/978-1-4939-3606-9_1, © Springer Science+Business Media New York 2016

This review provides an overview of mouse and human DC subsets and their specific features, with a special emphasis on their antigen presentation capacity.

2   Mouse Dendritic Cell Subsets

2.1   Resident Versus Migratory DCs

DCs can be broadly separated into two main groups: resident and migratory DCs. Resident DCs are present in secondary lymphoid organs during their entire lifespan. In the steady state, resident DCs display an immature phenotype and express low levels of co-stimulatory molecules [3]. DCs that are present in peripheral tissues and non-lymphoid organs constitutively migrate through the lymph to draining lymph nodes [4]. During their migration, they acquire a mature phenotype. Therefore, non-draining lymphoid organs, such as the spleen, only possess resident DCs, whereas lymph nodes contain both resident and migratory DCs, which can be distinguished by their differential expression of maturation markers (Fig. 1).

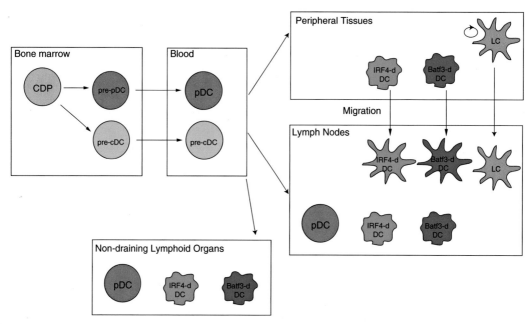

Fig. 1 Resident and migratory dendritic cells. Common dendritic cell progenitors (CDPs) give rise in the bone marrow to pre-pDCs and pre-cDCs. pre-cDCs and differentiated pDCs circulate in the blood and constantly repopulate lymphoid organs and peripheral tissues. Pre-cDCs differentiate into IRF4-dependent (IRF4-d) DCs and Batf3-dependent (Batf3-d) DCs. In some tissues (the skin, mucosa), self-renewing Langerhans cells (LC) are also present. Tissue DCs migrate to draining lymph nodes where they display a mature phenotype. Lymphoid organ-resident DCs display an immature phenotype

Table 1
Surface markers of mouse DC subsets

Marker	pDC	Batf3-dependent DC	IRF4-dependent DC	Langerhans cell	Monocyte-derived DC
CD11c	+	+	+	+	+
Siglec H	+	–	–	–	–
BST2/CD317	+	– *immature* + *mature*	– *immature* + *mature*	–	– *immature* + *mature*
B220	+	–	–	–	–
XCR1	–	+	–	–	–
Clec9A	Low	+	–	–	–
CD8	Heterogeneous	+ *resident* – *migratory*	–	–	–
CD103	–	+	– + *intestine*	–	–
CD11b	–	–	+	–	+
Sirp-α/CD172a	–	–	+	+	+
MR/CD206	–	–	– *resident* + *migratory*	–	+
Langerin/CD207	–	+	–	+	–
EpCAM/CD326	–	–	–	+	–
E-cadherin	–	–	–	+	–
FceRI	–	–	–	–	+
CD64	–	–	–	–	+
CD14	–	–	–	–	+

Resident and migratory DCs comprise several DC subpopulations that are best described according to their ontogeny rather than their surface phenotype (Table 1).

2.2 Plasmacytoid DCs

Plasmacytoid DCs (pDCs) are considered resident DCs; however, they can be recruited to peripheral tissues during inflammation [5]. pDCs derive from a pre-pDC precursor that originates from the common DC progenitor [6, 7]. pDC development is dependent on Fms-like tyrosine kinase 3-ligand (Flt3-L) and on the BCL11A and E2-2 transcription factors [8, 9].

pDCs are specialized for the recognition of virus-derived products and the production of type I interferon [10]. In vivo studies have shown that pDCs are essential for antiviral responses, but do not play a major role in antigen presentation [11–14].

2.3 Classical DCs

Classical DCs (cDCs) have a short half-life and are constantly replaced by precursors originating from the bone marrow [15]. cDCs derive from the pre-cDC progenitor that stems from the common DC progenitor and are dependent on Flt3-L [6, 7]. Transcriptomic analysis has shown that cDCs possess a specific molecular signature that distinguishes them from pDCs and other myeloid cell populations [16]. In particular, the transcription factor zbtb46 is specifically expressed by cDCs [17, 18]. In addition, fate mapping and genetic tracing using Clec9A expression have confirmed that cDCs represent a bona fide hematopoietic lineage [19].

cDCs can be further divided into two subsets that differ in their ontogeny: Batf3-dependent and IRF4-dependent DCs.

2.3.1 Batf3-Dependent DCs

Batf3-dependent DCs comprise resident CD8$^+$ DCs and migratory CD103$^+$ langerin$^+$ DCs. These subpopulations share a common ontogeny and molecular signature, including the specific expression of XCR1 and TLR3 [16]. Their development is dependent on the transcription factors IRF8 and Batf3 [20–22].

Although IRF4-dependent DCs can cross-present (i.e., present exogenous antigens on their MHC class I molecules) some forms of antigens, Batf3-dependent DCs are specialized for cross-presentation due to specific features of their endocytic pathway [23]. Furthermore, in vivo ablation of Batf3-dependent DCs abrogates cross-presentation of soluble or cell-associated model antigens [24–26] and of pathogen-derived or tumor antigens [21, 22, 27, 28].

2.3.2 IRF4-Dependent DCs

IRF4-dependent DCs comprise resident CD8$^-$ CD11b$^+$ DCs and migratory CD11b$^+$ DCs. Their development is dependent on the transcription factors RelB and IRF4 [29–31].

IRF4-dependent DCs are specialized for MHC II-restricted presentation of antigens [32, 33]. After allergen challenge or pathogen infection, IRF4-dependent DCs induce Th17 or Th2 responses in the draining lymph nodes [31, 34–37].

2.4 Langerhans Cells

Langerhans cells are present in the skin epidermis and in oral and vaginal mucosa. They differ markedly from other migratory DCs in their ontogeny. Langerhans cells are self-renewing and derive from embryonic monocytes that seed the tissues before birth [38]. Their development is dependent on CD115/MCSF-R and its ligand IL-34 [39, 40]. Because of these features, Langerhans cells have been proposed to be classified as macrophages rather than DCs [41].

Nevertheless, Langerhans cells present some of the functional characteristics of DCs. Langerhans cells are able to migrate to lymph nodes and present antigens to CD4 T cells. In skin infection with *Candida albicans*, Langerhans cells are essential for inducing Th17 responses [27]. After epicutaneous immunization with a model antigen, Langerhans cells induce Th2 responses [42]. By contrast, Langerhans cells do not cross-present antigens in vivo [27, 43, 44].

2.5 Monocyte-Derived DCs

During pathogen-induced or sterile inflammation, monocytes recruited to the site of inflammation can differentiate in situ into cells that express DC markers (CD11c and MHC class II) and display some functional characteristics of DCs (migration to the lymph nodes and antigen presentation capacity): these cells have therefore been identified as DC and are usually termed "inflammatory DCs" [45]. Their development is dependent on CD115/MCSF-R [46]. Monocyte-derived DCs can also be found in peripheral tissues in the absence of inflammation: in the intestine [47, 48], muscle [49], or skin [50]. Because they do not derive from the common DC progenitor, it was recently proposed that monocyte-derived DCs be classified as a separate lineage, distinct from pDCs and cDCs [41]. Of note, monocyte-derived DCs have been shown to express the transcription factor zbtb46 [17, 51].

"Inflammatory DCs" can perform both cross-presentation and MHC class II-restricted presentation and can induce Th1, Th2, or Th17 responses depending on the inflammatory environment [45]. In a vaccination setting, "inflammatory DCs" promote T follicular helper differentiation [52]. However, recent studies suggest that the primary role of "inflammatory DCs" would be to stimulate antigen-specific T cells, either effector or memory T cells, directly in the tissues rather than in the lymph nodes [36, 53–55].

3 Human Dendritic Cell Subsets

3.1 Blood Versus Tissue DCs

Our knowledge of human DCs has been hindered by sample accessibility constrains. Blood has been the dominant source of human DCs for ex vivo studies. However, recent studies suggest that blood cDCs actually represent a precursor form of cDCs and have not yet developed their full functional capacities, while blood pDCs would be terminally differentiated [56–59]. Consistent with this, it has recently been shown that blood DCs, but not tonsil DCs, retain the potential to differentiate in vitro into Langerhans cell-like DCs [60].

Human tissue DCs can be separated into resident and migratory DCs. Similar to the mouse, migratory DCs migrate through the lymph and display high levels of maturation markers when they reach the lymph nodes [57, 61, 62].

3.2 Plasmacytoid DCs

pDCs were initially described in humans [63]. pDC development seems to depend on Flt3-L, as injection of Flt3-L into healthy volunteers increases the number of circulating pDCs [64]. In vitro experiments and analysis of patients harboring monoallelic mutations of the gene coding for E2-2 have shown that the transcription factor E2-2 is essential for pDC development [8, 65].

Like their mouse counterparts, human pDCs are specialized for type I interferon secretion [10]. However, human pDCs also present antigens to T cells. Depending on the activation signals they receive, pDCs can induce Th1 polarization [66] or regulatory T cell differentiation [67, 68]. Multiple studies have shown that human pDCs can cross-present antigens efficiently, whether soluble [59, 69–71], viral [71–73], or cell-associated antigen [70]. However, pDCs are unable to cross-present necrotic cell-derived antigen [59].

3.3 Classical DCs

Human cDCs have historically been separated into two subsets that are found in the blood, spleen, tonsil, and lymph nodes: BDCA1/CD1c+ DCs and BDCA3/CD141+ DCs [57, 74–76]. DC populations with a similar phenotype have been described in the lung, liver, and intestine [61, 77, 78]. CD141+ DCs have also been found in the skin [61]. cDC development seems dependent on Flt3-L, as injection of Flt3-L into healthy volunteers increases blood cDC numbers [64]. The cDC-specific expression of the transcription factor zbtb46 is also conserved in humans [17]. Comparative transcriptomic analysis suggests that CD141+ DCs are homologue to mouse Batf3-dependent DCs, while CD1c+ DCs are homologue to mouse IRF4-dependent DCs [31, 78–80]. In addition, some phenotypic markers are conserved between mouse and human DC subsets (Table 2). In vitro experiments suggest that CD141+ DC development is dependent on Batf3 [81]. However, other transcription factors involved may be different from the mouse, as patients with IRF8 mutations lack both cDC subsets [82].

Steady-state lymph nodes CD1c+ and CD141+ DCs can induce Th1 and Th2 polarization [57]. When activated by *Aspergillus fumigatus*, lung CD1c+ DCs are potent inducers of Th17 responses [31]. Recent studies have shown that both resident cDC subsets can cross-present soluble antigens with overall similar efficiency [57, 59, 69]. When activated by signals to which they both respond, blood cDC subsets also display comparable capacity for cross-presentation [58, 69, 70]. In addition, both cDC subsets from humanized mice can cross-present a recall antigen to memory CD8 T cells [77].

3.4 CD1a+ and CD14+ Tissue DCs

In the skin and vaginal mucosa, additional DC subsets have been identified: CD1a+ DCs and CD14+ DCs [83, 84]. Their lineage origin and relationship to other DC subsets remain unclear. It has been proposed that skin CD1a+ DCs represent migratory equivalents of CD1c+ cDCs [61].

Functional analysis has shown that skin CD14+ DCs are potent inducers of T follicular helper cells, but are poor at cross-presentation [57, 84]. Mucosal CD14+ DCs can also induce Th1 polarization [83]. Skin CD1a+ DCs preferentially induce Th2 polarization and are able to cross-present antigens [57, 84].

Table 2
Surface markers of human DC subsets

Marker	pDC	CD141+ DC	CD1c+ DC	Tissue CD1a+ DC	Tissue CD14+ DC	Langerhans cell	Monocyte-derived DC
CD11c	–	+	+	+	+	Low	+
BDCA2/ CD303	+	–	–	–	–	–	–
BDCA4/ CD304	+	–	–	–	–	–	–
CD123	+	–	–	–	–	–	–
XCR1	–	+	–	–	–	–	–
Clec9A	–	+	–	–	–	–	–
BDCA3/ CD141	–	+	Low *immature* + *mature*	–	+	–	–
BDCA1/ CD1c	–	–	+	+	+	+	+
CD11b	–	–	+	+	+	+	+
Sirp-α/ CD172a	–	–	+	+	+	+	+
MR/CD206	–	–	–	+	+	–	+
Langerin/ CD207	–	–	–	–	–	+	–
EpCAM/ CD326	–	–	–	–	–	+	–
CD1a	–	–	–	+	–	+	+
DC-SIGN/ CD209	–	–	–	–	+	–	Heterogeneous
FcεRI	–	–	+ *blood* – *tissues*	–	–	–	+
CD64	–	–	Low	–	?	–	+
CD14	–	–	–	–	+	–	+

3.5 Langerhans Cells   Langerhans cells resembling murine Langerhans cells are found in human skin epidermis and mucosal tissues [83, 85]. Human Langerhans cells represent a lineage distinct from DCs and monocytes as patients affected with a mutation in *GATA2* retain normal numbers of epidermal Langerhans cells while lacking blood monocytes and all DC subsets [86].

Langerhans cells are potent activators of CD4 T cells and induce Th2 polarization [57, 83, 84]. Skin Langerhans cells can also perform cross-presentation [87] and efficiently induce effector cytotoxic CD8 T cells [88].

3.6 Monocyte-Derived DCs

Human equivalents of "inflammatory DCs" have been identified in psoriasis, atopic dermatitis, rheumatoid arthritis, and tumor ascites [89–91]. "Inflammatory DCs" from tumor ascites express a transcriptomic signature similar to that of in vitro-generated monocyte-derived DC, suggesting that they derive from monocytes rather than from a DC precursor [90]. Of note, "inflammatory DCs" express the transcription factor zbtb46.

Furthermore, transcriptomic analysis suggests that dermal CD14$^+$ DCs and intestinal CD103$^-$ CD172a$^+$ DCs are related to monocytes [61, 78] and could represent steady-state monocyte-derived DCs.

There is limited data on the functions of these monocyte-derived DCs. "Inflammatory DCs" from tumor ascites and rheumatoid arthritis synovial fluid can stimulate naive CD4 T cells and induce Th17 polarization [90].

Acknowledgments

This work is supported by INSERM, the European Research Council, LabEx DCBiol, and Ligue contre le Cancer.

References

1. Steinman RM, Nussenzweig MC (1980) Dendritic cells: features and functions. Immunol Rev 53:127–147

2. Merad M, Sathe P, Helft J, Miller J, Mortha A (2013) The dendritic cell lineage: ontogeny and function of dendritic cells and their subsets in the steady state and the inflamed setting. Annu Rev Immunol 31:563–604

3. Wilson NS, El-Sukkari D, Belz GT, Smith CM, Steptoe RJ, Heath WR, Shortman K, Villadangos JA (2003) Most lymphoid organ dendritic cell types are phenotypically and functionally immature. Blood 102(6):2187–2194

4. Wilson NS, Young LJ, Kupresanin F, Naik SH, Vremec D, Heath WR, Akira S, Shortman K, Boyle J, Maraskovsky E, Belz GT, Villadangos JA (2008) Normal proportion and expression of maturation markers in migratory dendritic cells in the absence of germs or Toll-like receptor signaling. Immunol Cell Biol 86(2):200–205

5. Reizis B, Bunin A, Ghosh HS, Lewis KL, Sisirak V (2011) Plasmacytoid dendritic cells: recent progress and open questions. Annu Rev Immunol 29:163–183

6. Naik SH, Sathe P, Park HY, Metcalf D, Proietto AI, Dakic A, Carotta S, O'Keeffe M, Bahlo M, Papenfuss A, Kwak JY, Wu L, Shortman K (2007) Development of plasmacytoid and conventional dendritic cell subtypes from single precursor cells derived in vitro and in vivo. Nat Immunol 8(11):1217–1226

7. Onai N, Obata-Onai A, Schmid MA, Ohteki T, Jarrossay D, Manz MG (2007) Identification of clonogenic common Flt3+M-CSFR+ plasmacytoid and conventional dendritic cell progenitors in mouse bone marrow. Nat Immunol 8(11):1207–1216

8. Cisse B, Caton ML, Lehner M, Maeda T, Scheu S, Locksley R, Holmberg D, Zweier C, den Hollander NS, Kant SG, Holter W, Rauch A, Zhuang Y, Reizis B (2008) Transcription

factor E2-2 is an essential and specific regulator of plasmacytoid dendritic cell development. Cell 135(1):37–48

9. Ippolito GC, Dekker JD, Wang YH, Lee BK, Shaffer AL III, Lin J, Wall JK, Lee BS, Staudt LM, Liu YJ, Iyer VR, Tucker HO (2014) Dendritic cell fate is determined by BCL11A. Proc Natl Acad Sci U S A 111(11): E998–E1006

10. Gilliet M, Cao W, Liu YJ (2008) Plasmacytoid dendritic cells: sensing nucleic acids in viral infection and autoimmune diseases. Nat Rev Immunol 8(8):594–606

11. Lund JM, Linehan MM, Iijima N, Iwasaki A (2006) Cutting edge: plasmacytoid dendritic cells provide innate immune protection against mucosal viral infection in situ. J Immunol 177(11):7510–7514

12. GeurtsvanKessel CH, Willart MA, van Rijt LS, Muskens F, Kool M, Baas C, Thielemans K, Bennett C, Clausen BE, Hoogsteden HC, Osterhaus AD, Rimmelzwaan GF, Lambrecht BN (2008) Clearance of influenza virus from the lung depends on migratory langerin+ CD11b− but not plasmacytoid dendritic cells. J Exp Med 205(7):1621–1634

13. Swiecki M, Gilfillan S, Vermi W, Wang Y, Colonna M (2010) Plasmacytoid dendritic cell ablation impacts early interferon responses and antiviral NK and CD8(+) T cell accrual. Immunity 33(6):955–966

14. Cervantes-Barragan L, Lewis KL, Firner S, Thiel V, Hugues S, Reith W, Ludewig B, Reizis B (2012) Plasmacytoid dendritic cells control T-cell response to chronic viral infection. Proc Natl Acad Sci U S A 109(8):3012–3017

15. Liu K, Waskow C, Liu X, Yao K, Hoh J, Nussenzweig M (2007) Origin of dendritic cells in peripheral lymphoid organs of mice. Nat Immunol 8(6):578–583

16. Miller JC, Brown BD, Shay T, Gautier EL, Jojic V, Cohain A, Pandey G, Leboeuf M, Elpek KG, Helft J, Hashimoto D, Chow A, Price J, Greter M, Bogunovic M, Bellemare-Pelletier A, Frenette PS, Randolph GJ, Turley SJ, Merad M, Gautier EL, Jakubzick C, Randolph GJ, Best AJ, Knell J, Goldrath A, Miller J, Brown B, Merad M, Jojic V, Koller D, Cohen N, Brennan P, Brenner M, Shay T, Regev A, Fletcher A, Elpek K, Bellemare-Pelletier A, Malhotra D, Turley S, Jianu R, Laidlaw D, Collins J, Narayan K, Sylvia K, Kang J, Gazit R, Rossi DJ, Kim F, Rao TN, Wagers A, Shinton SA, Hardy RR, Monach P, Bezman NA, Sun JC, Kim CC, Lanier LL, Heng T, Kreslavsky T, Painter M, Ericson J, Davis S, Mathis D, Benoist C (2012) Deciphering the transcriptional network of the dendritic cell lineage. Nat Immunol 13(9): 888–899

17. Satpathy AT, Kc W, Albring JC, Edelson BT, Kretzer NM, Bhattacharya D, Murphy TL, Murphy KM (2012) Zbtb46 expression distinguishes classical dendritic cells and their committed progenitors from other immune lineages. J Exp Med 209(6):1135–1152

18. Meredith MM, Liu K, Darrasse-Jeze G, Kamphorst AO, Schreiber HA, Guermonprez P, Idoyaga J, Cheong C, Yao KH, Niec RE, Nussenzweig MC (2012) Expression of the zinc finger transcription factor zDC (Zbtb46, Btbd4) defines the classical dendritic cell lineage. J Exp Med 209(6):1153–1165

19. Schraml BU, van Blijswijk J, Zelenay S, Whitney PG, Filby A, Acton SE, Rogers NC, Moncaut N, Carvajal JJ, Reis e Sousa C (2013) Genetic tracing via DNGR-1 expression history defines dendritic cells as a hematopoietic lineage. Cell 154(4):843–858

20. Ginhoux F, Liu K, Helft J, Bogunovic M, Greter M, Hashimoto D, Price J, Yin N, Bromberg J, Lira SA, Stanley ER, Nussenzweig M, Merad M (2009) The origin and development of nonlymphoid tissue CD103+ DCs. J Exp Med 206(13):3115–3130

21. Hildner K, Edelson BT, Purtha WE, Diamond M, Matsushita H, Kohyama M, Calderon B, Schraml BU, Unanue ER, Diamond MS, Schreiber RD, Murphy TL, Murphy KM (2008) Batf3 deficiency reveals a critical role for CD8alpha+ dendritic cells in cytotoxic T cell immunity. Science 322(5904):1097–1100

22. Edelson BT, Kc W, Juang R, Kohyama M, Benoit LA, Klekotka PA, Moon C, Albring JC, Ise W, Michael DG, Bhattacharya D, Stappenbeck TS, Holtzman MJ, Sung SS, Murphy TL, Hildner K, Murphy KM (2010) Peripheral CD103+ dendritic cells form a unified subset developmentally related to CD8alpha+ conventional dendritic cells. J Exp Med 207(4):823–836

23. Joffre OP, Segura E, Savina A, Amigorena S (2012) Cross-presentation by dendritic cells. Nat Rev Immunol 12(8):557–569

24. Desch AN, Randolph GJ, Murphy K, Gautier EL, Kedl RM, Lahoud MH, Caminschi I, Shortman K, Henson PM, Jakubzick CV (2011) CD103+ pulmonary dendritic cells preferentially acquire and present apoptotic cell-associated antigen. J Exp Med 208(9): 1789–1797

25. Kashiwada M, Pham NL, Pewe LL, Harty JT, Rothman PB (2011) NFIL3/E4BP4 is a key transcription factor for CD8alpha(+) dendritic cell development. Blood 117(23): 6193–6197

26. Yamazaki C, Sugiyama M, Ohta T, Hemmi H, Hamada E, Sasaki I, Fukuda Y, Yano T, Nobuoka M, Hirashima T, Iizuka A, Sato K, Tanaka T, Hoshino K, Kaisho T (2013) Critical roles of a dendritic cell subset expressing a chemokine receptor, XCR1. J Immunol 190(12): 6071–6082

27. Igyarto BZ, Haley K, Ortner D, Bobr A, Gerami-Nejad M, Edelson BT, Zurawski SM, Malissen B, Zurawski G, Berman J, Kaplan DH (2011) Skin-resident murine dendritic cell subsets promote distinct and opposing antigen-specific T helper cell responses. Immunity 35(2):260–272

28. Helft J, Manicassamy B, Guermonprez P, Hashimoto D, Silvin A, Agudo J, Brown BD, Schmolke M, Miller JC, Leboeuf M, Murphy KM, Garcia-Sastre A, Merad M (2012) Cross-presenting CD103+ dendritic cells are protected from influenza virus infection. J Clin Invest 122(11):4037–4047

29. Wu L, D'Amico A, Winkel KD, Suter M, Lo D, Shortman K (1998) RelB is essential for the development of myeloid-related CD8alpha– dendritic cells but not of lymphoid-related CD8alpha+ dendritic cells. Immunity 9(6):839–847

30. Suzuki S, Honma K, Matsuyama T, Suzuki K, Toriyama K, Akitoyo I, Yamamoto K, Suematsu T, Nakamura M, Yui K, Kumatori A (2004) Critical roles of interferon regulatory factor 4 in CD11bhighCD8alpha– dendritic cell development. Proc Natl Acad Sci U S A 101(24):8981–8986

31. Schlitzer A, McGovern N, Teo P, Zelante T, Atarashi K, Low D, Ho AW, See P, Shin A, Wasan PS, Hoeffel G, Malleret B, Heiseke A, Chew S, Jardine L, Purvis HA, Hilkens CM, Tam J, Poidinger M, Stanley ER, Krug AB, Renia L, Sivasankar B, Ng LG, Collin M, Ricciardi-Castagnoli P, Honda K, Haniffa M, Ginhoux F (2013) IRF4 transcription factor-dependent CD11b+ dendritic cells in human and mouse control mucosal IL-17 cytokine responses. Immunity 38(5):970–983

32. Dudziak D, Kamphorst AO, Heidkamp GF, Buchholz VR, Trumpfheller C, Yamazaki S, Cheong C, Liu K, Lee HW, Park CG, Steinman RM, Nussenzweig MC (2007) Differential antigen processing by dendritic cell subsets in vivo. Science 315(5808):107–111

33. Vander Lugt B, Khan AA, Hackney JA, Agrawal S, Lesch J, Zhou M, Lee WP, Park S, Xu M, DeVoss J, Spooner CJ, Chalouni C, Delamarre L, Mellman I, Singh H (2014) Transcriptional programming of dendritic cells for enhanced MHC class II antigen presentation. Nat Immunol 15(2):161–167

34. Persson EK, Uronen-Hansson H, Semmrich M, Rivollier A, Hagerbrand K, Marsal J, Gudjonsson S, Hakansson U, Reizis B, Kotarsky K, Agace WW (2013) IRF4 transcription-factor-dependent CD103(+)CD11b(+) dendritic cells drive mucosal T helper 17 cell differentiation. Immunity 38(5):958–969

35. Gao Y, Nish SA, Jiang R, Hou L, Licona-Limon P, Weinstein JS, Zhao H, Medzhitov R (2013) Control of T helper 2 responses by transcription factor IRF4-dependent dendritic cells. Immunity 39(4):722–732

36. Plantinga M, Guilliams M, Vanheerswynghels M, Deswarte K, Branco-Madeira F, Toussaint W, Vanhoutte L, Neyt K, Killeen N, Malissen B, Hammad H, Lambrecht BN (2013) Conventional and monocyte-derived CD11b(+) dendritic cells initiate and maintain T helper 2 cell-mediated immunity to house dust mite allergen. Immunity 38(2):322–335

37. Williams JW, Tjota MY, Clay BS, Vander Lugt B, Bandukwala HS, Hrusch CL, Decker DC, Blaine KM, Fixsen BR, Singh H, Sciammas R, Sperling AI (2013) Transcription factor IRF4 drives dendritic cells to promote Th2 differentiation. Nat Commun 4:2990

38. Hoeffel G, Wang Y, Greter M, See P, Teo P, Malleret B, Leboeuf M, Low D, Oller G, Almeida F, Choy SH, Grisotto M, Renia L, Conway SJ, Stanley ER, Chan JK, Ng LG, Samokhvalov IM, Merad M, Ginhoux F (2012) Adult Langerhans cells derive predominantly from embryonic fetal liver monocytes with a minor contribution of yolk sac-derived macrophages. J Exp Med 209(6):1167–1181

39. Wang Y, Szretter KJ, Vermi W, Gilfillan S, Rossini C, Cella M, Barrow AD, Diamond MS, Colonna M (2012) IL-34 is a tissue-restricted ligand of CSF1R required for the development of Langerhans cells and microglia. Nat Immunol 13(8):753–760

40. Greter M, Lelios I, Pelczar P, Hoeffel G, Price J, Leboeuf M, Kundig TM, Frei K, Ginhoux F, Merad M, Becher B (2012) Stroma-derived interleukin-34 controls the development and maintenance of Langerhans cells and the maintenance of microglia. Immunity 37(6): 1050–1060

41. Guilliams M, Ginhoux F, Jakubzick C, Naik SH, Onai N, Schraml BU, Segura E, Tussiwand R, Yona S (2014) Dendritic cells, monocytes and macrophages: a unified nomenclature based on ontogeny. Nat Rev Immunol 14(8):571–578

42. Nakajima S, Igyarto BZ, Honda T, Egawa G, Otsuka A, Hara-Chikuma M, Watanabe N, Ziegler SF, Tomura M, Inaba K, Miyachi Y, Kaplan DH, Kabashima K (2012) Langerhans

cells are critical in epicutaneous sensitization with protein antigen via thymic stromal lymphopoietin receptor signaling. J Allergy Clin Immunol 129(4):1048–1055, e1046

43. Bursch LS, Rich BE, Hogquist KA (2009) Langerhans cells are not required for the CD8 T cell response to epidermal self-antigens. J Immunol 182(8):4657–4664

44. Seneschal J, Jiang X, Kupper TS (2014) Langerin+ dermal DC, but not Langerhans cells, are required for effective CD8-mediated immune responses after skin scarification with vaccinia virus. J Invest Dermatol 134(3): 686–694

45. Segura E, Amigorena S (2013) Inflammatory dendritic cells in mice and humans. Trends Immunol 34(9):440–445

46. Greter M, Helft J, Chow A, Hashimoto D, Mortha A, Agudo-Cantero J, Bogunovic M, Gautier EL, Miller J, Leboeuf M, Lu G, Aloman C, Brown BD, Pollard JW, Xiong H, Randolph GJ, Chipuk JE, Frenette PS, Merad M (2012) GM-CSF controls nonlymphoid tissue dendritic cell homeostasis but is dispensable for the differentiation of inflammatory dendritic cells. Immunity 36(6):1031–1046

47. Bogunovic M, Ginhoux F, Helft J, Shang L, Hashimoto D, Greter M, Liu K, Jakubzick C, Ingersoll MA, Leboeuf M, Stanley ER, Nussenzweig M, Lira SA, Randolph GJ, Merad M (2009) Origin of the lamina propria dendritic cell network. Immunity 31(3):513–525

48. Varol C, Vallon-Eberhard A, Elinav E, Aychek T, Shapira Y, Luche H, Fehling HJ, Hardt WD, Shakhar G, Jung S (2009) Intestinal lamina propria dendritic cell subsets have different origin and functions. Immunity 31(3): 502–512

49. Langlet C, Tamoutounour S, Henri S, Luche H, Ardouin L, Gregoire C, Malissen B, Guilliams M (2012) CD64 expression distinguishes monocyte-derived and conventional dendritic cells and reveals their distinct role during intramuscular immunization. J Immunol 188(4):1751–1760

50. Tamoutounour S, Guilliams M, Montanana Sanchis F, Liu H, Terhorst D, Malosse C, Pollet E, Ardouin L, Luche H, Sanchez C, Dalod M, Malissen B, Henri S (2013) Origins and functional specialization of macrophages and of conventional and monocyte-derived dendritic cells in mouse skin. Immunity 39(5): 925–938

51. Zigmond E, Varol C, Farache J, Elmaliah E, Satpathy AT, Friedlander G, Mack M, Shpigel N, Boneca IG, Murphy KM, Shakhar G, Halpern Z, Jung S (2012) Ly6C(hi) monocytes in the inflamed colon give rise to proinflammatory

effector cells and migratory antigen-presenting cells. Immunity 37(6):1076–1090

52. Chakarov S, Fazilleau N (2014) Monocyte-derived dendritic cells promote T follicular helper cell differentiation. EMBO Mol Med 6(5):590–603

53. Wakim LM, Waithman J, van Rooijen N, Heath WR, Carbone FR (2008) Dendritic cell-induced memory T cell activation in nonlymphoid tissues. Science 319(5860):198–202

54. Iijima N, Mattei LM, Iwasaki A (2011) Recruited inflammatory monocytes stimulate antiviral Th1 immunity in infected tissue. Proc Natl Acad Sci U S A 108(1):284–289

55. Aldridge JR Jr, Moseley CE, Boltz DA, Negovetich NJ, Reynolds C, Franks J, Brown SA, Doherty PC, Webster RG, Thomas PG (2009) TNF/iNOS-producing dendritic cells are the necessary evil of lethal influenza virus infection. Proc Natl Acad Sci U S A 106(13): 5306–5311

56. Ziegler-Heitbrock L, Ancuta P, Crowe S, Dalod M, Grau V, Hart DN, Leenen PJ, Liu YJ, MacPherson G, Randolph GJ, Scherberich J, Schmitz J, Shortman K, Sozzani S, Strobl H, Zembala M, Austyn JM, Lutz MB (2010) Nomenclature of monocytes and dendritic cells in blood. Blood 116(16):e74–e80

57. Segura E, Valladeau-Guilemond J, Donnadieu MH, Sastre-Garau X, Soumelis V, Amigorena S (2012) Characterization of resident and migratory dendritic cells in human lymph nodes. J Exp Med 209(4):653–660

58. Nizzoli G, Krietsch J, Weick A, Steinfelder S, Facciotti F, Gruarin P, Bianco A, Steckel B, Moro M, Crosti MC, Romagnani C, Stolzel K, Torretta S, Pignataro L, Scheibenbogen C, Neddermann P, Defrancesco R, Abrignani S, Geginat J (2013) Human CD1c+ dendritic cells secrete high levels of IL-12 and potently prime cytotoxic T cell responses. Blood 122(6): 932–942

59. Segura E, Durand M, Amigorena S (2013) Similar antigen cross-presentation capacity and phagocytic functions in all freshly isolated human lymphoid organ-resident dendritic cells. J Exp Med 210(5):1035–1047

60. Martinez-Cingolani C, Grandclaudon M, Jeanmougin M, Jouve M, Zollinger R, Soumelis V (2014) Human blood BDCA-1 dendritic cells differentiate into Langerhans-like cells with thymic stromal lymphopoietin and TGF-beta. Blood 124(15):2411–2420

61. Haniffa M, Shin A, Bigley V, McGovern N, Teo P, See P, Wasan PS, Wang XN, Malinarich F, Malleret B, Larbi A, Tan P, Zhao H, Poidinger M, Pagan S, Cookson S, Dickinson R, Dimmick I, Jarrett RF, Renia L, Tam J,

Song C, Connolly J, Chan JK, Gehring A, Bertoletti A, Collin M, Ginhoux F (2012) Human tissues contain CD141(hi) cross-presenting dendritic cells with functional homology to mouse CD103(+) nonlymphoid dendritic cells. Immunity 37(1):60–73

62. Morandi B, Bonaccorsi I, Mesiti M, Conte R, Carrega P, Costa G, Iemmo R, Martini S, Ferrone S, Cantoni C, Mingari MC, Moretta L, Ferlazzo G (2013) Characterization of human afferent lymph dendritic cells from seroma fluids. J Immunol 191(9):4858–4866

63. Cella M, Jarrossay D, Facchetti F, Alebardi O, Nakajima H, Lanzavecchia A, Colonna M (1999) Plasmacytoid monocytes migrate to inflamed lymph nodes and produce large amounts of type I interferon. Nat Med 5(8):919–923

64. Pulendran B, Banchereau J, Burkeholder S, Kraus E, Guinet E, Chalouni C, Caron D, Maliszewski C, Davoust J, Fay J, Palucka K (2000) Flt3-ligand and granulocyte colony-stimulating factor mobilize distinct human dendritic cell subsets in vivo. J Immunol 165(1):566–572

65. Nagasawa M, Schmidlin H, Hazekamp MG, Schotte R, Blom B (2008) Development of human plasmacytoid dendritic cells depends on the combined action of the basic helix-loop-helix factor E2-2 and the Ets factor Spi-B. Eur J Immunol 38(9):2389–2400

66. Cella M, Facchetti F, Lanzavecchia A, Colonna M (2000) Plasmacytoid dendritic cells activated by influenza virus and CD40L drive a potent TH1 polarization. Nat Immunol 1(4):305–310

67. Manches O, Munn D, Fallahi A, Lifson J, Chaperot L, Plumas J, Bhardwaj N (2008) HIV-activated human plasmacytoid DCs induce Tregs through an indoleamine 2,3-dioxygenase-dependent mechanism. J Clin Invest 118(10):3431–3439

68. Palomares O, Ruckert B, Jartti T, Kucuksezer UC, Puhakka T, Gomez E, Fahrner HB, Speiser A, Jung A, Kwok WW, Kalogjera L, Akdis M, Akdis CA (2012) Induction and maintenance of allergen-specific FOXP3+ Treg cells in human tonsils as potential first-line organs of oral tolerance. J Allergy Clin Immunol 129(2):510–520, 520 e511–519

69. Mittag D, Proietto AI, Loudovaris T, Mannering SI, Vremec D, Shortman K, Wu L, Harrison LC (2011) Human dendritic cell subsets from spleen and blood are similar in phenotype and function but modified by donor health status. J Immunol 186(11):6207–6217

70. Tel J, Schreibelt G, Sittig SP, Mathan TS, Buschow SI, Cruz LJ, Lambeck AJ, Figdor CG, de Vries IJ (2013) Human plasmacytoid dendritic cells efficiently cross-present exogenous Ags to CD8+ T cells despite lower Ag uptake than myeloid dendritic cell subsets. Blood 121(3):459–467

71. Hoeffel G, Ripoche AC, Matheoud D, Nascimbeni M, Escriou N, Lebon P, Heshmati F, Guillet JG, Gannage M, Caillat-Zucman S, Casartelli N, Schwartz O, De la Salle H, Hanau D, Hosmalin A, Maranon C (2007) Antigen crosspresentation by human plasmacytoid dendritic cells. Immunity 27(3):481–492

72. Di Pucchio T, Chatterjee B, Smed-Sorensen A, Clayton S, Palazzo A, Montes M, Xue Y, Mellman I, Banchereau J, Connolly JE (2008) Direct proteasome-independent cross-presentation of viral antigen by plasmacytoid dendritic cells on major histocompatibility complex class I. Nat Immunol 9(5):551–557

73. Lui G, Manches O, Angel J, Molens JP, Chaperot L, Plumas J (2009) Plasmacytoid dendritic cells capture and cross-present viral antigens from influenza-virus exposed cells. PLoS One 4(9):e7111

74. Dzionek A, Fuchs A, Schmidt P, Cremer S, Zysk M, Miltenyi S, Buck DW, Schmitz J (2000) BDCA-2, BDCA-3, and BDCA-4: three markers for distinct subsets of dendritic cells in human peripheral blood. J Immunol 165(11):6037–6046

75. Summers KL, Hock BD, McKenzie JL, Hart DN (2001) Phenotypic characterization of five dendritic cell subsets in human tonsils. Am J Pathol 159(1):285–295

76. McIlroy D, Troadec C, Grassi F, Samri A, Barrou B, Autran B, Debre P, Feuillard J, Hosmalin A (2001) Investigation of human spleen dendritic cell phenotype and distribution reveals evidence of in vivo activation in a subset of organ donors. Blood 97(11):3470–3477

77. Yu CI, Becker C, Wang Y, Marches F, Helft J, Leboeuf M, Anguiano E, Pourpe S, Goller K, Pascual V, Banchereau J, Merad M, Palucka K (2013) Human CD1c(+) dendritic cells drive the differentiation of CD103(+) CD8(+) mucosal effector T cells via the cytokine TGF-beta. Immunity 38(4):818–830

78. Watchmaker PB, Lahl K, Lee M, Baumjohann D, Morton J, Kim SJ, Zeng R, Dent A, Ansel KM, Diamond B, Hadeiba H, Butcher EC (2014) Comparative transcriptional and functional profiling defines conserved programs of intestinal DC differentiation in humans and mice. Nat Immunol 15(1):98–108

79. Robbins SH, Walzer T, Dembele D, Thibault C, Defays A, Bessou G, Xu H, Vivier E, Sellars M, Pierre P, Sharp FR, Chan S, Kastner P,

Dalod M (2008) Novel insights into the relationships between dendritic cell subsets in human and mouse revealed by genome-wide expression profiling. Genome Biol 9(1):R17

80. Crozat K, Guiton R, Contreras V, Feuillet V, Dutertre CA, Ventre E, Vu Manh TP, Baranek T, Storset AK, Marvel J, Boudinot P, Hosmalin A, Schwartz-Cornil I, Dalod M (2010) The XC chemokine receptor 1 is a conserved selective marker of mammalian cells homologous to mouse CD8alpha+ dendritic cells. J Exp Med 207(6):1283–1292

81. Poulin LF, Reyal Y, Uronen-Hansson H, Schraml B, Sancho D, Murphy KM, Hakansson UK, Moita LF, Agace WW, Bonnet D, Reis ESC (2012) DNGR-1 is a specific and universal marker of mouse and human Batf3-dependent dendritic cells in lymphoid and non-lymphoid tissues. Blood 119(25): 6052–6062

82. Hambleton S, Salem S, Bustamante J, Bigley V, Boisson-Dupuis S, Azevedo J, Fortin A, Haniffa M, Ceron-Gutierrez L, Bacon CM, Menon G, Trouillet C, McDonald D, Carey P, Ginhoux F, Alsina L, Zumwalt TJ, Kong XF, Kumararatne D, Butler K, Hubeau M, Feinberg J, Al-Muhsen S, Cant A, Abel L, Chaussabel D, Doffinger R, Talesnik E, Grumach A, Duarte A, Abarca K, Moraes-Vasconcelos D, Burk D, Berghuis A, Geissmann F, Collin M, Casanova JL, Gros P (2011) IRF8 mutations and human dendritic-cell immunodeficiency. N Engl J Med 365(2):127–138

83. Duluc D, Gannevat J, Anguiano E, Zurawski S, Carley M, Boreham M, Stecher J, Dullaers M, Banchereau J, Oh S (2013) Functional diversity of human vaginal APC subsets in directing T-cell responses. Mucosal Immunol 6(3): 626–638

84. Klechevsky E, Morita R, Liu M, Cao Y, Coquery S, Thompson-Snipes L, Briere F, Chaussabel D, Zurawski G, Palucka AK, Reiter Y, Banchereau J, Ueno H (2008) Functional specializations of human epidermal Langerhans cells and CD14+ dermal dendritic cells. Immunity 29(3):497–510

85. Nestle FO, Zheng XG, Thompson CB, Turka LA, Nickoloff BJ (1993) Characterization of dermal dendritic cells obtained from normal human skin reveals phenotypic and functionally distinctive subsets. J Immunol 151(11): 6535–6545

86. Bigley V, Haniffa M, Doulatov S, Wang XN, Dickinson R, McGovern N, Jardine L, Pagan S, Dimmick I, Chua I, Wallis J, Lordan J, Morgan C, Kumararatne DS, Doffinger R, van der Burg M, van Dongen J, Cant A, Dick JE, Hambleton S, Collin M (2011) The human syndrome of dendritic cell, monocyte, B and NK lymphoid deficiency. J Exp Med 208(2): 227–234

87. Polak ME, Newell L, Taraban VY, Pickard C, Healy E, Friedmann PS, Al-Shamkhani A, Ardern-Jones MR (2012) CD70-CD27 interaction augments CD8+ T-cell activation by human epidermal Langerhans cells. J Invest Dermatol 132(6):1636–1644

88. Banchereau J, Thompson-Snipes L, Zurawski S, Blanck JP, Cao Y, Clayton S, Gorvel JP, Zurawski G, Klechevsky E (2012) The differential production of cytokines by human Langerhans cells and dermal CD14(+) DCs controls CTL priming. Blood 119(24): 5742–5749

89. Wollenberg A, Mommaas M, Oppel T, Schottdorf EM, Gunther S, Moderer M (2002) Expression and function of the mannose receptor CD206 on epidermal dendritic cells in inflammatory skin diseases. J Invest Dermatol 118(2):327–334

90. Segura E, Touzot M, Bohineust A, Cappuccio A, Chiocchia G, Hosmalin A, Dalod M, Soumelis V, Amigorena S (2013) Human inflammatory dendritic cells induce Th17 cell differentiation. Immunity 38(2):336–348

91. Guttman-Yassky E, Lowes MA, Fuentes-Duculan J, Whynot J, Novitskaya I, Cardinale I, Haider A, Khatcherian A, Carucci JA, Bergman R, Krueger JG (2007) Major differences in inflammatory dendritic cells and their products distinguish atopic dermatitis from psoriasis. J Allergy Clin Immunol 119(5):1210–1217

Part II

In Vitro Culture of Dendritic Cells

Chapter 2

In Vitro Generation of Human XCR1⁺ Dendritic Cells from CD34⁺ Hematopoietic Progenitors

Sreekumar Balan and Marc Dalod

Abstract

Dendritic cells (DCs) are a heterogeneous population of professional antigen-presenting cells which play a key role in orchestrating immune defenses. Most of the information gained on human DC biology was derived from studies conducted with DCs generated in vitro from peripheral blood CD14⁺ monocytes (MoDCs) or from CD34⁺ hematopoietic progenitors. Recent advances in the field revealed that these types of in vitro-derived DCs strikingly differ from the DC subsets that are naturally present in human lymphoid organs, in terms of global gene expression, of specialization in the sensing of different types of danger signals, and of the ability to polarize T lymphocytes toward different functions. Major efforts are being made to better characterize the biology and the functions of lymphoid organ-resident DC subsets in humans, as an essential step for designing innovative DC-based vaccines against infections or cancers. However, this line of research is hampered by the low frequency of certain DC subsets in most tissues, their fragility, and the complexity of the procedures necessary for their purification. Hence, there is a need for robust procedures allowing large-scale in vitro generation of human DC subsets, under conditions allowing their genetic or pharmacological manipulation, to decipher their functions and their molecular regulation. Human CD141⁺CLEC9A⁺XCR1⁺ DCs constitute a very interesting DC subset for the design of immunotherapeutic treatments against infections by intracellular pathogens or against cancer, because these cells resemble mouse professional cross-presenting CD8α⁺Clec9a⁺Xcr1⁺ DCs. Human XCR1⁺ DCs have indeed been reported by several teams to be more efficient than other human DC subsets for cross-presentation, in particular of cell-associated antigens but also of soluble antigens especially when delivered into late endosomes or lysosomes. However, human XCR1⁺ DCs are the rarest and perhaps the most fragile of the human DC subsets and hence the most difficult to study ex vivo. Here, we describe a protocol allowing simultaneous in vitro generation of human MoDCs and XCR1⁺ DCs, which will undoubtedly be extremely useful to better characterize the functional specialization of human XCR1⁺ DCs and to identify its molecular bases.

Key words Human immune system, CD34⁺ hematopoietic stem cells, Differentiation, XCR1⁺ dendritic cells, Monocyte-derived dendritic cells, Cross-presentation

Elodie Segura and Nobuyuki Onai (eds.), *Dendritic Cell Protocols*, Methods in Molecular Biology, vol. 1423,
DOI 10.1007/978-1-4939-3606-9_2, © Springer Science+Business Media New York 2016

1   Introduction

The development of better vaccines or immunotherapies against cancer or intracellular pathogens is one of the major current challenges of immunological research. Innovative approaches are needed to induce efficient and long-lasting memory cytotoxic CD8 T cell responses. Conversely, in the case of inflammatory or autoimmune diseases, better treatments are needed to selectively dampen deleterious inflammatory reactions or to specifically inactivate autoimmune lymphocytes, without compromising overall immune responses of the patients to avoid increasing sensitivity to infections or cancers. Dendritic cells (DCs) are professional antigen-presenting cells which are exquisitely efficient for the activation of naïve T lymphocytes upon their primary encounter with their cognate antigen [1]. During this interaction, DCs instruct the functional polarization of T lymphocytes toward different activities depending on the physiopathological context, including promoting either immunity or tolerance. DCs exert this function by delivering to T cells the three types of output signals necessary for their priming: the triggering of the T cell receptor by peptide-MHC-I complexes (signal 1), the triggering of co-stimulation receptors by the co-stimulation molecules induced on mature DCs (signal 2), and cytokines which can contribute to promote the proliferation or survival of T cells and which instruct their differentiation toward specific functions (signal 3). The nature of the output signals delivered to T lymphocytes by DCs is determined by the integration by DCs of a variety of input signals that they can detect in their environment. Indeed, DCs are equipped with a variety of innate immune recognition receptors (I2R2s) allowing them on the one hand to detect pathogens, infections, transformed cells, and cytokines and on the other hand to engulf molecules, microorganisms, or cellular debris [2]. The combinations of I2R2s which are engaged on DCs during their activation control the type of maturation that DCs undergo, either tolerogenic or immunogenic, with the expression of specific combinations of positive or negative co-stimulation molecules and the production of particular patterns of immunoactivating or immunosuppressive cytokines. Hence, DC functions are highly plastic, allowing to induce the type of immune responses needed depending upon the physiopathological context [3]. The existence of distinct DC subsets specialized in different functions is another key feature contributing to promote a diversity of DC responses to match the diversity of the threats which the immune system has to face [4]. Indeed, several DC subsets exist which express different arrays of I2R2s [2] and have different potentials for exerting distinct functions such as the production of various cytokines or the activation of CD8 T cells [5].

A particular subset of mouse DCs excels at inducing protective CD8 T cell responses, in particular through uptake and processing of exogenous antigens for their presentation in association with major histocompatibility class-I (MHC-I) molecules, a process called cross-presentation [6]. Cross-presentation is critical for the induction of CD8 T cell responses against cancer, since most tumors are not derived from APCs and are inefficient for priming. Cross-presentation is also critical for the induction of CD8 T cell responses against intracellular pathogens that do not infect DCs or that escape direct antigen presentation in infected DCs, for example, by downregulating the expression of MHC or activating co-stimulation molecules or by enhancing the expression of inhibitory co-stimulation molecules [7]. Besides selectively expressing CD8α or CD103, the professional cross-presenting mouse DC subset is characterized by its unique expression of the endocytic receptor Clec9a, of the Toll-like receptor Tlr3, and of the chemokine receptor Xcr1 [8]. These three molecules have been shown or are thought to play a major role in endowing mouse Xcr1⁺ DCs with their high efficiency for CD8 T cell activation. Clec9a promotes the cross-presentation of dead cell-associated antigens, by allowing recognition of filamentous actin on dying cells for uptake by Xcr1⁺ DCs and intracellular routing into proper endosomes [9–11]. Tlr3 allows XCR1⁺ DCs to sense the abnormal presence of double-stranded RNA in the materials that they have engulfed in their endosomes, which triggers their production of interferons-β (IFN-β) and IFN-λ, which, in turn, at least for IFN-β, can enhance antigen cross-presentation by, and boost immunogenic maturation of, Xcr1⁺ DCs [12]. The ligand for Xcr1, Xcl1, is selectively produced by activated natural killer (NK) and CD8 T cells and might promote their physical encounters with Xcr1⁺ DCs as a mechanism amplifying the activation of cytotoxic effector lymphocytes by Xcr1⁺ DCs [13–15]. In vivo, targeting of mouse Xcr1⁺ DCs for vaccination, by co-administration of adjuvants and antigen-fused specific antibodies, yielded very encouraging results [16–19]. However, the efficacy of the immune responses induced and the generation of protective long-term memory require improvement. Moreover, how to translate this vaccination strategy for human health was not obvious.

In 2008, we discovered that the previously identified human CD141/BDCA3⁺ DC subset is equivalent to mouse Xcr1⁺ DCs, based on the similarities of their gene expression programs including unique expression of XCR1, CLEC9A, and TLR3 [20, 21]. Since 2010, we and several other teams have confirmed that human XCR1⁺ DCs are more efficient than other human DC subsets for the cross-presentation of cell-associated antigens or even of soluble antigens in specific experimental settings, in particular when delivered to late endosomes or lysosomes [13, 22–25]. However, in other experimental settings, all human DC subsets from the blood, spleen, or tonsils were found to be equally efficient for cross-presentation of soluble or particulate antigens [23, 26, 27]. Human XCR1+ DCs represent in average only 0.02 % of human peripheral blood mononuclear cells, thus constituting the rarest human DC subset. They also seem to be the most fragile human DC subset upon ex vivo isolation. Hence, human XCR1+ DCs are the most difficult human DC subset to study ex vivo. Thus, alternative approaches are needed to study human XCR1+ DCs. This is all the more important as most of the information gained on human DC biology was derived from studies conducted with DCs generated in vitro from peripheral blood CD14+ monocytes (MoDCs) or from CD34+ hematopoietic progenitors [21]. MoDCs are the most frequently used human DC subset for immunotherapeutic treatments in cancer patients or in individuals infected with human immunodeficiency virus type 1, but only with moderately encouraging results [28, 29] at least until recently [30].

Based on the combination and optimization of procedures previously reported by others [31, 32], we developed a protocol allowing simultaneous in vitro generation of high numbers of human MoDCs and XCR1+ DCs from CD34+ hematopoietic stem cells [33]. This system uniquely enables rigorous side-by-side functional comparison of XCR1+ DCs and MoDCs from the same culture [33], which will undoubtedly be extremely useful to better characterize their respective functional specializations and to identify how they are molecularly regulated.

2 Materials

2.1 Enrichment of CD34+ Hematopoietic Progenitor Stem Cells

1. A source of human CD34+ hematopoietic cells (*see* **Note 1**).
2. CD34+ cell selection kit (*see* **Note 2**).
3. Ficoll-Paque PLUS (GE healthcare). Store away from light at around +4 °C.
4. Dilution buffer (DB): Phosphate-buffered saline (PBS), 2 % fetal calf serum (FCS), 1 mM EDTA.
5. T 75 mL flask.
6. 50 mL Falcon tubes.
7. Sterile scissors.
8. Serological pipettes 5, 10, and 25 mL.

2.2 Expansion of Hematopoietic Precursors

1. StemSpan (StemSpan™ SFEM, serum-free medium for expansion of hematopoietic cells, Stem cell technology). This medium can be stored at −20 °C in aliquots.
2. FCS (*see* **Note 3**).

3. Recombinant human cytokines: FLT3-L, SCF, IL-3, TPO (Peprotech).

4. Amplification medium: StemSpan, FCS 10 %, FLT3-L (100 ng/mL), SCF (100 ng/mL), IL-3 (20 ng/mL), and TPO (50 ng/mL), to be prepared extemporaneously.

5. U-bottom 96-well tissue culture-treated plates.

6. Roswell Park Memorial Institute medium (RPMI).

7. 15 or 50 mL polypropylene tissue culture Falcon tubes.

2.3 Cryopreservation and Revival of Expanded Hematopoietic Precursors

1. Iscove's Modified Dulbecco's Medium (IMDM).

2. DMSO.

3. Deoxyribonuclease I from bovine pancreas (Nalgene, Sigma Aldrich).

4. FCS.

5. Cryotubes, e.g., Nunc® CryoTubes®, cryogenic vial, 1.8 mL, internal thread, round-bottomed, starfoot, free standing (Sigma).

6. Isopropanol.

7. Freezing container (e.g., Mr. Frosty, Nalgene).

8. Freezing medium#1 (FM1): IMDM, 30 % FCS.

9. Freezing medium#2 (FM2): IMDM, 30 % FCS, 20 % DMSO, to be prepared extemporaneously.

10. 15 or 50 mL polypropylene tissue culture Falcon tubes.

11. Water bath adjustable to 37 °C.

2.4 Differentiation of DCs from Expanded Hematopoietic Precursors

1. RPMI.

2. FCS.

3. Recombinant human cytokines: FLT3-L, SCF, GM-CSF, IL-4 (Peprotech).

4. Medium#2: RPMI, 10 % FCS, 10 mM HEPES, 1 mM sodium pyruvate, penicillin, streptomycin, 2 mM L-glutamine, 50 μM β mercaptoethanol.

5. Differentiation medium#1: medium#2, 100 ng/mL FLT3-L, 20 ng/mL SCF, 2.5 ng/mL IL-4, 2.5 ng/mL GM-CSF, to be prepared extemporaneously.

6. Differentiation medium#2: medium#2, 200 ng/mL FLT3-L, 40 ng/mL SCF, 5 ng/mL IL-4, 5 ng/mL GM-CSF, to be prepared extemporaneously (*see* **Note 4**).

7. U-bottom 96-well tissue culture-treated plates.

8. 15 or 50 mL polypropylene tissue culture Falcon tubes.

2.5 Staining for Flow Cytometry Analysis

Staining for flow cytometry can be used to phenotypically identify the different cell populations at the end of the culture, for sorting DC subsets, or for evaluation of DC subset maturation following their stimulation.

1. Fluorochrome-coupled monoclonal antibodies depending on the intended cell populations or biological process to study. The manufacturers, molecular targets, fluorochrome conjugation, hybridoma clones, and dilutions of use in our experimental settings (vol:vol) are given in Table 1 for the different antihuman antibodies used.

Table 1
List of the different fluorochrome-conjugated antihuman antibodies used for phenotyping

Antigen	Fluorochrome	Clone	Suppliers	Final dilution (v/v)
CD141	APC/FITC	AD5-14H12	Miltenyi Biotec	1:30
CD141	PE	VI E013	BD Bioscience	1:30
CLEC9A	APC	683409	R&D Systems	1:30
CLEC9A	APC/PE	8F9	Miltenyi Biotec	1:30
CADM1	Purified	3E1	MBL	1:1000
CD11C	V450	B-ly6	BD Biosciences	1:100
CD11B	FITC	ICRF44	eBioscience	1:100
CD23	PE	EBVCS2	eBioscience	1:50
CD32	APC	6C4	eBioscience	1:50
CD206	PE-Cy7	19.2	eBioscience	1:100
CD209	APC	eB-h209	eBioscience	1:100
CD86	AF700	2331 (FUN-1)	BD Biosciences	1:50
CD83	APC	HB15e	BD Biosciences	1:30
HLA-DR	AF700	LN3	eBioscience	1:100
CD40	PE		Beckman Coulter	1:10
TLR3	PE	34A3	Innate Pharma	1:30
TLR4	PE	HTA125	eBioscience	1:100
CD1B	APC	SN13 K5-1B8	eBioscience	1:50
CD103	PE-Cy7	B-ly7	eBioscience	1:100
CD14	FITC	M5E2	BD Biosciences	1:30
IgG (H+L)	FITC	Chicken	Southern Biotech	1:400
CLEC9A	PE	8F9	BioLegend	1:30

2. U-bottom 96-well tissue culture-treated plates.

3. FACS buffer: PBS, 1 mM EDTA, 10 mM HEPES.

4. Staining buffer (SB): FACS buffer, 2 % FCS.

5. Human TruStain FcX™ (Fc Receptor Blocking Solution, BioLegend).

6. Blocking buffer (BB): SB complemented 1:20, vol:vol, with Human TruStain FcX™ (e.g., 50 μL TruStain FcX™ for 1 mL SB).

7. LIVE/DEAD® Fixable Aqua Dead Cell Stain Kit (Invitrogen).

8. 0.5 % paraformaldehyde working solution: prepare 4 % (w/v) stock solution in PBS, adjusted to pH 7, according to manufacturer's instructions. Stock solution should be aliquoted in 10 mL volumes in 15 mL polypropylene tubes and frozen at −20 °C. Extemporaneously prepare 0.5 % working solution by diluting stock solution 1/8 in PBS.

9. OneComp eBeads (eBioscience) for compensation control.

10. Fluorescence-activated cell sorter for analysis of cells.

2.6 Sort of DC Subsets at the End of the Culture

1. Antibodies, FACS buffer, SB, and BB (*see* Subheading 2.5 above), all sterile.

2. 15 or 50 mL polypropylene tissue culture Falcon tubes.

3. FACS tubes (5 mL polystyrene round-bottomed 12 mm × 75 mm snap-cap tissue culture tubes).

4. Fluorescence-activated cell sorter for isolation of cells.

5. PBS containing 2 % BSA for coating the FACS tubes, sterile.

6. Collection medium: RPMI, 10 % FCS, sterile.

7. 70 μM cell strainers.

8. SYTOX® Blue Dead Cell Stain Kit (Invitrogen).

9. OneComp eBeads (eBioscience) for compensation control.

2.7 Stimulation of Sorted DC Subsets with Different Adjuvants

1. Adjuvants: Poly(I:C) (high molecular weight), R848 (imidazoquinoline compound), and LPS (ultrapure LPS from *Salmonella minnesota*).

2. 15 mL polypropylene Falcon tissue culture tubes.

3. U-bottom 96-well tissue culture-treated plates.

4. Medium#2.

5. Differentiation medium#2.

6. Round-bottomed 96-well tissue culture-treated plates.

3   Methods

The in vitro generation of XCR1+ DCs consists in a two-step culture system (*see* Fig. 1). The first phase is a 7-day expansion of enriched CD34+ cells. During this step, the CD34+ cells proliferate under the instruction of a combination of cytokines (FLT3-L, SCF, TPO, and IL-3). This step allows amplifying the cells up to 100-fold. The expanded cells can be directly differentiated or cryopreserved for future use. The second phase is the differentiation of

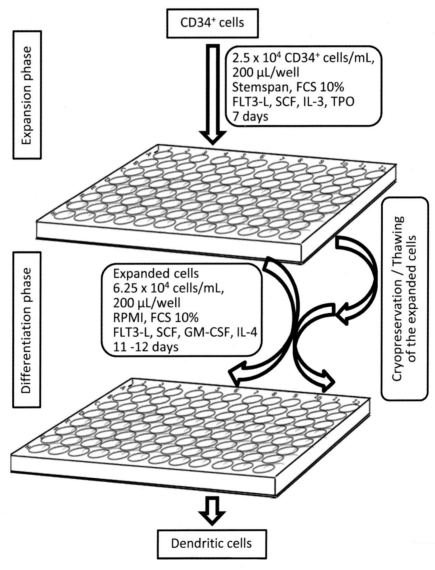

Fig. 1 General overview of the procedure for in vitro generation of XCR1+ DCs from CD34+ hematopoietic stem cell progenitors

the expanded cells, for 11–12 days, under the instruction of a different combination of cytokines (FLT3-L, SCF, IL-4, and GM-CSF). At the end of the culture, the different cell populations can be identified phenotypically by flow cytometry and eventually sorted for functional characterization as exemplified here by characterization of their responses to the stimulation by adjuvants.

All the experimental procedure should be performed under sterile cell culture conditions and with the appropriate safety measures required for handling human samples.

3.1 Isolation of the Mononuclear Cells (MNCs) from Cord Blood, Bone Marrow Aspirate, or G-CSF-Mobilized Peripheral Blood Apheresis

1. Bring the Ficoll-Paque from +4 °C to room temperature (RT).

2. Open the cord blood/bone marrow aspirate/peripheral blood apheresis collection (with sterile scissors in the case of collection bags), and transfer content to a T 75 mL tissue culture flask (for bags) or to 50 mL tissue culture tubes. Measure the volume.

3. Dilute the cord blood at 1:1 (v/v) with DB or dilute the bone marrow aspirate or peripheral blood apheresis at 1:2 (v/v).

4. Mix well.

5. Prepare one 50 mL tube for each 20–30 mL of the diluted cord blood or bone marrow aspirate or peripheral blood apheresis.

6. Dispense 15 mL of Ficoll-Paque to each of the 50 mL tube.

7. Carefully overlay 20–30 mL of blood on 15 mL of Ficoll-Paque (*see* **Note 5**).

8. Centrifuge the tubes at $800 \times g$ for 25 min at 20 °C without break.

9. Collect and pool the mononuclear cells at the interphase in new 50 mL tubes.

10. Centrifuge at $800 \times g$ for 5 min.

11. Resuspend the pellets in 10 mL of DB.

12. Centrifuge at $450 \times g$ for 6 min.

13. Resuspend the pellets in 5 mL of DB.

14. Determine the viable cell count with trypan blue. The cells are ready to use for CD34⁺ cell enrichment.

3.2 Enrichment of CD34⁺ Hematopoietic Progenitor Stem Cells

Enrich CD34⁺ cells using commercially available CD34⁺ enrichment kits according to the manufacturer's protocols. Enriched CD34⁺ cells can be directly used for the culture or cryopreserved for future use.

3.3 Expansion of Hematopoietic Precursors

1. Prepare the amplification medium as described in Subheading 2.

2. Wash the CD34⁺cells and resuspend them in the amplification medium at a cell density of 2.5×10^4 CD34⁺ cells/mL.

3. Plate 200 μL/well of the cell suspension, in U-bottom 96-well tissue culture-treated plates.

4. Harvest the cells on the seventh day: transfer the cells into 15 or 50 mL tubes and centrifuge at $450 \times g$ for 5 min.

5. Resuspend the cells in RPMI, 10 % FCS, and determine the viable cell count using trypan blue. Expanded progenitors can be directly used for the culture or cryopreserved for future use (*see* **Note 6**).

3.4 Cryopreservation of Expanded Hematopoietic Precursors

1. The day before, prepare the freezing container by replenishing with fresh isopropanol according to the manufacturer's instructions. Precool it overnight at around +4 °C (*see* **Note 7**).

2. Prepare FM1 and FM2 and incubate them on ice for a time long enough to allow them to cool to +4 °C (for ≥10 min, depending on the volume) (*see* **Note 7**).

3. Label the appropriate number of cryotubes with sample name, cell number, date, etc.

4. Cool the cryotubes in ice for >10 min.

5. Harvest the cell culture and determine the viable count.

6. Resuspend the cells in FM1, in half of the final volume of cell suspension to be frozen (*see* **Note 8**).

7. Keep the cell suspension in ice for a time long enough to allow it to cool to +4 °C.

8. Add drop by drop to the cell suspension an identical volume of FM2, to achieve a 1:1 mixture of cell suspension and FM2, with continuous gentle agitation of the cell suspension tube. The tubes must be kept cold, on ice, during the entire procedure.

9. Transfer the cells to cryotubes, on ice.

10. Transfer the vials to the precooled freezing container.

11. Cool the freezing container at −80 °C overnight.

12. The day after, transfer the vials to liquid nitrogen for long-term storage.

3.5 Revival of Frozen Expanded Hematopoietic Precursors

1. Set the water bath at 37 °C.

2. Transfer the vials to the water and thaw the cells rapidly until only a small piece of ice is left in the tube (*see* **Note 9**).

3. Transfer the cells to a 15 mL polypropylene tissue culture tube.

4. Dilute the cell suspension fivefold in cold IMDM, 5 % FCS, 20 U/mL DNase I.

5. Gently mix the cell suspension, on ice.

6. Centrifuge the cell at $450 \times g$ for 5 min at low break.

7. Resuspend the cells in medium#2.

3.6 Differentiation of DCs from Expanded Hematopoietic Precursors

1. Count the cells and resuspend them in differentiation medium#1 at a cell density of 6.25×10^4 cells/mL.

2. Plate 200 μL/well of the cell suspension in U-bottom 96-well tissue culture-treated plates.

3. On day 6: remove half of the medium from each well (100 μL) and replenish with differentiation medium#2.

4. The cells can be harvested on day 11 or 12 (see **Note 10**).

3.7 Phenotypic Identification of the Different Cell Populations at the End of the Culture

The in vitro culture consists in three different populations based on the expression of CD11c and CD141 (see **Note 11**) (see Fig. 2). The CD11c^{high}CD141$^+$ (XCR1^{neg}) fraction specifically expresses CD11b, CD206, CD209, and TLR4. These cells are the equivalents of MoDCs as classically derived in cultures of monocytes with GM-CSF and IL-4. The CD11c^{low}CD141^{high} (XCR1$^+$) fraction specifically expresses CLEC9A, CADM1, TLR3, and XCR1. These cells are equivalent to the bona fide blood XCR1$^+$ DC (see **Note 12**). The third fraction is low or negative for CD141 and CD11c (DN cells).

1. Prepare the staining buffer and blocking buffer as detailed in Subheading 2.

2. Resuspend the cells in blocking buffer (2–10×10^5 cells/50 μL) (see **Note 13**).

3. Incubate at +4 °C for 5–10 min.

4. Prepare the desired antibody cocktails in blocking buffer, at twice the final dilution of use (see **Note 14**).

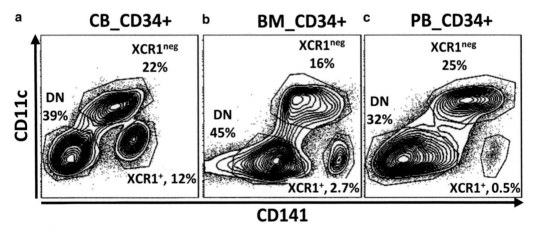

Fig. 2 Phenotypic identification of cell subsets in the cultures. The dot plot shows the different cell populations in the in vitro culture of CD34$^+$ cells from cord blood (**a**), bone marrow (**b**), and G-CSF-mobilized peripheral blood (**c**). All culture generates three different populations based on the expression of CD11c and CD141. The XCR1$^+$ cells correspond to the bona fide CD141$^+$CLEC9A$^+$ cells present in human blood. The XCR1^{neg} cells are equivalent to MoDCs

5. Accordingly to the staining design, add to each cell suspension well 50 μL of the adequate antibody combination and aqua dead cell stain (*see* **Note 15**).

6. Incubate for 30 min at +4 °C.

7. Add 100 μL of SB to each well and centrifuge at $450 \times g$ for 4 min.

8. Wash the cells pellets with 200 μL of SB and centrifuge at $450 \times g$ for 4 min.

9. Resuspend the cell pellets in 100 μL of 0.5 % paraformaldehyde and incubate for 10–15 min for fixing the cells.

10. Spin the cells at $450 \times g$ for 4 min.

11. Resuspend cell pellets in 200 μL of SB. The cells are ready for the FACS analysis.

12. Prepare single staining tubes for compensation with adequate compensation beads according to the manufacturer recommendations (*see* **Note 16**).

3.8 Sorting of the DC Subsets at the End of the Differentiation Culture

Purifying the different cell subsets in the culture is essential for their functional characterization and transcriptomic analysis (*see* **Note 17**). DCs can be sorted in two major populations, CD11c^{high}CD141$^+$ cells (XCR1^{neg} fraction) and CD11c^{low}CD141^{high}CLEC9A$^+$ cells (XCR1$^+$ fraction) (*see* Fig. 2).

1. Coat the 5 mL snap-cap collection tubes overnight with PBS, 2 % BSA.

2. Extemporaneously fill the collection tubes with 250 μL/tube of collection medium.

3. Harvest the culture by collecting the cells in 15 or 50 mL polypropylene tissue culture tubes.

4. Centrifuge the cells at $450 \times g$ for 5 min.

5. Resuspend the cell pellet in 5 mL of sterile SB and determine the viable cell number.

6. Prepare the single staining tubes with individual markers for compensation setup.

7. Prepare the antibody cocktails for phenotypic identification of subsets as described in Subheading 3.7.

8. Suspend ten million cells/200 μL of SB, and incubate the desired combination of antibodies for 30 min on ice in 15 mL tubes or 5 mL polystyrene round-bottomed snap-cap FACS tubes.

9. Wash the cells with 500 μL of SB.

10. Resuspend the cells in 1 mL of staining buffer containing SYTOX® Blue Dead Cell Stain.

11. Filter the cell suspension through a 70 μM cell strainer into 5 mL polystyrene round-bottomed snap-cap FACS tubes.

12. The live single cells should be plotted in CD11c vs. CD141 dot plot (*see* Fig. 2). Check that CLEC9A is specifically expressed on all CD11c^{low}CD141^{high} cells and that CD209, CD206, and CD11b are specifically expressed on CD11c^{high}CD141$^+$ cells.

13. Sort the XCR1$^+$ DC as CD11c^{low}CD141^{high} cells and the XCR1^{neg} DCs as CD11c^{high}CD141$^+$ cells and if required the DN cells as CD11c$^{neg/low}$CD141^{neg} cells (*see* **Note 18**).

14. Collect the sorted cells into the collection tubes.

15. Transfer a small fraction of sorted cell populations (~1:20 of the total volume) each into a properly labeled FACS tube, and complete it with SB to reach a total volume of 150 μL. Analyze these tubes by flow cytometry for measuring the purity achieved by the sorting procedure (*see* **Note 19**).

3.9 Stimulation of Sorted DC Subsets with Different Adjuvants and Evaluation of Their Activation

1. Transfer the sorted cells to 15 mL Falcon tubes and spin down at $450 \times g$ for 5 min.

2. Resuspend the cells in a volume of medium#2 that should theoretically yield a cell concentration of 3×10^6 cells/mL based on the number of cells sorted for each population.

3. Determine the viable cell count.

4. Adjust the cell density to 10^6 cells/mL in medium#2.

5. Transfer 100 μL/well in U-bottom 96-well plate.

6. Add to each well 100 μL of differentiation medium#2 alone or supplemented with the selected TLR ligands, poly(I:C) at 5 μg/mL final, R848 at 10 μg/mL, or LPS at 1 μg/mL (*see* **Note 20**).

7. Incubate the cells for 16 h in a cell culture incubator.

8. Spin the culture plates at $450 \times g$ for 5 min.

9. Transfer 150 μL of cell supernatant into a round-bottomed tissue culture-treated polypropylene 96-well plate, and freeze it at –20 °C for eventual later measurement of cytokine or chemokine titers by ELISA, Luminex, or any suitable method of your choice.

10. Resuspend cell pellets in 180 μL of staining buffer, centrifuge plates at $450 \times g$ for 5 min, and discard supernatant.

11. Stain cells for flow cytometry as explained in Subheading 3.7, for the evaluation of their maturation as assessed by their enhanced expression of CD83, CD86, or HLA-DR under the different stimulation conditions.

4 Notes

1. The protocol works best with cord blood CD34⁺ cells. This protocol also allows generating XCR1⁺ DCs by using CD34⁺ cells from bone marrow or from G-CSF-mobilized peripheral blood. Alternatively, CD34⁺ cells can be purchased from different commercial suppliers. The frequency of XCR1⁺ DCs varies from donor to donor but is generally higher (3–10 %) when using CD34⁺ cells from cord blood as compared to bone marrow or to G-CSF-mobilized peripheral blood.

2. CD34⁺ cells selection kits are available from different manufacturers (EasySep™ Human Cord Blood CD34 Positive Selection Kit, Stem cell technology; Dynal® CD34 Progenitor Cell Selection System, Dynal, Invitrogen; CD34 MicroBead Kit, human, Miltenyi Biotec). All the kits are efficient for the enrichment of CD34⁺ cells. In our studies, we mainly utilized the Dynal or EasySep kits.

3. The selection of FCS is one of the most critical factors to ensure obtaining the highest possible frequencies of XCR1⁺ DCs in the cultures, which can vary for the same CD34⁺ cells from 0 to >10 % depending on the FCS. The frequency of the other cell subsets does not seem to be strongly affected by the FCS batches. Hence, a preliminary but very important step for establishing the culture system is to screen for a proper batch of FCS. Practically, small-scale cultures from two to three different donors must be seeded in parallel with different batches of FCS and compared for the frequency and absolute yields of XCR1⁺ DCs obtained at the end of the culture.

4. This medium is used for feeding the cultures by replacement of half of their volume with fresh medium on the sixth day of the differentiation culture. Accordingly, it contains twice the concentrations of cytokines as compared to differentiation medium#1.

5. Preparation of the overlay of diluted samples on Ficoll-Paque is a critical step for the recovery of MNCs. To prepare the separate layers of Ficoll-Paque and diluted samples, it is crucial to hold the Ficoll-Paque containing tubes in a slanting position and to add the samples very gently in a very slow but continuous flow on the walls of the tubes by handling the pipette at a vertical angle with the tube walls. The loaded tubes should be handled very gently to avoid any mixing of Ficoll-Paque and samples which may reduce the recovery of MNCs.

6. Cryopreservation of expanded cells prior to their differentiation does not affect their propensity at generating XCR1⁺ DCs. This step provides flexibility, including by allowing screening

CD34+ samples in small-scale cultures to select those yielding a higher frequency of XCR1+ DCs for later experiments requiring large-scale cultures.

7. All the items used in this step (FM1, FM2, cryotubes, freezing container, cell suspension, etc.) must be precooled by keeping them on ice, or at about +4 °C, for a minimum of 10–15 min or more depending upon the total volume of the cell suspension to be cryopreserved.

8. Cells are frozen in FM2. $1–5 \times 10^6$ cells can be cryopreserved in 1 mL per vial. The cells should be suspended at a 2× concentration in FM1 and diluted with FM2 at 1:1 ratio (i.e., if planning to cryopreserve 2×10^6 cells in 1 mL, suspend the cells in 500 μL of freezing medium 1, in one 50 mL polypropylene tissue culture tube, and add 500 μL of precooled freezing medium 2, drop by drop, under gentle mixing, on ice).

9. This is to ensure that the cell suspension does not warm above +4 °C.

10. The maximum yield of XCR1+ DCs is at days 11–12. Culturing cells longer does not improve the frequency or yield of XCR1+ DCs.

11. The combination of CD11c and CD141 allows discrimination between XCR1+ DCs (CD141^{high}CD11c^{low}) and MoDCs (CD141+CD11c^{high}) versus other cell types (CD11c-CD141-) present in the culture. However, to ensure rigorous identification of cell types, it is highly recommended to examine their expression of additional markers, in particular CLEC9A or CADM1 which are specifically expressed on XCR1+ DCs versus CD206 and CD209 which are specifically expressed on MoDC subsets [33].

12. Other studies recently reported different protocols for in vitro generation of human XCR1+ DCs subsets [34, 35]. These protocols always use FLT3-L but with different combinations of cytokines including SCF, IL-3, IL-6, TPO, GM-CSF, or IL-4 and in one case using an antagonist of the aryl hydrocarbon receptor (StemRegenin 1) [35]. These protocols are especially interesting as they were reported to also simultaneously yield cells equivalent to the two other human blood DC subsets, pDCs and CD1c+ DCs. However, further characterization of the cell populations obtained with these protocols is required to ensure that each of them as identified corresponds to a single, homogeneous, cell type. The extent to which these in vitro-generated cell subsets are similar to human blood DC subsets also needs to be investigated in greater details.

13. Cells can be stained in U-bottom 96-well plate or in 5 mL round-bottomed FACS tubes.

Table 2
Differential expression of selected markers between the XCR1⁺
and XCR1⁻ subsets of DCs generated from CD34⁺ cells

Marker	XCR1⁺ DCs	XCR1ⁿᵉᵍ DCs
CD11c	Low	High
CD141	High	+
CLEC9A	+	−
XCR1	+	−
CADM1	+	−
TLR3	+	−
TLR4	−	+
CD209	−	+
CD206	−	+
CD11b	−	+
CD14	−	+
CD1b	+	++
CD23	−	+
CD32	−/+	++
CD103	−/+	++

14. For characterization of the cell populations by multiparameter flow cytometry analysis, a combination of CD11c and CD141 staining is required in each tube, in combination with the staining of other markers such as listed in Table 2 to ensure the identity of the cell populations.

15. Dead cell discrimination dyes must be used, according to the recommendations from the manufacturer, to ensure proper discrimination of cell subsets without the confounding factors of nonspecific binding of antibodies to dying cells or autofluorescence of these cells.

16. For compensation of the spectral overlaps between the different fluorochromes selected for the phenotyping, it is necessary to prepare single staining tubes, one for each of the individual antibodies used. Single stainings for compensation can be achieved with adequate compensation beads according to the manufacturer's recommendations (OneComp eBeads, eBioscience). However, it is highly recommended to additionally prepare and acquire single stainings performed with the cell suspension, in order to be able to replay compensation after data acquisition using this set of single stainings if needed.

17. Sorted cells at the end of Subheadings 3.8 and 3.9 can be processed for gene expression profiling as described previously [33] and detailed in another chapter of this book (Chapter 16).

18. The cells should be sorted with 100 μm nozzle and under low pressure (and hence low speed) to limit mechanical stress, using sterile FACS buffer instead of sheath fluid and keeping the collection tubes at 4 °C to preserve the best cell viability possible.

19. It is necessary to measure the purity achieved after each sort, by flow cytometry analysis of a small fraction of each sorted cell population.

20. XCR1⁺ DCs specifically and strongly respond to poly(I:C) and are also activated by R848 but not by LPS. MoDCs in these cultures exhibit a strong response to LPS and are also activated R848 but not strongly by poly(I:C) [33].

Acknowledgments

We thank Caetano Reis e Sousa and Lionel Poulin (Immunobiology Laboratory, Cancer Research UK, London, UK) for sharing with us the details of their protocol for in vitro generation of human CLEC9A⁺ DCs, an achievement they were the first to report to the best of our knowledge. This work was performed in the frame of the I₂HD collaborative project between CIML, AVIESAN, and SANOFI. It received additional funding from Inserm, CNRS, Agence Nationale de Recherches sur le SIDA et les hépatites virales (ANRS to M.D.), Institut National du Cancer (INCa grant #2011-155), FRM (Equipe labellisée to M.D.), and the European Research Council under the European Community's Seventh Framework Programme (FP7/2007–2013 Grant Agreement no. 281225, to M.D.). S.B. was supported through the Agence Nationale de la Recherche (EMICIF, ANR-08-MIEN-008-02 to M.D.) and the I₂HD project.

References

1. Mellman I, Steinman RM (2001) Dendritic cells: specialized and regulated antigen processing machines. Cell 106(3):255–258

2. Crozat K, Vivier E, Dalod M (2009) Crosstalk between components of the innate immune system: promoting anti-microbial defenses and avoiding immunopathologies. Immunol Rev 227(1):129–149

3. Reis e Sousa C (2006) Dendritic cells in a mature age. Nat Rev Immunol 6(6):476–483

4. Ueno H, Klechevsky E, Schmitt N, Ni L, Flamar AL, Zurawski S, Zurawski G, Palucka K, Banchereau J, Oh S (2011) Targeting human dendritic cell subsets for improved vaccines. Semin Immunol 23(1):21–27

5. Guilliams M, Henri S, Tamoutounour S, Ardouin L, Schwartz-Cornil I, Dalod M, Malissen B (2010) From skin dendritic cells to a simplified classification of human and mouse dendritic cell subsets. Eur J Immunol 40(8):2089–2094

6. Joffre OP, Segura E, Savina A, Amigorena S (2012) Cross-presentation by dendritic cells. Nat Rev Immunol 12(8):557–569

7. Alexandre YO, Cocita CD, Ghilas S, Dalod M (2014) Deciphering the role of DC subsets in MCMV infection to better understand immune protection against viral infections. Front Microbiol 5:378

8. Crozat K, Tamoutounour S, Vu Manh TP, Fossum E, Luche H, Ardouin L, Guilliams M, Azukizawa H, Bogen B, Malissen B, Henri S, Dalod M (2011) Cutting edge: expression of XCR1 defines mouse lymphoid-tissue resident and migratory dendritic cells of the CD8alpha+ type. J Immunol 187(9):4411–4415

9. Zhang JG, Czabotar PE, Policheni AN, Caminschi I, Wan SS, Kitsoulis S, Tullett KM, Robin AY, Brammananth R, van Delft MF, Lu J, O'Reilly LA, Josefsson EC, Kile BT, Chin WJ, Mintern JD, Olshina MA, Wong W, Baum J, Wright MD, Huang DC, Mohandas N, Coppel RL, Colman PM, Nicola NA, Shortman K, Lahoud MH (2012) The dendritic cell receptor Clec9A binds damaged cells via exposed actin filaments. Immunity 36(4): 646–657

10. Ahrens S, Zelenay S, Sancho D, Hanc P, Kjaer S, Feest C, Fletcher G, Durkin C, Postigo A, Skehel M, Batista F, Thompson B, Way M, Reis e Sousa C, Schulz O (2012) F-actin is an evolutionarily conserved damage-associated molecular pattern recognized by DNGR-1, a receptor for dead cells. Immunity 36(4): 635–645

11. Sancho D, Joffre OP, Keller AM, Rogers NC, Martinez D, Hernanz-Falcon P, Rosewell I, Reis e Sousa C (2009) Identification of a dendritic cell receptor that couples sensing of necrosis to immunity. Nature 458(7240): 899–903

12. Tomasello E, Pollet E, Vu Manh T-P, Uzé G, Dalod M (2014) Harnessing mechanistic knowledge on beneficial versus deleterious IFN-I effects to design innovative immunotherapies targeting cytokine activity to specific cell types. Front Immunol 5:526

13. Crozat K, Guiton R, Contreras V, Feuillet V, Dutertre CA, Ventre E, Vu Manh TP, Baranek T, Storset AK, Marvel J, Boudinot P, Hosmalin A, Schwartz-Cornil I, Dalod M (2010) The XC chemokine receptor 1 is a conserved selective marker of mammalian cells homologous to mouse CD8alpha+ dendritic cells. J Exp Med 207(6):1283–1292

14. Dorner BG, Dorner MB, Zhou X, Opitz C, Mora A, Guttler S, Hutloff A, Mages HW, Ranke K, Schaefer M, Jack RS, Henn V, Kroczek RA (2009) Selective expression of the chemokine receptor XCR1 on cross-presenting dendritic cells determines cooperation with CD8+ T cells. Immunity 31(5):823–833

15. Kroczek RA, Henn V (2012) The role of XCR1 and its ligand XCL1 in antigen cross-presentation by murine and human dendritic cells. Front Immunol 3:14

16. Idoyaga J, Lubkin A, Fiorese C, Lahoud MH, Caminschi I, Huang Y, Rodriguez A, Clausen BE, Park CG, Trumpfheller C, Steinman RM (2011) Comparable T helper 1 (Th1) and CD8 T-cell immunity by targeting HIV gag p24 to CD8 dendritic cells within antibodies to Langerin, DEC205, and Clec9A. Proc Natl Acad Sci U S A 108(6):2384–2389

17. Caminschi I, Proietto AI, Ahmet F, Kitsoulis S, Shin Teh J, Lo JC, Rizzitelli A, Wu L, Vremec D, van Dommelen SL, Campbell IK, Maraskovsky E, Braley H, Davey GM, Mottram P, van de Velde N, Jensen K, Lew AM, Wright MD, Heath WR, Shortman K, Lahoud MH (2008) The dendritic cell subtype-restricted C-type lectin Clec9A is a target for vaccine enhancement. Blood 112(8):3264–3273

18. Bonifaz LC, Bonnyay DP, Charalambous A, Darguste DI, Fujii S, Soares H, Brimnes MK, Moltedo B, Moran TM, Steinman RM (2004) In vivo targeting of antigens to maturing dendritic cells via the DEC-205 receptor improves T cell vaccination. J Exp Med 199(6): 815–824

19. Sancho D, Mourao-Sa D, Joffre OP, Schulz O, Rogers NC, Pennington DJ, Carlyle JR, Reis e Sousa C (2008) Tumor therapy in mice via antigen targeting to a novel, DC-restricted C-type lectin. J Clin Invest 118(6): 2098–2110

20. Robbins SH, Walzer T, Dembele D, Thibault C, Defays A, Bessou G, Xu H, Vivier E, Sellars M, Pierre P, Sharp FR, Chan S, Kastner P, Dalod M (2008) Novel insights into the relationships between dendritic cell subsets in human and mouse revealed by genome-wide expression profiling. Genome Biol 9(1):R17

21. Crozat K, Guiton R, Guilliams M, Henri S, Baranek T, Schwartz-Cornil I, Malissen B, Dalod M (2010) Comparative genomics as a tool to reveal functional equivalences between human and mouse dendritic cell subsets. Immunol Rev 234(1):177–198

22. Bachem A, Guttler S, Hartung E, Ebstein F, Schaefer M, Tannert A, Salama A, Movassaghi K, Opitz C, Mages HW, Henn V, Kloetzel PM, Gurka S, Kroczek RA (2010) Superior antigen cross-presentation and XCR1 expression define human CD11c+CD141+ cells as homologues of mouse CD8+ dendritic cells. J Exp Med 207(6):1273–1281

23. Cohn L, Chatterjee B, Esselborn F, Smed-Sorensen A, Nakamura N, Chalouni C, Lee BC, Vandlen R, Keler T, Lauer P, Brockstedt D, Mellman I, Delamarre L (2013) Antigen delivery to early endosomes eliminates the superiority of human blood BDCA3+ dendritic cells at cross presentation. J Exp Med 210(5):1049–1063

24. Haniffa M, Shin A, Bigley V, McGovern N, Teo P, See P, Wasan PS, Wang XN, Malinarich F, Malleret B, Larbi A, Tan P, Zhao H, Poidinger M, Pagan S, Cookson S, Dickinson R, Dimmick I, Jarrett RF, Renia L, Tam J, Song C, Connolly J, Chan JK, Gehring A, Bertoletti A, Collin M, Ginhoux F (2012) Human tissues contain CD141hi cross-presenting dendritic cells with functional homology to mouse CD103+ nonlymphoid dendritic cells. Immunity 37(1):60–73

25. Jongbloed SL, Kassianos AJ, McDonald KJ, Clark GJ, Ju X, Angel CE, Chen CJ, Dunbar PR, Wadley RB, Jeet V, Vulink AJ, Hart DN, Radford KJ (2010) Human CD141+ (BDCA-3)+ dendritic cells (DCs) represent a unique myeloid DC subset that cross-presents necrotic cell antigens. J Exp Med 207(6):1247–1260

26. Segura E, Durand M, Amigorena S (2013) Similar antigen cross-presentation capacity and phagocytic functions in all freshly isolated human lymphoid organ-resident dendritic cells. J Exp Med 210(5):1035–1047

27. Mittag D, Proietto AI, Loudovaris T, Mannering SI, Vremec D, Shortman K, Wu L, Harrison LC (2011) Human dendritic cell subsets from spleen and blood are similar in phenotype and function but modified by donor health status. J Immunol 186(11):6207–6217

28. Adema GJ, de Vries IJ, Punt CJ, Figdor CG (2005) Migration of dendritic cell based cancer vaccines: in vivo veritas? Curr Opin Immunol 17(2):170–174

29. Anguille S, Smits EL, Lion E, van Tendeloo VF, Berneman ZN (2014) Clinical use of dendritic cells for cancer therapy. Lancet Oncol 15(7):e257–e267

30. Garcia F, Climent N, Guardo AC, Gil C, Leon A, Autran B, Lifson JD, Martinez-Picado J, Dalmau J, Clotet B, Gatell JM, Plana M, Gallart T (2013) A dendritic cell-based vaccine elicits T cell responses associated with control of HIV-1 replication. Sci Transl Med 5(166):166ra2

31. Poulin LF, Salio M, Griessinger E, Anjos-Afonso F, Craciun L, Chen JL, Keller AM, Joffre O, Zelenay S, Nye E, Le Moine A, Faure F, Donckier V, Sancho D, Cerundolo V, Bonnet D, Reis e Sousa C (2010) Characterization of human DNGR-1+ BDCA3+ leukocytes as putative equivalents of mouse CD8alpha+ dendritic cells. J Exp Med 207(6):1261–1271

32. Chen W, Antonenko S, Sederstrom JM, Liang X, Chan AS, Kanzler H, Blom B, Blazar BR, Liu YJ (2004) Thrombopoietin cooperates with FLT3-ligand in the generation of plasmacytoid dendritic cell precursors from human hematopoietic progenitors. Blood 103(7):2547–2553

33. Balan S, Ollion V, Colletti N, Chelbi R, Montanana-Sanchis F, Liu H, Vu Manh TP, Sanchez C, Savoret J, Perrot I, Doffin AC, Fossum E, Bechlian D, Chabannon C, Bogen B, Asselin-Paturel C, Shaw M, Soos T, Caux C, Valladeau-Guilemond J, Dalod M (2014) Human XCR1+ dendritic cells derived in vitro from CD34+ progenitors closely resemble blood dendritic cells, including their adjuvant responsiveness, contrary to monocyte-derived dendritic cells. J Immunol 193(4):1622–1635

34. Proietto AI, Mittag D, Roberts AW, Sprigg N, Wu L (2012) The equivalents of human blood and spleen dendritic cell subtypes can be generated in vitro from human CD34(+) stem cells in the presence of fms-like tyrosine kinase 3 ligand and thrombopoietin. Cell Mol Immunol 9(6):446–454

35. Thordardottir S, Hangalapura BN, Hutten T, Cossu M, Spanholtz J, Schaap N, Radstake TR, van der Voort R, Dolstra H (2014) The aryl hydrocarbon receptor antagonist StemRegenin 1 promotes human plasmacytoid and myeloid dendritic cell development from CD34+ hematopoietic progenitor cells. Stem Cells Dev 23(9):955–967

Chapter 3

Derivation and Utilization of Functional CD8⁺ Dendritic Cell Lines

Matteo Pigni, Devika Ashok, and Hans Acha-Orbea

Abstract

It is notoriously difficult to obtain large quantities of non-activated dendritic cells ex vivo. For this reason, we produced and characterized a mouse model expressing the large T oncogene under the CD11c promoter (Mushi mice), in which $CD8\alpha^+$ dendritic cells transform after 4 months. We derived a variety of stable cell lines from these primary lines. These cell lines reproducibly share with freshly isolated dendritic cells most surface markers, mRNA and protein expression, and all tested biological functions. Cell lines can be derived from various strains and knockout mice and can be easily transduced with lentiviruses. In this article, we describe the derivation, culture, and lentiviral transduction of these dendritic cell lines.

Key words Dendritic cell, Cell line, Tissue culture, Ex vivo derivation

1 Introduction

It is very difficult to obtain large numbers of the different subsets of dendritic cells (DCs) that can have unique or overlapping functions. Many DC subsets have been described in mouse. The major lymphoid tissue-resident DC subsets can be distinguished by their expression of $CD8\alpha$ (these DCs are major cross-presenter and are better for CD8 T cell priming), CD11b (these DCs are better for CD4 T cell priming), or CD45RA (these DCs are termed plasmacytoid DCs (pDC) and are the major type I interferon producers after viral infection) [1, 2]. In other tissues, several migratory subsets are found such as epidermal Langerhans cells, dermal $CD103^+$ DC (these are equivalent to the lymphoid tissue-resident $CD8^+$ DCs), and $CD11b^+CD103^-$ DC subsets. These migratory DCs migrate to secondary lymphoid organs upon microbial infection and inflammation, leading to their activation and differentiation, and present encountered antigens to T cells. A few millions of total DCs, including 100,000–400,000 DCs of the major subsets, are obtained per mouse spleen after sorting ex vivo [3]. An alternative is the derivation of bone-marrow-derived DCs with growth and

Elodie Segura and Nobuyuki Onai (eds.), *Dendritic Cell Protocols*, Methods in Molecular Biology, vol. 1423,
DOI 10.1007/978-1-4939-3606-9_3, © Springer Science+Business Media New York 2016

differentiation factors such as GM-CSF, with or without IL-4 (yielding mostly CD11b⁺ DC subsets, up to 100 million per mouse depending on the procedure, with 40 % contaminating non-DCs, including macrophages), or Flt3 ligand (Flt3L) (yielding mainly CD8-like DCs which do not express CD8 but are closely related to this subset [4], but also CD11b⁺ DC and pDC subsets, a few millions per mouse). However, further purification steps are required to obtain pure DC subsets using these isolation and in vitro differentiation procedures [4–6]. Isolated cells are often stressed or activated after the long isolation procedures, and, in addition, many cells die within 24–48 h of in vitro culture. The in vitro life span can be slightly extended by adding growth and differentiation factors or by using Bcl-2 overexpression or cells deficient for proapoptotic molecules [7].

Injection of Flt3L into mice, adoptive transfer of tumors secreting Flt3L, and more recently, generating transgenic mice with Flt3L overexpression allowed generation of different DC subsets in the spleen in about 30-fold higher numbers than from normal mice [8, 9].

Alternatively, several groups have established cell lines after retroviral transduction with oncogenes, transgenesis, or from primary cultures of splenocytes with growth factors [10–17]. All these lines have been cultured for many passages and have lost many of the initially described functions typical of DCs.

Despite the advances in method development, it is still difficult to obtain sufficient numbers of DCs to perform large-scale drug screenings, subcellular fractionations, proteomics, etc.

For these reasons, we generated transgenic mice, which spontaneously develop DC tumors, and derived stable cell lines with controlled low passage numbers [18–21]. These mice (called Mushi mice) are transgenic for the SV-40 large T oncogene under the CD11c promoter and generate spontaneous DC tumors. Cell lines derived from these tumors are surprisingly similar to freshly isolated DCs. They grow up to one-two million per mL with divisions every 1.5 days, require no additional growth factors, grow slightly adherent allowing the removal of dead cells before isolation, and share all tested functions with freshly isolated DCs such as protein and mRNA expression, activation-induced cytokine and chemokine secretion, co-stimulatory molecule expression (CD40, CD70, CD80, CD86), antigen (cross)-presentation, etc.

Mushi mice develop exclusively CD8⁺ DC tumors spontaneously between 4 and 4.5 months of age. The reason for the preferential outgrowth of CD8⁺ DC tumors is most likely due to the threefold higher expression of the large T oncogene in this subset compared to CD11b⁺ DCs with the CD11c promoter fragment used to generate the transgenic mice [21]. Tumors first develop in spleens, but also infiltrate the liver, mesenteric lymph nodes, thymus, and bone marrow. From these tumors, DC lines (termed MuTu DC) can be derived. They have very similar functions to freshly isolated splenic

DCs with the advantage that they are not activated during the isolation procedure and keep being viable. In addition, the cell lines can be easily transduced with lentiviruses for overexpression or shRNA to knockdown gene expression [19, 20]. Knockout or transgenic DC lines can be obtained by crossing the Mushi mice with the corresponding knockout or transgenic mice (*see* Table 1). Here, we describe the derivation, culture, and lentiviral transduction of the MuTu DC lines. We recently were able to derive the first CD11b and pDC lines from BatF3 KO large T transgenic mice (*see* Table 1).

2 Materials

2.1 Derivation of CD8+ DC Lines

1. Microcapillary tubes.
2. Critoseal Capillary Tube Sealant (Fisher Scientific).
3. Microhematocrit centrifuge with a hematocrit rotor.
4. Microhematocrit tube capillary reader (McCormick Scientific).
5. Scissors, forceps and scalpel.
6. Mesh or cell strainer and syringe plunger.
7. PBS.
8. Complete medium: IMDM-glutamax, 8–10% heat-inactivated fetal calf serum (FCS), 10 mM Hepes, 50 μM 2-mercaptoethanol (*see* **Note 1**), and, if required, 50 U/mL penicillin and 50 μg/mL streptomycin. Adjust with $NaHCO_3$ to 308 mOsm, if required (*see* **Note 2**). The medium is not supplemented with additional growth factors (*see* **Notes 3** and **4**). Prepare fresh medium and keep at 4 °C for a maximum of 2 weeks.
9. Culture vessels (*see* **Note 5**).
10. Humidified incubator set at 37 °C with 5% CO_2 (*see* **Note 6**).

2.2 Freezing

1. CryoTubes.
2. Containers for slow freezing filled with fresh ice-cold isopropanol. Precool at 4 °C for a minimum of 16 h, and replace the isopropanol after using the container five times.
3. Freezing medium: complete medium, 50% FCS, 10% DMSO. Keep at 4 °C.

2.3 Cell Culture

1. Complete medium. Pre-warm before use.
2. Cell dissociation buffer: PBS, 5 mM EDTA (non-enzymatic). Pre-warm at room temperature before use.
3. Culture vessels (*see* **Note 5**).

2.4 Lentiviral Transduction

1. 293T cells.
2. 293T culture medium: DMEM, 4.5% glucose, 10% FCS, 2 mM glutamine, penicillin/streptomycin. Prepare fresh medium every week.

Table 1
List of currently available MuTu DC lines

Cell line	DC type	Genotype (all but the Langerhans cell line are GFP+ large T+)	Strain background	Lentiviral transduction	References
MuTu1940	CD8	WT	C57BL/6	–	[19, 20]
MuTu BALB/c	CD8	WT	BALB/c	–	Not yet published
MuTu1 NOD	CD8	WT	NOD	–	Not yet published
MuTu IFNI R−/−	CD8	Type I IFN receptor KO	C57BL/6	–	[20]
MuTu TLR3−/−	CD8	TLR3 KO	C57BL/6	–	[20]
MuTu TLR9−/−	CD8	TLR9 KO	C57BL/6	–	[20]
MuTu MAVS−/−	CD8	MAVS/Cardiff KO	C57BL/6	–	Not yet published
MuTu NOX2−/−	CD8	NADPH Oxidase 2 KO	C57BL/6	–	Not yet published
MuTu H-2 K$^{b-/-}$	CD8	H-2 K^b KO	C57BL/6	–	Not yet published
MuTu CD11b	CD11b	BatF3 KO	C57BL/6	–	Not yet published
MuTu pDC	pDC?	BatF3 KO	C57BL/6	–	Not yet published
MuTu LC	Langerhans cell?	Generated from Langerin promoter-large T transgenic mouse. Has human CD2 as reporter	C57BL/6	–	Not yet published
MuTu IFNβ, IFNI R−/−	CD8	Interferon β secreting, type I IFNR KO	C57BL/6	+	[20]

MuTu IL-2	CD8	IL-2 producing	C57BL/6	+	Not yet published
MuTu IL-10	CD8	IL-10 producing	C57BL/6	+	Submitted
MuTu IL-12	CD8	IL-12 producing	C57BL/6	+	Not yet published
MuTu IL-15	CD8	IL-15 producing	C57BL/6	+	Not yet published
MuTu IL-35	CD8	IL-35 producing	C57BL/6	+	Submitted
MuTu TGFβ	CD8	TGFβ producing	C57BL/6	+	Submitted
MuTu Active TGFβ	CD8	Constitutively active TGFβ producing	C57BL/6	+	Submitted
MuTu Arginine	CD8	Arginase producing	C57BL/6	+	Not yet published
MuTu Indo	CD8	Indoleamine 2,3-dioxygenase producing	C57BL/6	+	Not yet published
MuTu CTLA-4	CD8	Surface CTLA-4 expressing	C57BL/6	+	Not yet published
MuTu PD2L	CD8	Surface PD2-L expressing	C57BL/6	+	Not yet published
MuTu luciferase	CD8	Luciferase expressing	C57BL/6	+	[18]

The MuTu1940 DC line is the best characterized, already distributed to many labs. From several KO mice, we have independently derived cell lines with similar phenotype and function. All the lentiviral-transduced cell lines are originally MuTu1940 cell lines at low passage numbers. They generally keep the expression of the transduced gene for at least ten passages, but it is recommended to test them from time to time. If they lose expression, we thaw another sample or lentivirally re-transduce a MuTu dendritic cell line

3. Plasmid psPAX2 (Addgene).

4. Plasmid pENV (Addgene).

5. Plasmid containing the lentivector of interest.

6. 0.5 M $CaCl_2$.

7. 2× HeBS (HEPES-buffered saline): dissolve 1.64 g NaCl, 1.19 g HEPES free acid, and 0.02 g Na_2HPO_4 in 80 mL of water. Adjust pH to 7.05 and filter sterilize. Keep at 4 °C.

8. 0.45 μm syringe filters (Millex-HV PVDF, Durapore).

9. 10 cm Petri dishes.

10. Puromycin.

11. Humidified incubator set at 37 °C with 5% CO_2.

3   Methods

3.1 Derivation of CD8+ DC Lines from Tumor-Bearing Mushi Mice

1. When mice start looking pale, measure their hematocrit (*see* **Note 7**). Bleed the mice.

2. Collect the blood in a microcapillary.

3. Seal the microcapillaries with Critoseal Capillary Tube Sealant.

4. Spin the microcapillaries ($15,000 \times g$ for 3 min) in a microhematocrit centrifuge with a hematocrit rotor.

5. Determine the hematocrit by placing the tube in a microhematocrit tube capillary reader.

6. If the hematocrit is below 0.3 (*see* **Note 8**), sacrifice the mouse.

7. Remove the organs from which you want to isolate MuTu DC.

8. Dissociate the organs by pressing them through a fine mesh, or cell strainer, using a syringe plunger (*see* **Note 9**).

9. Wash the cells in PBS by centrifugating at $360 \times g$ for 5–10 min.

10. Culture the cells in 6- or 24-well plates in at least ten replicates. Prepare twofold serial dilutions from 10^7 cells/mL to 10^6 cells/mL (*see* **Note 10**).

11. Replace the medium after overnight culture, discarding overgrown cultures. This step will remove many dead cells as the DC lines already start adhering.

12. Replace half the medium at least weekly until confluent adherent cultures are observed. This may take several weeks.

13. Once the medium starts changing color, split the cells very gently into culture vessels of the same size in two to three serial dilutions (*see* **Notes 11** and **12**).

14. Repeat **step 13** until the lower dilutions start growing (this may take several months) (*see* **Notes 13** and **14**).

15. Keep track of passage numbers (*see* **Note 15**).

16. Expand the culture by diluting the cells at 10^5 cells/mL in the largest possible volume. Go to **step 6** of Subheading 3.3 and/or.

17. Freeze many tubes of cells at early passage.

3.2 Freezing MuTu DC Lines

1. Place the freezing vials at 4 °C 1 h before use.

2. Dissociate the cells by incubating them for 10 min in cold cell dissociation buffer.

3. Centrifuge the cells at $360 \times g$ for 5 min.

4. Label the precooled freezing vials (*see* **Note 16**).

5. Gently dissociate the dry pellet.

6. Resuspend the cells at 3×10^6/mL in ice-cold freezing medium.

7. Transfer the cells to the vials (1 mL per vial). Close the tubes well.

8. Place the tubes quickly in the freezing container and place the container at –70 °C (*see* **Note 17**).

9. After a minimum of 2 days, transfer the tubes into liquid nitrogen for long-term storage.

3.3 Culture of MuTu DC Lines

1. Thaw the cells by warming up the tubes quickly until most ice crystals have melted and dilute dropwise with ice cold complete medium.

2. Centrifuge the cells at $360 \times g$ for 5 min.

3. Count the cells.

4. Seed the cells in culture vessels in twofold serial dilutions from 10^6 cells/mL to 10^5 cells/mL in complete medium.

5. Once the cells reach confluence ($1–2 \times 10^6$ cells/mL), you need to dilute them for expansion. Cells will be seeded into new culture vessels at 10^5/mL - 5×10^5/mL (*see* **Note 18**).

6. Remove the medium supernatant.

7. Dissociate the cells by adding pre-warmed PBS containing 5 mM EDTA. Just cover the culture vessel bottom.

8. Incubate for 5–10 min at room temperature.

9. Gently tap the culture vessels or pipette the cells up and down to detach them from the culture vessels.

10. Spin the cells at $360 \times g$ for 5 min.

11. Remove the supernatant.

12. Dissociate the dry pellet by gently tapping the tube.

13. Resuspend the dissociated pellet in the desired volume of complete medium.

14. Discard the culture when passage 40 is reached (*see* **Notes 19** and **20**).

3.4 Lentiviral Transduction

1. On day 1, seed 293T cells in a 10 cm Petri dish (two million cells in 8 mL of 293T culture medium per dish).

2. On day 2, refresh the medium with pre-warmed 293T culture medium. Allow transfection solutions to equilibrate at room temperature.

3. In a 15 mL Falcon tube, add to 250 μL of sterile distilled water 5 μg of psPAX2 plasmid, 15 μg of pENV plasmid, and 20 μg of the lentivector of interest.

4. Add 250 μL of 0.5 M CaCl$_2$. Mix well by vortexing.

5. Add the above 500 μL dropwise (one drop per second) to 50 μL of 2× HeBS by vortexing at full speed the 15 mL Falcon tube continuously.

6. Incubate for 20 min at room temperature, without the cap, under the hood.

7. Add the precipitate of DNA/calcium phosphate (total volume of 1 mL) to the 293T cells dropwise, slightly tilting the plates for mixing slowly.

8. Incubate the plates into the incubator for a maximum of 18 h.

9. On day 3, check on the microscope for the formation of a precipitate between the cells. Remove the 293T cells supernatant carefully (total volume of 9 mL) and refresh the medium with 9 mL of pre-warmed 293T culture medium.

10. Split the MuTu DC at 10^5 cells/mL in 6-well plates (25,000 cells/cm^2).

11. On day 4, filter the 293T cell supernatant containing viral particles using 0.45 μm filters.

12. The viral supernatants can be stored for a few days at 4 °C or frozen at −70 °C. However, viral titers may drop.

13. On day 5, remove the medium of the MuTu DC and replace it with lentivirus-containing medium (*see* **Note 21**).

14. 48 h after lentiviral infection, add puromycin to the MuTu DC medium (0.5 μg/mL).

15. Allow 3 days for efficient lentiviral transduction.

4 Notes

1. In our experience, the cells will not survive if 2-mercaptoethanol is omitted.

2. In some instances, when human osmolarity medium (290 mOsm) is used, cells may suffer. We then adjust osmolarity to 318 mOsm by adding 7.5 % sodium bicarbonate solution at 1/100 dilution to increase the osmolarity.

3. The MuTu DC lines require rich media. We recommend RPMI1640 or IMDM. The cells do not grow in DMEM.

4. The batch of fetal calf serum is important: about one in three serum batches quickly kill the cells in our experience. In general, survival of cells correlates with low LPS in the serum, but we do not know whether LPS or other contaminants are toxic for the lines. It is advised to test several batches for growth of the MuTu DC lines.

5. All types of culture vessels we tested work well: culture flasks of any size, Petri dishes, 96-, 48-, 24-, 6-well plates, etc. Non-tissue culture-coated plates or non-coated Petri dishes also work, but cells adhere less. However, as the cells grow adherent, roller bottles are not recommended.

6. Some groups grow DC in 10% CO_2. Check if your medium is at the right pH at this CO_2 concentration.

7. When MuTu tumors heavily infiltrate the bone marrow, hematopoiesis is disturbed and the mice become pale. We usually measure the hematocrit when they start becoming pale.

8. Usually, mice have a hematocrit between 0.4 and 0.5.

9. Alternatively, you can digest the organs with collagenase using the methods provided in Chapter 5 from this book.

10. In parallel to the primary culture, we often transfer tumor splenocytes into T cell-deficient Rag2 KO or CD3ε KO mice. They can also be transferred into CD8β T cell-depleted mice using 1 mg of H35-17.2 monoclonal antibody just before intravenous injection of one million tumor splenocytes [18]. Within a month, much more aggressive tumors grow in which DCs make up for more than 40% of splenocytes. This gives (1) a second chance to derive cell lines, and (2) it is generally easier to derive cell lines after adoptive transfer.

11. It can help to add 10% of conditioned medium from an established DC line.

12. If you plate multiple wells, ensure that you add the medium before distributing the cells and gently mix after distribution. Otherwise, the cells may adhere quickly at high local concentrations.

13. During this phase, fibroblasts may overgrow which will terminate the culture. To avoid this, we prepare many replicates and discard the ones in which fibroblasts overgrow.

14. It can take up to 6 months and up to ten passages to obtain easy growing functional DC lines.

15. It is important to keep track of the passage numbers to avoid selecting for lines that have lost some of their functions.

16. Use a pencil, not ink, to avoid removing the label if isopropanol spills over the tubes.

17. Keep the cells for as short as possible at 4°C before moving them to –70 °C.

18. Established cell lines are seeded above 10^5 cells/mL. Seeding at lower concentrations leads to a prolonged lag phase from which they usually slowly recover. If too diluted, the cells may die. Using 0.2 μm-filtered conditioned medium from parallel confluent DC line cultures helps to overcome this lag phase and even allows cloning. Let the cells grow to near confluence (not more than 1×10^6/mL), and split them before the medium is too consumed (slightly orange), otherwise they may lose functions quickly, as you will select for variants.

19. In general, established DC lines keep their functions for at least 40 passages if they are maintained properly. We do not let them overgrow and discard them if they do. They can remain functional much longer, but at late passages we often see decreased cytokine production upon stimulation and decreased cross-presentation capacity. It is worth testing the cells from time to time for cytokine production and cross-presentation, as these are the first functions they lose after prolonged culture.

20. We have observed that after adoptive transfer of aged cell lines with reduced functionality into Rag KO, CD3 KO, or CD8 T cell-depleted C57BL/6 mice (as they are rejected by a CD8 T cell response to the strong large T transgene [18]), they may regain their functions. Some of the cell lines increase their CD11b expression in culture. After such an adoptive transfer, they become CD11b^{low} again.

21. Infection may be increased by adding 7 μg/mL Polybrene.

References

1. Colonna M, Trinchieri G, Liu YJ (2004) Plasmacytoid dendritic cells in immunity. Nat Immunol 5(12):1219–1226

2. Merad M, Sathe P, Helft J, Miller J, Mortha A (2013) The dendritic cell lineage: ontogeny and function of dendritic cells and their subsets in the steady state and the inflamed setting. Annu Rev Immunol 31:563–604

3. Vremec D, Segura E (2013) The purification of large numbers of antigen presenting dendritic cells from mouse spleen. Methods Mol Biol 960:327–350

4. Naik SH, Proietto AI, Wilson NS, Dakic A, Schnorrer P, Fuchsberger M, Lahoud MH, O'Keeffe M, Shao QX, Chen WF, Villadangos JA, Shortman K, Wu L (2005) Cutting edge: generation of splenic CD8+ and CD8– dendritic cell equivalents in Fms-like tyrosine kinase 3 ligand bone marrow cultures. J Immunol 174(11):6592–6597

5. Brasel K, De Smedt T, Smith JL, Maliszewski CR (2000) Generation of murine dendritic cells from flt3-ligand-supplemented bone marrow cultures. Blood 96(9):3029–3039

6. Inaba K, Swiggard WJ, Steinman RM, Romani N, Schuler G, Brinster C (2009) Isolation of dendritic cells. Curr Protoc Immunol Chapter 3:Unit 3 7

7. Vremec D, Hansen J, Strasser A, Acha-Orbea H, Zhan Y, O'Keeffe M, Shortman K (2015) Maintaining dendritic cell viability in culture. Mol Immunol 63(2):264–267

8. Maraskovsky E, Brasel K, Teepe M, Roux ER, Lyman SD, Shortman K, McKenna HJ (1996) Dramatic increase in the numbers of functionally mature dendritic cells in Flt3 ligand-treated mice: multiple dendritic cell subpopulations identified. J Exp Med 184(5):1953–1962

9. Tsapogas P, Swee LK, Nusser A, Nuber N, Kreuzaler M, Capoferri G, Rolink H, Ceredig

R, Rolink A (2014) In vivo evidence for an instructive role of fms-like tyrosine kinase-3 (FLT3) ligand in hematopoietic development. Haematologica 99(4):638–646

10. Ebihara S, Endo S, Ito K, Ito Y, Akiyama K, Obinata M, Takai T (2004) Immortalized dendritic cell line with efficient cross-priming ability established from transgenic mice harboring the temperature-sensitive SV40 large T-antigen gene. J Biochem 136(3):321–328

11. Mohty M, Gaugler B, Olive D (2003) Generation of leukemic dendritic cells from patients with acute myeloid leukemia. Methods Mol Biol 215:463–471

12. Mortellaro A, Urbano M, Citterio S, Foti M, Granucci F, Ricciardi-Castagnoli P (2009) Generation of murine growth factor-dependent long-term dendritic cell lines to investigate host-parasite interactions. Methods Mol Biol 531:17–27

13. Ruiz S, Beauvillain C, Mevelec MN, Roingeard P, Breton P, Bout D, Dimier-Poisson I (2005) A novel CD4–CD8alpha+CD205+CD11b– murine spleen dendritic cell line: establishment, characterization and functional analysis in a model of vaccination to toxoplasmosis. Cell Microbiol 7(11):1659–1671

14. Santegoets SJ, van den Eertwegh AJ, van de Loosdrecht AA, Scheper RJ, de Gruijl TD (2008) Human dendritic cell line models for DC differentiation and clinical DC vaccination studies. J Leukoc Biol 84(6):1364–1373

15. Shen Z, Reznikoff G, Dranoff G, Rock KL (1997) Cloned dendritic cells can present exogenous antigens on both MHC class I and class II molecules. J Immunol 158(6):2723–2730

16. van Helden SF, van Leeuwen FN, Figdor CG (2008) Human and murine model cell lines for dendritic cell biology evaluated. Immunol Lett 117(2):191–197

17. Winzler C, Rovere P, Rescigno M, Granucci F, Penna G, Adorini L, Zimmermann VS, Davoust J, Ricciardi-Castagnoli P (1997) Maturation stages of mouse dendritic cells in growth factor-dependent long-term cultures. J Exp Med 185(2):317–328

18. Duval A, Fuertes Marraco SA, Schwitter D, Leuenberger L, Acha-Orbea H (2014) Large T antigen-specific cytotoxic T cells protect against dendritic cell tumors through perforin-mediated mechanisms independent of CD4 T cell help. Front Immunol 5:338

19. Fuertes Marraco SA, Grosjean F, Duval A, Rosa M, Lavanchy C, Ashok D, Haller S, Otten LA, Steiner QG, Descombes P, Luber CA, Meissner F, Mann M, Szeles L, Reith W, Acha-Orbea H (2012) Novel murine dendritic cell lines: a powerful auxiliary tool for dendritic cell research. Front Immunol 3:331

20. Fuertes Marraco SA, Scott CL, Bouillet P, Ives A, Masina S, Vremec D, Jansen ES, O'Reilly LA, Schneider P, Fasel N, Shortman K, Strasser A, Acha-Orbea H (2011) Type I interferon drives dendritic cell apoptosis via multiple BH3-only proteins following activation by PolyIC in vivo. PLoS One 6(6):e20189

21. Steiner QG, Otten LA, Hicks MJ, Kaya G, Grosjean F, Saeuberli E, Lavanchy C, Beermann F, McClain KL, Acha-Orbea H (2008) In vivo transformation of mouse conventional CD8alpha+ dendritic cells leads to progressive multisystem histiocytosis. Blood 111(4):2073–2082

Part III

Purification and Isolation of Dendritic Cell Subsets

Chapter 4

Isolation of Dendritic Cell Progenitor and Bone Marrow Progenitor Cells from Mouse

Nobuyuki Onai and Toshiaki Ohteki

Abstract

Dendritic cells (DCs) comprise two major subsets, conventional DC (cDC) and plasmacytoid DC (pDC) in the steady-state lymphoid organ. These cells have a short half-life and therefore, require continuous generation from hematopoietic stem cells and progenitor cells. Recently, we identified DC-restricted progenitors called common DC progenitors (CDPs) in the bone marrow of mouse. The CDPs can be isolated from mouse bone marrow based on the hematopoietic cytokine receptors, such as Flt3 (Fms-related tyrosine kinase 3) (CD135), c-kit (CD117), M-CSF (macrophage colony-stimulating factor) receptor (CD115), and IL-7 (interleukin-7) receptor-α (CD127). The CDPs comprise of two progenitors, CD115$^+$ CDPs and CD115$^-$ CDPs, and give rise to only DC subsets in both in vitro and in vivo. The former CDPs are the main source of cDC, while the later CDPs are the main source of pDC in vivo. Here, we provide a protocol for the isolation of dendritic cell progenitor and bone marrow progenitor cells from mouse.

Key words Conventional dendritic cells (cDCs), Plasmacytoid DCs (pDCs), Common DC progenitors (CDPs), Cytokine receptor

1 Introduction

Dendritic cells (DCs) are professional antigen-presenting cells and are essential for the induction and maintenance of immunity [1, 2]. Several subsets of DCs have been identified in the lymphoid and nonlymphoid tissues. There are two major DC subsets in the lymphoid tissue such as conventional DCs (cDCs) and plasmacytoid DCs (pDCs).

Recently, a new nomenclature has been proposed, such as pDC1 and pDC2, based on the ontogeny of the cells and their functions. cDC1 is a classical CD8α^+ cDCs which is a dependent Batf3, while cDC2 is inclusive of CD8α^- cDCs and CD4$^-$CD8α^- cDCs [3].

All DC subsets have a short half-life and do not proliferate in the surrounding environment; therefore, it is essential to continuously generate them from hematopoietic stem cells via progenitors [4, 5].

Elodie Segura and Nobuyuki Onai (eds.), *Dendritic Cell Protocols*, Methods in Molecular Biology, vol. 1423,
DOI 10.1007/978-1-4939-3606-9_4, © Springer Science+Business Media New York 2016

Recently, we identified DC-restricted progenitors, such as common DC progenitors, CDPs, which give rise to only DC subsets both in vitro and in vivo, and not to other cell lineages; they also process high-proliferation capacity. The CDPs comprise of two progenitors, such CD115+ and CD115- CDPs. The former is a main source of cDCs, while the later is the main source of pDCs in the steady state [6, 7]. CDPs differentiate into cDC subsets via pre-cDCs [8, 9] and mature CCR9+ pDCs via CCR9- pDCs [7, 10]. It has been shown that the CDPs are derived from macrophage and DC progenitors (MDPs), which give rise to monocytes, macrophages, and DC subsets [9, 11]. Recently, we found lymphoid-primed multipotent progenitor (LMPPs) directly giving rise to and MDPs in vivo and revised the load map for DC development [7]. Here, we provide a protocol for the isolation of dendritic cell progenitor and bone marrow progenitor cells from mouse.

2   Materials

2.1   Preparation of Bone Marrow Cell (BMCs)

1. C57BL/6 mice, 8–12 weeks old.
2. 70% ethanol.
3. Phosphate-buffered saline (PBS).
4. 10 ml syringes with 19 G needles.
5. Mortar and pestle.
6. Nylon meshes (150 μm pore size).
7. Histopague-1077 (Sigma-Aldrich).
8. 15 and 50 ml Falcon tubes.

2.2   Isolation of Lineage Negative Cells from BMCs

1. PE-Cy5-conjugated antibodies against lineage antigens. For DC and BM progenitors isolation: CD3ε (145-2C11); CD4 (GK1.5); CD8α (53-6.7); B220 (RA3-6B2); CD19 (MB19-1); CD11b (M1/70); CD11c (N418); I-A/I-E (M-15/114.15.2);Gr-1 (RB6-8C5); TER119 (TER119); NK1.1 (PK136). For pre-cDC isolation: CD3ε (145-2C11); CD4 (GK1.5); CD8α (53-6.7); B220 (RA3-6B2); CD19 (MB19-1); CD11b (M1/70);Gr-1 (RB6-8C5); TER119 (TER119); NK1.1 (PK136).
2. Staining buffer: PBS 1% fetal calf serum (FCS), 2 mM EDTA.
3. Anti-Cy5/Anti-Alexa Flour 647 microbeads (Miltenyi Biotec).
4. AutoMACS Pro Separator (Miltenyi Biotec).

2.3   Antibody Staining and Cell Sorting for DC and BM Progenitors

1. Staining buffer. Store at 4 °C.
2. Primary antibodies: FITC-conjugated anti-CD34 (RAM34), PE-conjugated anti-CD135 (A2F10.1), APC-conjugated anti-CD117 (ACK2), Brilliant Violet 421-conjugated anti-CD127 (A7R34), and biotin-conjugated anti-CD115 (AFS-98).

3. Streptavidin-APC-Cy7.

4. Propidium iodide solution (1000×). Dissolve at 10 mg/ml in PBS and store at 4 °C in the dark (*see* **Note 1**).

5. FCS-IMDM: Iscove's Modified Dulbecco's Medium (IMDM) supplemented with 10% FCS, 100 U/ml penicillin, 100 μg/ml streptomycin.

6. Cell sorter: BD FACSAria III (Becton Dickinson Immuno-cytometry Systems).

2.4 Antibody Staining and Cell Sorting for Pre-cDC

1. Staining buffer. Store at 4 °C.

2. Primary antibodies: FITC-conjugated anti-I-A/I-E (M-15/114.15.2), PE-conjugated CD135 (A2F10.1), APC-conjugated anti-CD11c (N418), PE/Cy7-conjugated anti-CD172a (P84).

3. Propidium iodide solution (1000×).

4. FCS-IMDM.

5. Cell sorter: BD FACSAria III (Becton Dickinson Immuno-cytometry Systems).

3 Methods

All procedures should be performed under sterile condition.

3.1 Preparation of Bone Marrow Cells (BMCs)

1. Wet the whole body of the mouse with 70% ethanol for sterilization.

2. Remove the femurs, tibias, ilium, and backbone from five mice, and place them into ice-cold PBS (*see* Fig. 1a).

3. Remove the muscles from the femurs, tibias, ilium, and backbone using scissors and forceps, and transfer them into a new Petri dish containing PBS (*see* Fig. 1b).

4. Add 10 ml of ice-cold PBS into mortar and crush the bones (the femurs, tibias, and ilium) using pestle (*see* Fig. 1c) or add 10 ml of ice-cold PBS into dish, and flush out marrow using syringe with 19 G needle to obtain the bone marrow cell suspension from bone shaft (*see* Fig. 1d) (*see* **Note 2**).

5. Pass the cell suspension through a nylon mesh to remove debris.

6. Add 10 ml of ice-cold PBS into mortar and transfer cleaned backbone. Crush and grind the backbone using the pestle to obtain the spinal marrow (*see* **Note 3**).

7. Pass the cell suspension through a nylon mesh to remove debris.

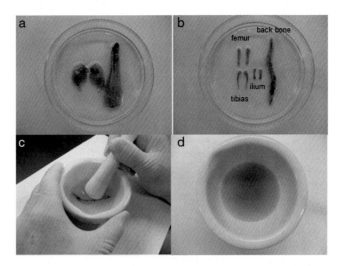

Fig. 1 Preparation of cell suspension from the femur, tibias, ilium, and the backbone. (**a**) Isolated legs and backbone from mouse. (**b**) Isolated femurs, tibias, ilium, and backbone after removal of excess muscle and fat. (**c**) Crush and grind the backbone using pestle to obtain the spinal marrow. (**d**) Cell suspension from the backbone

8. Mix bone marrow and spinal marrow cell suspensions, and centrifuge 5 min at $400 \times g$ at room temperature.

9. During centrifugation, add 5 ml of room temperature histopaque-1077 into a 15 ml tube.

10. Remove the supernatant and resuspend the cells in 5 ml of PBS at room temperature.

11. Carefully overlay the 5 ml of cell suspension onto histopaque-1077.

12. Centrifuge for 30 min at 18 °C at $900 \times g$ with acceleration and brakes set to "zero."

13. After centrifugation, carefully aspirate the uppermost layer. Subsequently transfer the intermediate mononuclear cell layer into a new tube.

14. Wash the cells with an excess of ice-cold PBS (5 ~ 10× volume). and centrifuge for 5 min at 4 °C at $400 \times g$.

15. Resuspend the cells in PBS, and count them.

3.2 Isolation of Lineage-Negative Cells from BMCs

1. Centrifuge cell suspension at $400 \times g$ for 5 min at 4 °C, and aspirate the supernatant.

2. Add to the cells the appropriate PE-Cy5-conjugated antibody cocktail against lineage antigens, mix well.

3. Incubate for 30 min at 4 °C in the dark.

4. Wash the cells with ice-cold staining buffer in excess (5~10× of volume), centrifuge at $400 \times g$ for 5 min at 4 °C, and aspirate the supernatant.

5. Resuspend the cell in staining buffer, and add appropriate volume of anti-Cy5/Anti-Alexa Flour 647 microbeads according to manufacturer's instructions.

6. Incubate for 15 min at 4 °C in the dark.

7. Wash the cells with ice-cold staining buffer in excess, centrifuge at $400 \times g$ for 5 min at 4 °C, and aspirate the supernatant.

8. Resuspend the cells in staining buffer. Proceed with magnetic separation to obtain lineage-negative cell fraction using AutoMACSPro Separator according to manufacturer's instructions.

9. Proceed to Subheading 3.3 or 3.4.

3.3 Antibody Staining and Cell Sorting for DC and BM Progenitors

1. Centrifuge the lineage-negative cell suspension at $400 \times g$ for 5 min at 4 °C, and aspirate the supernatant.

2. Add primary antibody mix to the cell suspension, mix well.

3. Incubate for 30 min at 4 °C in the dark.

4. Wash the cells with ice-cold staining buffer in excess and centrifuge for 5 min at $400 \times g$, and aspirate the supernatant.

5. Add the streptavidin to the cells, mix well.

6. Incubate for 30 min at 4 °C in the dark.

7. Wash the cells with ice-cold staining buffer in excess, centrifuge for 5 min at $400 \times g$, and aspirate the supernatant.

8. Resuspend the cells in staining buffer containing Propidium iodide (final concentration 10 µg/ml) to stain and exclude dead cells.

9. Prepare tubes containing 1 ml of FCS-IMDM for collecting the sorted target cells.

10. Sort CD115+ CDPs as lin−CD117intCD135+CD115+CD127−, CD115− CDPs as lin−CD117intCD135+CD115−CD127−, MDPs as in−CD117+CD135+CD115+Sca-1−, and LMPPs as in−CD117+CD135+CD34+Sca-1+ by using a cell sorter (*see* Fig. 2) (*see* **Note 4**).

3.4 Antibody Staining and Cell Sorting for Pre-cDC

1. Centrifuge the lineage-negative cell suspension at $400 \times g$ for 5 min at 4 °C, and aspirate the supernatant.

2. Add the primary antibody mix to the cell suspension, mix well.

3. Incubate for 30 min at 4 °C in the dark.

4. Wash the cells with ice-cold staining buffer in excess and centrifuge for 5 min at $400 \times g$, and aspirate the supernatant.

Figure 2. Onai *et al.*

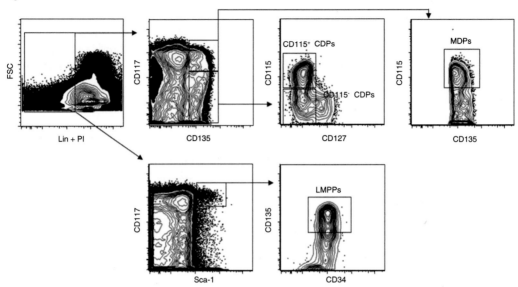

Fig. 2 Isolation of DC progenitors and BM progenitors from BM Lin⁻ cells were divided by CD117, CD135, and Sca-1 expression. Lin⁻CD117⁺CD135⁺ cells and lin⁻CD117^intCD135⁺ cells were further divided by CD115 and CD127 expression. Lin⁻CD117⁺Sca-1⁺ cells were divided by CD34 and CD135 expression. CD115⁺ CDPs, CD115⁻ CDPsMDPs, and LMPPs were defined as lin⁻ CD117^intCD135⁺CD115⁺CD127⁻, lin⁻CD117^intCD135⁺CD115⁻CD127⁻, lin⁻CD117⁺CD135⁺CD115⁺Sca-1⁻, and lin⁻CD117⁺Sca-1⁺CD135⁺CD34⁺, respectively

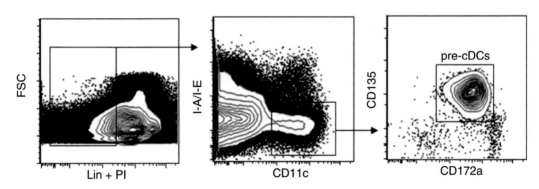

Fig. 3 Isolation of pre-cDC from BM. Lin⁻ cells were divided by CD11c, I-A/I-E, CD135, and CD172a. pre-cDCs were defined as lin⁻CD11c⁺I-A/I-E⁻CD135⁺CD172a^int cells

5. Resuspend the cells in staining buffer containing Propidium iodide (final concentration 10 μg/ml) to stain and exclude dead cells.

6. Prepare tubes containing 1 ml of FCS-IMDM for collecting the sorted target cells.

7. Sort pre-cDCs as lin⁻CD11c⁺I-A/I-E⁻CD135⁺CD172a^int cells (*see* Fig. 3).

4 Notes

1. Propidium iodide is light sensitive.

2. In this protocol, we introduced bone marrow cell preparation from the femurs, tibias, ilium, and backbone using mortar and pestle for crushing and grinding the bones. Using this method, the total number of bone marrow cells from the mice is increased. There is no functional difference between the progenitors isolated from the femurs, tibias, ilium, and backbone.

3. Remove and discard the white funiculus that will be extracted as well during the crushing.

4. Sorted CD115+ CDPs, CD115− CDPs, MDPs, and LMPPs are cultured in FCS-IMDM supplemented with Flt3-ligand (10 ng/ml). The progenies derived from these progenitors are analyzed on day 8 after culture.

References

1. Banchereau J, Steinman RM (1998) Dendritic cells and the control of immunity. Nature 392: 245–252

2. Shortman K, Naik SH (2007) Steady-state and inflammatory dendritic-cell development. Nat Rev Immunol 7:19–30

3. Guilliams M, Ginhoux F, Jakubzick C, Naik SH, Onai N, Schraml BU, Segura E, Tussiwand R, Yona S (2014) Dendritic cells, monocytes and macrophages: a unified nomenclature based on ontogeny. Nat Rev Immunol 14:571–578

4. Liu YJ (2005) IPC: professional type 1 interferon-producing cells and plasmacytoid dendritic cell precursors. Annu Rev Immunol 23:275–306

5. Kondo M, Wagers AJ, Manz MG, Prohaska SS, Scherer DC, Beilhack GF, Shizuru JA, Weissman IL (2003) Biology of hematopoietic stem cells and progenitors: implications for clinical application. Annu Rev Immunol 21: 759–806

6. Onai N, Obata-Onai A, Schmid MA, Ohteki T, Jarrossay D, Manz MG (2007) Identification of clonogenic common Flt3+M-CSFR+ plasmacytoid and conventional dendritic cell progenitors in mouse bone marrow. Nat Immunol 8:1207–1216

7. Onai N, Kurabayashi K, Hosoi-Amaike M, Toyama-Sorimachi N, Matsushima K, Inaba K, Ohteki T (2013) A clonogenic progenitor with prominent plasmacytoid dendritic cell developmental potential. Immunity 38:943–957

8. Naik SH, Metcalf D, van Nieuwenhuijze A, Wicks I, Wu L, O'Keeffe M, Shortman K (2006) Intrasplenic steady-state dendritic cell precursors that are distinct from monocytes. Nat Immunol 7:663–671

9. Liu K, Victora GD, Schwickert TA, Guermonprez P, Meredith MM, Yao K, Chu FF, Randolph GJ, Rudensky AY, Nussenzweig M (2009) In vivo analysis of dendritic cell development and homeostasis. Science 324: 392–397

10. Schlitzer A, Loschko J, Mair K, Vogelmann R, Henkel L, Einwacher H, Schiemann M, Niess JH, Reindl W, Krug A (2011) Identification of CCR9− murine plasmacytoid DC precursors with plasticity to differentiate into conventional DCs. Blood 117:6562–6570

11. Fogg DK, Sibon C, Miled C, Jung S, Aucouturier P, Littman DR, Cumano A, Geissmann F (2006) A clonogenic bone marrow progenitor specific for macrophages and dendritic cells. Science 311:83–87

Chapter 5

The Isolation and Enrichment of Large Numbers of Highly Purified Mouse Spleen Dendritic Cell Populations and Their In Vitro Equivalents

David Vremec

Abstract

Dendritic cells (DCs) form a complex network of cells that initiate and orchestrate immune responses against a vast array of pathogenic challenges. Developmentally and functionally distinct DC subtypes differentially regulate T-cell function. Importantly it is the ability of DC to capture and process antigen, whether from pathogens, vaccines, or self-components, and present it to naive T cells that is the key to their ability to initiate an immune response. Our typical isolation procedure for DC from murine spleen was designed to efficiently extract all DC subtypes, without bias and without alteration to their in vivo phenotype, and involves a short collagenase digestion of the tissue, followed by selection for cells of light density and finally negative selection for DC. The isolation procedure can accommodate DC numbers that have been artificially increased via administration of fms-like tyrosine kinase 3 ligand (Flt3L), either directly through a series of subcutaneous injections or by seeding with an Flt3L secreting murine melanoma. Flt3L may also be added to bone marrow cultures to produce large numbers of in vitro equivalents of the spleen DC subsets. Total DC, or their subsets, may be further purified using immunofluorescent labeling and flow cytometric cell sorting. Cell sorting may be completely bypassed by separating DC subsets using a combination of fluorescent antibody labeling and anti-fluorochrome magnetic beads. Our procedure enables efficient separation of the distinct DC subsets, even in cases where mouse numbers or flow cytometric cell sorting time is limiting.

Key words Dendritic cell, DC purification, DC expansion, DC subtypes, Conventional DCs, Plasmacytoid DCs

1 Introduction

Dendritic cells (DCs) were identified in mouse spleen in the 1970s by Ralph Steinman [1]. They are bone marrow-derived cells that are found at low frequency in all tissues and are indispensable to the immune system. DCs have the ability to recognize bacteria, virus, fungi, and other immunogenic agents via numerous pattern recognition receptors, taking up and processing the antigen into peptide form and presenting it to naive T cells in order to initiate an antigen-specific immune response [2]. Mice deficient in DC are

Elodie Segura and Nobuyuki Onai (eds.), *Dendritic Cell Protocols*, Methods in Molecular Biology, vol. 1423,
DOI 10.1007/978-1-4939-3606-9_5, © Springer Science+Business Media New York 2016

incapable of mounting an immune response to pathogens [3–5]. DCs also play a pivotal role in controlling reactivity to self-antigen in the absence of pathogens or inflammation. They maintain self-tolerance in the thymus by presenting self-antigen and either deleting developing self-reactive T cells or rendering them unresponsive and in the periphery deleting mature T cells [6, 7] and inducing the expansion of regulatory T cells [8].

By the 1990s, it was evident that murine lymphoid organ DC could be divided into two subsets based on expression of CD8α [9, 10]. These two subsets were subsequently shown to have different immune functions and were the first indication of the vast functional diversity of DC [11]. Spleen DCs are a heterogeneous group which differ in the antigen processing and presenting pathways they use and the role they play in immunity. The surface phenotype and abilities of DC vary between subtypes [12, 13] and change as they undergo the process of maturation [14].

The vast majority of splenic DCs are in an immature state as indicated by their low-level expression of major histocompatibility complex class II (MHCII) molecules and co-stimulatory molecules. In this state they are extremely efficient at mediating endocytosis and processing of antigen and its traffic to the cell surface for presentation to T cells. Upon exposure to microbial patterns or other "danger" signals, DCs undergo a process of maturation in which they cease uptake and processing of antigen. MHC and various membrane-associated co-stimulatory molecules are up-regulated, as is the production of many cytokines and chemokines. In this form they are able to induce an effector T-cell response [15].

All mature DCs are very efficient at presenting endogenous antigen; however, presentation of exogenous antigen is largely restricted to MHC class II molecules [16]. Cross presentation is a specialized function which allows particular DC subsets to present exogenous antigen via MHC class I [17] in a role which is vital to antiviral and antitumor T-cell responses and in inducing and maintaining tolerance [18].

Due to the vast heterogeneity of the DC network, our aim when devising an isolation and purification procedure was not only to ensure that DC from lymphoid tissue could be adequately and efficiently extracted and purified regardless of their maturation state, surface phenotype, and function, but to do so in a way that conserved their in vivo state. Accidental induction of DC maturation gives an inaccurate view of their in vivo steady-state form.

Early protocols depended on overnight culture at 37 °C to separate transiently adherent splenic DC from adherent macrophages. This inadvertently induced a maturation process resulting in isolated DCs that were no longer equivalent to their in vivo steady-state form, but rather resembled mature DC [14]. Subsequently even short periods at 37 °C were shown to induce maturation of DC, particularly under conditions of high DC concentration, due to these very rare cells coming into close proximity of each other [19].

We designed our isolation procedure with this limitation in mind. Our collagenase digestion of tissue is a mild treatment performed for a relatively short time at room temperature, and has been shown not to activate DC based on expression of MHCII and co-stimulatory molecules, which remain at similar levels to those found on the small number of DC that can be isolated without collagenase at 4 °C [19]. The digestion extracts all DC subtypes without any observed bias. After the digestion, DCs constitute less than 1 % of the extracted splenocytes. A short treatment with EDTA ensures that all multicellular complexes between DC and T cells are dissociated, before the first enrichment step is performed. Selection for light-density cells is the first DC enrichment step. It results in the removal of erythrocytes and dead cells as well as many other contaminating lineages. DCs account for 10–15 % of light-density cells, and it is possible to isolate them directly at this stage using either flow cytometric cell sorting or positive selection using immunomagnetic beads. However, it is more economical to enrich further via negative selection. Non-DC lineage cells are coated with a cocktail of monoclonal antibodies and depleted with anti-immunoglobulin coated immunomagnetic beads. Care must be taken when selecting the monoclonal antibodies included in the cocktail, in order to avoid losing DC subtypes that bear molecules found more commonly on T cells, B cells, and macrophages. Residual contaminating autofluorescent macrophages and NK cells are often a problem [20, 21]. Ideally, inclusion of monoclonal antibodies specific for these cells in the depletion cocktail would be the preferred solution. Unfortunately, the candidate macrophage molecules that could potentially be targeted are also expressed on some DC subtypes and so cannot be used, and we have not been able to find an effective NK cell-specific monoclonal antibody that is also a rat IgG. Nevertheless, both can be removed during fluorescence activated cell sorting or analysis by gating out autofluorescent cells and CD49b+ cells, respectively. Alternatively NK cells may be removed by a second negative selection step. After depletion, spleen DCs account for approximately 90 % of the recovered cells (*see* Fig. 1) and are a mixture of migratory plasmacytoid DC (pDC) and lymphoid resident conventional (or classical) DC (cDC). These two subtypes are easily distinguished and separated using multicolor immunofluorescent staining and flow cytometry.

In the steady state, mouse pDCs are produced in the bone marrow and migrate to lymphoid organs including the spleen, via the blood. They are characterized by low expression of the integrin CD11c and high expression of either of the pDC-specific molecules siglec H or, in the steady state, CD317 and are enriched during DC isolation if the appropriate cocktail of monoclonal antibodies is used in the depletion. CD45R and CD45RA have been used in

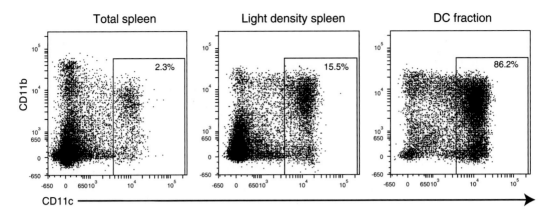

Fig. 1 Sequential enrichment of spleen cDC. The proportion of cDC is increased from approximately 2 % of total leucocytes after the digestion up to approximately 90 % of all cells after depletion of non-DC lineage cells. cDCs are identified as CD11c^{high} cells

conjunction with CD11c to identify pDC in the past, but this may be problematic if no additional NK cell depletion is planned, as some NK cells express significant levels of these molecules and may inadvertently be included in any pDC gate that is set (*see* Fig. 2). pDCs account for approximately 15 % of DC in mouse spleen and express high levels of toll-like receptors 7 and 9, which when ligated by viral antigens lead to production of high levels of IFN-α and IFN-λ; upregulation of MHC class II, CD40, CD69, CD80, and CD86; acquisition of typical DC morphology; and the ability to present foreign antigen [22].

cDCs populate most lymphoid and nonlymphoid tissue. All cDCs in the spleen are tissue resident and have developed in the spleen from a blood-borne precursor [23]. They are found in the marginal zone where they constantly acquire tissue and blood antigen. Spleen cDCs are distinguished from, and can be separated from, pDC by higher expression of CD11c and the absence of siglec H, CD317, CD45R, and CD45RA [24].

Early observation of CD8 on the surface of spleen cDC made it possible to begin dividing spleen cDC into two subsets: CD8$^+$ and CD8$^-$. Although CD8 was adequate for cDC identification in mouse lymphoid tissue, it did not translate well to use in cDC in nonlymphoid tissue or in humans. The introduction of new markers provides a more complete characterization of the two populations. The CD8$^+$ cDC can be defined as CD8$^+$CD11b$^-$CD24^{high}CD205$^+$CD4$^-$CD172a^{low}Clec9A$^+$, while the CD8$^-$ cDC are CD8$^-$CD11b$^+$CD24^{low}CD205$^-$CD4$^{-/+}$CD172a^{high}Clec9A$^-$ [25] (*see* Fig. 3a). Current terminology refers to these two populations as CD8$^+$ and CD11b$^+$ cDC.

CD8$^+$ cDCs account for approximately 25 % of mouse spleen DC, respond to TLR3 stimulation [26], secrete IL-12p70 [27] and IFN-λ [28], and are essential for cross-presenting antigen

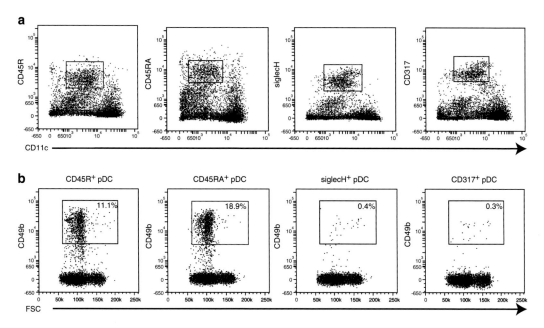

Fig. 2 NK cell contamination of isolated pDC. (**a**) cDCs are identified as CD11c^{high} and negative for either of CD45R, CD45RA, siglec H, or CD317, while pDCs are CD11c^{low} and positive for either of CD45R, CD45RA, siglec H, or CD317. (**b**) NK cells, identified as CD49b$^+$, are significant contaminants of pDC preparations separated using CD45R or CD45RA, contributing up to 19 % of recovered cells. NK cells may be gated out during flow cytometric sorting or analysis, or by depleting using a combination of anti-CD49b conjugated to biotin and anti-biotin-coated magnetic beads. Siglec H and CD317 are better alternatives for separating pDC from NK cells with very low NK cell contamination detected in pDC preparations isolated using either

[29, 30]. Expression of Clec9A on CD8$^+$ cDC is of great importance (*see* Fig. 3b). Clec9A is a receptor for necrotic tissue which binds and directs tissue antigens associated with this tissue to the cross-presentation pathway [31, 32]. Further, Clec9A allows us to align the CD8$^+$ cDC subtype in mouse with its functional human equivalent, which does not express CD8 [33]. Terminal differentiation of CD8$^+$ cDC is dependent on GM-CSF and results in expression of CD103 and the acquisition of cross-presentation functions [34, 35].

The CD11b$^+$ cDCs, which account for up to 60 % of DC in the mouse spleen, are a heterogeneous population and less well characterized. They may be further divided into two populations using CD4, the endothelial cell-selective adhesion molecule (ESAM) [36] and the C-type lectin Clec12A (*see* Fig. 3c). The CD11b$^+$CD4$^+$ subtype is Clec12A$^-$ and largely ESAM$^+$ and is essential for presentation of MHCII-antigen complexes to CD4$^+$ T cells [37]. The CD11b$^+$CD4$^-$ subtype is Clec12A$^+$ and ESAM$^-$ and is the superior producer of inflammatory cytokines such as CCL3, CCL4, and CCL5 after TLR engagement [38].

The rarity of splenic DC and the labor-intensive process required to isolate them in sufficient numbers have limited their

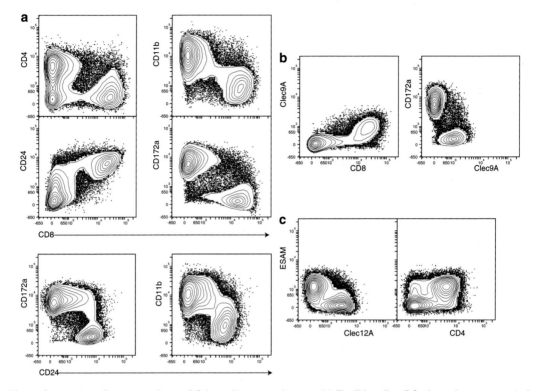

Fig. 3 Separation of mouse spleen cDC into discrete subtypes. (**a**) Traditionally cDCs have been separated using CD8 and CD4, but alignment of spleen cDC using neither CD8 or CD4, is best accomplished using CD24 and either CD11b or CD172a. The CD24^{high}CD11b$^-$CD172a^{low} and the CD24^{low}CD11b$^+$CD172a^{high} populations correspond to the CD8$^+$CD4$^-$ and CD8$^-$CD4$^+$ subtypes, respectively. (**b**) Clec9A can also be used to distinguish the traditional CD8$^+$ and CD8$^-$ cDC. Clec9A$^+$ cDCs are CD8$^+$CD172a^{low}, and Clec9A$^-$ cDCs are CD8$^-$CD172a^{high}. (**c**) The CD8$^-$ cDC subtype can be divided further using CD4, ESAM, and Clec12A. The traditional CD4$^+$CD8$^-$ cDCs are ESAM$^+$Clec12A^{low}, and the CD4$^-$CD8$^-$ cDCs are largely ESAM$^-$Clec12A$^+$

availability for study. The cytokine fms-like tyrosine kinase 3 ligand (Flt-3L) and its receptor (Flt-3) play an essential role in the commitment of hematopoietic precursors to the DC lineage and their subsequent development [39]. Flt-3L is ubiquitously produced by multiple tissue stroma, endothelial cells, and activated T cells and drives DC differentiation from both mouse and human precursors [40]. Mice deficient in Flt-3 and Flt-3L have reduced numbers of both DC precursors and lymphoid tissue pDC and cDC [41–43].

Administration of Flt-3L to mice acts on the Flt-3$^+$ fraction of bone marrow precursors [44] to produce large numbers of the equivalents of the steady-state mouse DC populations either in vivo or in vitro.

In vivo methods involve the administration of daily intravenous Flt-3L injections for a period of 10 days or subcutaneous injection of the Flt-3L secreting melanoma B16FLT3L. In both cases, total spleen cellularity increases up to threefold, but spleen DC numbers can be boosted up to 20-fold. DC subtypes in the spleen of mice treated in this way are identified and separated using the same markers used in spleen from untreated mice; however, there is a larger expansion of the CD8$^+$ subtype [45, 46].

Large numbers of pDC and cDC can also be produced in in vitro cultures of bone marrow treated with Flt-3L [47, 48]. cDCs generated in these cultures do not express CD8 or CD4, so these markers are inadequate for alignment of the culture-generated cDC with the subtypes found in vivo. Equivalents of the CD8$^+$ and CD11b$^+$ cDC can be identified using the markers CD11b, CD24, CD172a, and Clec9A. The numbers of DC generated in these cultures can be boosted further by an initial addition of a small amount of GM-CSF. Addition of a second larger dose of GM-CSF after 6 days of culture preferentially expands the CD103$^+$ cDC population which is the population in these cultures capable of cross presentation [34] (*see* Fig. 4).

Subpopulations of highly purified cDCs have traditionally been obtained by flow cytometric sorting. Sorting, however, is both expensive and time consuming. An alternative method, resulting in the isolation of highly purified cDC subtypes, involves positive selection using anti-fluorochrome-conjugated magnetic beads. DCs of one subtype are stained with a specific fluorochrome-conjugated monoclonal antibody, allowed to bind to anti-fluorochrome coated magnetic beads, and then selected using a magnet. If the other subtype is also required, the negative fraction may be stained with a monoclonal antibody, conjugated to another fluorochrome, that is specific for this subtype and the positive selection process repeated. In this way, it is possible to sequentially select the two cDC subtypes [49, 50] (*see* Fig. 5).

A similar process may be used to remove contaminating NK cells. Biotin-conjugated anti-CD49b and anti-biotin magnetic beads can be used in combination to remove and discard NK cells. Alternatively, NK cells may be removed during flow cytometric sorting.

The ability to isolate purified DC populations in an immature state and unaffected by the isolation procedure, the ability to expand their numbers using Flt-3L, and the design of economical methods to purify their subtypes, are all crucial in the ongoing study of DC function.

2 Materials

2.1 In Vivo Administration of Flt3L

1. Donor mice.

2. 27G needle.

3. MTPBS: Mouse tonicity (308 mOsm/kg) phosphate-buffered saline. Filter through a 0.2 μM filter unit to sterilize and store at 4 °C.

4. Flt3L: Recombinant murine FLAG-tagged fms-like tyrosine kinase 3 ligand (constructs provided by N. Nicola, WEHI, Australia) was expressed in FreeStyle 293F cells (Invitrogen,

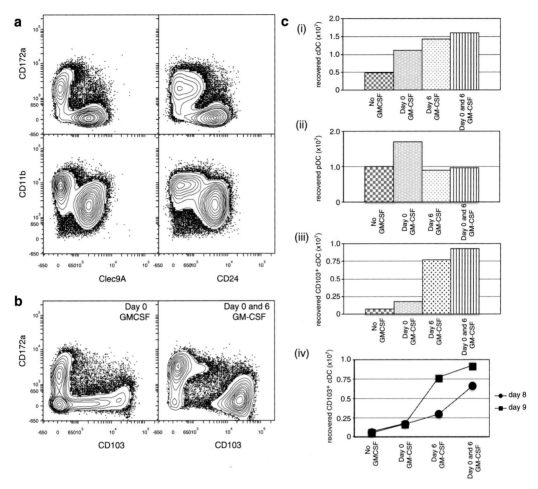

Fig. 4 Generation, separation, and modification of cDC in Flt-3L-supplemented bone marrow cultures. (**a**) cDCs produced in Flt-3L-supplemented bone marrow cultures do not express CD8 or CD4 but may be separated into CD8+CD4− and CD8−CD4+ equivalents using a combination of other molecules, usually CD24 or Clec9A and CD11b or CD172a. (**b**) Flt-3L-generated cDCs are modified by the addition of GM-CSF on day 6 of culture resulting in an increase in both the total number of cells produced and the proportion of CD103+ cross-presenting cDC after 9 days of culture. (**c**) Addition of GM-CSF at day 0, day 6, or both has an effect on numbers and the phenotype of DC recovered after 9 days of culture. (*i*) Numbers of cDC are increased two- to threefold by addition of GM-CSF at day 0 or day 6, but only marginally more if both doses are administered. (*ii*) pDC numbers are increased by addition of GM-CSF at day 0 but not at day 6. pDC production is slightly inhibited if GM-CSF is added on both days. (*iii*) The number of CD103+ cDC is increased fivefold by the addition of GM-CSF at day 6. (*iv*) A similar pattern of results is obtained when analysis is performed after 8 days of culture, but with a lower recovery of total DC (not shown) and CD103+ cDC

Victoria, Australia) by transient transfection using Freestyle Max (Invitrogen) and cultured in protein-free/serum-free media (FreeStyle Expression Media, Invitrogen) for 5 days. Media containing the secreted recombinant protein was concentrated using a 10,000 molecular weight cutoff centrifugal device (Millipore, Billerica, MA). Recombinant Flt3L-FLAG

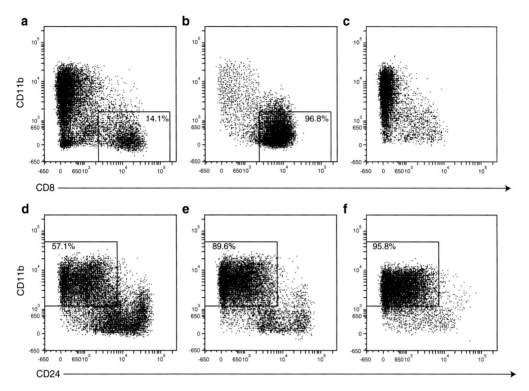

Fig. 5 The sequential separation of CD8$^+$ and CD8$^-$ cDC using fluorochrome-conjugated antibodies and anti-fluorochrome immunomagnetic beads. (**a**) Two major subsets of spleen DC can be visualized using the markers CD8 and CD11b: CD8$^+$CD11b$^-$ and CD8$^-$CD11b$^+$. (**b**) The first round of selection for CD8$^+$ DC results in a highly purified CD8$^+$CD11b$^-$ population. (**c**) The flow through contains a greatly reduced proportion of CD8$^+$ cells. (**d**) The flow through contains a mixture of CD11b$^+$CD24^{low} and CD11b$^-$CD24^{hi} cells. (**e**) A second round of depletion targets CD24 and removes many of the remaining CD8$^+$CD24^{high}CD11b$^-$ but also CD24^{high}CD8$^-$ precursors of the CD8$^+$ cDC lineage. The second flow through contains an enriched CD8$^-$CD24^{low}CD11b$^+$ population. (**f**) The final round of selection utilizes CD11b and results in a greatly enriched CD8$^-$CD24^{low}CD11b$^+$ population

was purified by affinity chromatography using an anti-FLAG M2 agarose resin (Sigma, Castle Hill, Australia) and elution with 100 mg/ml FLAG peptide (Auspep, Victoria, Australia) and further purified by size-exclusion chromatography using a prepacked Superdex 200 column (GE Healthcare, Rydalmere, Australia) (*see* **Note 1**).

5. MSA: Mouse serum albumin (Sigma Aldrich). Dissolve in MTPBS to give a 10 μg/ml stock solution. Filter through a 0.2 μM syringe filter unit to sterilize and store at 4 °C.

6. Flt3L/MSA: Sterile MTPBS containing Flt3L at 100 μg/ml and MSA at 1 μg/ml. Store at 4 °C for a maximum of 10 days.

7. B16FLT3L melanoma: Retroviral-mediated gene transfer generated B16-F10 melanoma line secreting murine Flt3L (J. Villadangos, University of Melbourne, Australia) (*see* **Note 2**).

2.2 Organ Removal

1. FCS: Fetal calf serum. Aliquot and store at −20 °C, or for short periods of time at 4 °C.

2. RPMI-FCS: Modify RPMI-1640 to mouse osmolarity (308 mOsm/kg), and add pH 7.2 HEPES buffering to reduce dependence on CO_2 concentration. Adjust to ~pH 7 with CO_2, sterilize by running through a 0.2 μM filter unit, and store at 4 °C. Add FCS to a final concentration of 2 % before use.

3. Dissecting instruments (scissors and forceps).

2.3 Digestion of Spleen and Release of DC

1. Enzyme digestion mix: Stock solution (7×) prepared by dissolving collagenase type III (Worthington Biochemicals) at 7 mg/ml and Dnase I (Boehringer Mannheim) at 140 μg/ml in RPMI-FCS. Ensure the collagenase used is free of trypsin and other trypsin-like proteases (*see* **Note 3**). Divide into 1 ml aliquots and store frozen as a stock solution at −20 °C. Add each 1 ml aliquot to 6 ml RPMI-FCS before use. Run through a 0.2 μM filter unit to sterilize if required. Use immediately.

2. EDTA solution: 0.1 M ethylenediamine tetra-acetic acid disodium salt adjusted to pH 7.2. Run through a 0.2 μM filter unit to sterilize and store at 4 °C.

3. EDTA-FCS: Add 1 ml of 0.1 M EDTA per 10 ml FCS before use.

2.4 Selection of the Light-Density Fraction of Spleen

1. BSS-EDTA: Modified salt solution containing 150 mM NaCl and 3.75 mM KCl (no Ca^{2+} or Mg^{2+}) and 5 mM EDTA. Adjust to pH 7.2 and mouse osmolarity (308 mOsm/kg). Filter to sterilize using a 0.2 μM filter unit and store at 4 °C.

2. BSS-EDTA-FCS: BSS-EDTA containing 2 % EDTA-FCS.

3. Nycodenz-EDTA: Nycodenz AG powder (Nycomed Pharma AS) is dissolved in water to produce a 0.372 M stock solution and then diluted and adjusted to the desired density of 1.077 g/cm^3 at 4 °C and anosmolarity of 308 mOsm/kg using BSS-EDTA (*see* **Note 4**). Sterilize using a 0.2 μM filter unit and store in 10 ml aliquots at −20 °C. Thaw at room temperature, mix thoroughly, and cool to 4 °C prior to use (*see* **Note 5**).

4. Polypropylene tubes: 14 ml polypropylene round bottom tubes (Becton Dickinson Labware).

2.5 Depletion of Non-DC Lineages

1. Monoclonal antibody depletion cocktail (*see* **Note 6**):
   Combine pre-titrated amounts of rat monoclonal antibodies specific for non-DC lineage cells.

(a) For purification of cDC only: Add KT3-1.1 (anti-CD3ε), T24/31.7 (anti-CD90), TER119 (anti-erythroid lineage), 1A8 (anti-Ly6G), and RA3-6B2 (anti-CD45R).

(b) For purification of both cDC and pDC: Add KT3-1.1 (anti-CD3ε), T24/31.7 (anti-CD90), TER119 (anti-erythroid lineage), 1A8 (anti-Ly6G), and 1D3 (anti-CD19).

Dilute to the appropriate volume with BSS-EDTA-FCS. Saturating levels of all monoclonal antibodies are used (*see* **Note 7**). Run through a 0.2 μM filter to sterilize, aliquot, and store at –20 °C.

2. Immunomagnetic beads: BioMag goat anti-rat IgG coated beads (Qiagen).

3. Spiral rotator: Spiramix 10 (Denley).

2.6 In Vitro Administration of Flt3L

1. Red cell lysis medium (RCLM): 0.168 M NH$_4$Cl. Run through a 0.2 μM filter unit to sterilize and store at 4 °C.

2. Flt3L DC culture medium: Modified RPMI-1640, isoosmotic with mouse serum, with additional HEPES buffering at pH 7.2, supplemented with 10 % FCS, 50 μM 2-mercaptoethanol, and 2 mM l-glutamine. Sterilize by running through a 0.2 μM filter unit and store frozen at –70 °C. Add 200 ng/ml murine Flt3L immediately prior to use (*see* **Note 8**).

3. GM-CSF: Recombinant murine granulocyte macrophage colony-stimulating factor (R&D Systems, Inc.).

2.6.1 Immunofluorescent Staining and Purification via Flow Cytometric Sorting

1. The majority of monoclonal antibodies are purified from hybridoma culture supernatant using Protein G Sepharose (Amersham Biosciences) and subsequently conjugated to fluorochromes in-house. Anti-siglec H (clone eBio440c)-FITC and anti-ESAM (clone 1G8)-PE conjugates are purchased from eBioscience. They are all titrated to determine saturating levels.

2. mAbs are conjugated to fluorochromes and biotin following the manufacturer's instructions:

(a) Conjugate P84 (anti-CD172a), 14.8 (anti-CD45RA), RA36B2 (anti-CD45R), and 120G8 (anti-CD317) to FITC (Molecular Probes, Inc.).

(b) M2/90 (anti-CD103) to phycoerythrin (PE) (ProZyme).

(c) M1/70 (anti-CD11b) to AlexaFluor680 (Molecular Probes, Inc.).

(d) N418 (anti-CD11c) and YTS169.4 (anti-CD8) to allophycocyanin (APC) (ProZyme).

(e) M1/69 (anti-CD24) to Pacific Blue (Molecular Probes, Inc.).

(f) N418 (anti-CD11c) to PerCp.Cy5.5 (Innova Biosciences).

(g) GK1.5 (anti-CD4), 10B4 (anti-Clec9A), and 5D3 (anti-Clec12A) to biotin (Molecular Probes, Inc.).

Add FCS to 1% and NaN_3 to a final concentration of 10 mM (*see* **Note 9**). Titrate to determine saturating levels. Aliquot stocks of FITC, biotin, AlexaFluor680, and Pacific Blue conjugates, and store at −70 °C. Stocks of PE, APC, and PerCp. Cy5.5 conjugates (*see* **Note 10**) and working stocks of FITC, biotin, AlexaFluor680, and Pacific Blue conjugates are stored at 4 °C, protected from light. Dilute to their final working concentration immediately prior to use.

3. PI: 100 μg/ml propidium iodide (Calbiochem) stock solution in normal saline. Aliquot and store at 4 °C protected from light (*see* **Note 11**).

4. BSS-EDTA-FCS-PI: Working solution of PI made by diluting the PI stock in BSS-EDTA-FCS to a final working concentration of 500 ng/ml before addition to cells.

5. FACSAria (BD Biosciences).

2.6.2 Immunofluorescent Staining and Purification via Immunomagnetic Beads

1. Fluorochrome-conjugated mAb:

   (a) Conjugate YTS169.4 and M1/69 to FITC (Molecular Probes, Inc.).

   (b) Conjugate M1/70 to phycoerythrin (PE) (ProZyme).

2. Anti-fluorochrome beads (Miltenyi Biotec):

   (a) Anti-FITC microbeads.

   (b) Anti-PE microbeads.

3. BSS-EDTA-0.5%FCS: BSS-EDTA   containing   0.5% EDTA-FCS.

4. MACS column (Miltenyi Biotec):

   (a) MACS LS column (×2).

   (b) MACS LD column.

5. Magnet and stand (Miltenyi Biotec):

   (a) Mini MACS magnet.

   (b) MACS multistand.

2.6.3 Staining and Purification via Immunomagnetic Beads: Removal of NK Cells

1. Conjugated mAb: Conjugate DX5 to biotin (Molecular Probes, Inc.).

2. Immunomagnetic beads (Miltenyi Biotec): anti-biotin microbeads.

3. BSS-EDTA-0.5%FCS: BSS-EDTA containing 0.5% EDTA-FCS.

4. MACS column (Miltenyi Biotec): MACS LD column.

5. Magnet and stand (Miltenyi Biotec):

    (a) Mini MACS magnet.

    (b) MACS multistand.

3 Methods

3.1 In Vivo Administration of Flt3L

3.1.1 Soluble Flt3L

1. Inject each mouse daily for 10 days with 100 μl of Flt3L/MSA subcutaneously into the nape of the neck using a 27G needle (*see* **Note 12**).

3.1.2 B16Flt3L

1. Maintain the B16Flt3L melanoma in in vitro culture in RPMI-FCS.

2. Pour off the supernatant from the culture flasks, and wash with 5 ml of MTPBS.

3. Pour off the MTPBS, add 2 ml of trypsin, and treat at 37 °C for 2 min.

4. Shake the flask to dislodge adherent cells and remove and pool cells from multiple flasks.

5. Centrifuge, resuspend the pellet in MTPBS, and count.

6. Dilute cells at 25×10^6/ml in MTPBS.

7. Inject 200 μl (5×10^6) cells per mouse subcutaneously into the nape of the neck using a 27G needle (*see* **Note 13**).

8. Monitor injected mice daily (*see* **Note 14**).

9. Allow melanoma to grow for up to 10 days (*see* **Note 15**).

3.2 Organ Removal

1. Remove spleens from eight untreated mice or from two Flt-3L-treated mice (*see* **Note 16**) into cold RPMI-FCS, taking care to remove the organs with as little fat and connective tissue as possible.

3.3 Digestion of Spleen and Release of DC

1. Prepare the enzyme digestion mix slightly ahead of time, and allow it to warm to room temperature before use.

2. Remove any remaining fat and/or connective tissue (*see* **Note 17**) from the spleens using two 20G needles, and transfer spleens to a small Petri dish containing 7 ml of enzyme digestion mix. Use a sharp pair of scissors or a single-sided razor blade to cut the tissue into very small fragments (*see* **Note 18**). Transfer the fragments to a 10 ml tube using a wide-bore Pasteur pipette. Mix frequently, using the same pipette, while digesting the tissue for 20–25 min at room temperature (~22 °C) (*see* **Note 19**).

3. Add 600 μl of EDTA solution to the digestion mix, and continue the incubation for a further 5 min (*see* **Note 20**).

4. Run the digestion mix through a sieve to remove any remaining undigested tissue. Discard anything caught in the sieve. Dilute the digestion mix to 9 ml with RPMI-FCS, underlay with 1 ml of cold FCS-EDTA, and centrifuge to recover the cells (*see* **Note 21**).

3.4 Selection of the Light-Density Fraction of Spleen

1. Thaw two 10 ml aliquots of Nycodenz-EDTA at room temperature. Once thawed, mix thoroughly and keep at 4 °C until required (*see* **Note 22**).

2. Resuspend the cell pellet in 10 ml of Nycodenz-EDTA (*see* **Note 23**).

3. Transfer 5 ml of the remaining Nycodenz-EDTA into the bottom of each of two polypropylene tubes.

4. Gently layer 5 ml of the cell suspension over the Nycodenz-EDTA in each of the two tubes (*see* **Note 24**). Add a 1–2 ml layer of EDTA-FCS over the cell suspension.

5. Disrupt the interface gently by inserting the tip of a Pasteur pipette, swirling and removing it (*see* **Note 25**).

6. Perform the density cut in a swing-out head, refrigerated centrifuge, set at 4 °C, for 10 min at $1700 \times g$ with the brake set on low.

7. Collect the light-density fraction in the upper zones down to the 4 ml mark using a Pasteur pipette (*see* **Note 26**). Discard the bottom 4 ml and the cell pellet.

8. Transfer the light-density fraction to a 50 ml tube and dilute up to 50 ml with BSS-EDTA. Mix thoroughly and centrifuge to wash and recover the cells (*see* **Note 27**).

9. Resuspend the cells in 5 ml BSS-EDTA-FCS and count (*see* **Note 28**). Calculate the total cell number recovered.

3.5 Depletion of Non-DC Lineages

1. Calculate the volume of monoclonal antibody depletion cocktail required if 10 µl is needed per 10^6 cells.

2. Add the required volume of the appropriate monoclonal antibody depletion cocktail (*see* **Note 29**) to the cell pellet and resuspend and incubate at 4 °C for 30 min.

3. Calculate the required volume of immunomagnetic beads required (*see* **Note 30**) and transfer them to 5 ml polypropylene tubes. Wash the beads by diluting with BSS-EDTA-FCS (*see* **Note 31**), placing the tubes into the magnet, allowing the beads to move to the magnet and removing the supernatant. Repeat the washing step three to four times. After the final wash, pellet the beads at the bottom of the tube in a small amount of BSS-EDTA-FCS and place the tube at 4 °C until required.

4. Dilute the cells up to 9 ml with BSS-EDTA-FCS and underlay with 1 ml of FCS-EDTA. Centrifuge the cells and remove the

supernatant from the top, leaving the FCS layer over the cells (*see* **Note 32**). Pulse in the centrifuge for 15 s to force any remaining supernatant down the wall of the tube. Then remove any supernatant and the FCS (*see* **Note 33**). Resuspend the cells in 400–500 μl of BSS-EDTA-FCS.

5. Remove the BSS-EDTA-FCS from the pellet of immunomagnetic beads and add the cells. Vortex the tube very briefly to produce a bead-cell slurry (*see* **Note 34**). Seal the tube and mix the slurry continuously for 20 min at 4 °C at an angle of 30° on a spiral mixer (*see* **Note 35**).

6. Dilute the slurry with 3 ml of BSS-EDTA-FCS, mix very gently, and attach the tube to the magnet for 2 min.

7. Recover the supernatant containing unbound DC with a Pasteur pipette, and transfer to a second 5 ml polypropylene tube. Discard the tube containing magnetic beads bound to non-DC (*see* **Note 36**). Place the tube containing the supernatant into the magnet for a further 2 min to remove any remaining beads. Transfer the supernatant to a 10 ml tube.

8. Layer 1 ml of FCS-EDTA under the cell suspension and centrifuge to recover the DC fraction. Resuspend the cells in 2 ml of BSS-EDTA-FCS and count.

9. Maintain the cells at 4 °C until they are required for immunofluorescent labeling.

3.6 In Vitro Administration of Flt3L

1. Remove the femurs and tibiae from the desired number of mice, removing as much tissue and sinew as possible, and place in RPMI-FCS at 4 °C (*see* **Note 37**).

2. Use scissors to remove the top of each bone to allow access to the bone marrow. Control each bone with a pair of sterile forceps, and flush the bone with RPMI-FCS using a 23G needle and syringe to remove the bone marrow. Pool the bone marrow from individual bones. Mix the suspension up and down in a syringe and 26G needle to create a single cell suspension. Underlay with FCS and centrifuge.

3. Remove the supernatant and gently resuspend the cells in 500 μl of RCLM per mouse used. Expose the cells to RCLM for 30 s (*see* **Note 38**). Dilute the cells immediately with RPMI-FCS, pass them into a new tube through a sieve to remove clumps, underlay with FCS, and wash by centrifugation. Repeat the washing step twice more. Resuspend the cells and count.

4. Centrifuge the cells and resuspend the pellet at 1.5×10^6/ml in Flt3L culture medium. If desired add 300 pg/ml GM-CSF to boost numbers at days 8 and 9 of culture (*see* **Note 39**). Culture the cells for 8–9 days at 37 °C in 10 % CO_2 in air (*see* **Note 40**).

5. If required, add 5 ng/ml GM-CSF at day 6 of culture, to increase the proportion of CD103$^+$ cDC at day 8 or 9.

6. At the completion of the culture period, harvest the cells by gently washing the flasks several times with cold BSS-EDTA-FCS (*see* **Note 41**). Centrifuge to recover the cells and resuspend the pellet in BSS-EDTA-FCS. Count to determine recovery.

7. Keep the cell suspension at 4 °C until the cells are required for immunofluorescent labeling.

3.7 Immunofluorescent Staining and Purification

3.7.1 Via Flow Cytometric Sorting

1. Prepare a cocktail of pre-titrated fluorochrome-conjugated monoclonal antibodies at the appropriate concentration immediately prior to use.

2. Centrifuge to recover the cells and remove the supernatant.

3. Add the fluorochrome-conjugated antibody cocktail at 10 μl per 10^6 cells, resuspend by flicking the tube, and incubate at 4 °C for 30 min (*see* **Note 42**).

4. Resuspend up to a larger volume with BSS-EDTA-FCS and underlay with FCS-EDTA. Centrifuge to wash the cells.

5. Remove the supernatant leaving the FCS-EDTA layer above the cells. Pulse the tube in the centrifuge for 15 s to force any remaining media to wash down the wall of the tube. Remove the remaining media and the FCS-EDTA layer.

6. Resuspend the cells in BSS-EDTA-FCS-PI and keep cells at 4 °C until they are required for flow cytometry (*see* **Note 43**).

7. Immediately prior to sorting, add the cell suspension to a syringe fitted with a 26G needle. Gently force the cells through the needle and into a cell strainer cap fitted on a 5 ml polystyrene round bottom tube. Allow the cells to run through the strainer and collect in the tube.

8. Sort the DC using a FACSAria (*see* **Note 44**). The flow cytometer should be set up with standard lasers in place: a blue 488 nm emitting laser for the detection of FITC, PE, PI, PerCp.Cy5.5, and PE.Cy7; a violet/near UV laser, emitting wavelengths of 375 and 405 nm, to detect Pacific Blue; and a red laser emitting a wavelength of 640 nm to detect APC, as well as the appropriate filters and dichroic mirrors. Select DC on the basis of high forward and side light scatter, excluding dead cells with high PI fluorescence. Contaminating macrophages should be removed by gating out autofluorescent cells using the PI channel in combination with another unused fluorescence channel (*see* **Note 45**). Identify CD11c^{int} siglec-H$^+$ pDC and CD11c^{hi}siglec-H$^-$ cDC (*see* **Note 46**). Use combinations of conjugated antibodies to identify and sort other dendritic cell subtypes (*see* **Note 47**).

1. Centrifuge cells to a pellet.

2. Add YTS169.4-FITC (anti-CD8) (*see* **Note 48**) at 10 µl/10^6 cells (*see* **Note 49**) and resuspend and incubate at 4 °C for 20 min (*see* **Note 50**).

3. Dilute up to a larger volume with BSS-EDTA-FCS, underlay with FCS-EDTA, and centrifuge to recover the cells.

4. Remove the supernatant leaving the FCS-EDTA layer above the cells and pulse in the centrifuge for 15 s to force any remaining media down the wall of the tube. Remove the remaining media and the FCS-EDTA layer.

5. Prepare anti-FITC microbeads (*see* **Note 51**) at 1 µl/4×10^6 cells in a final volume of 2.5 µl/10^6 cells BSS-EDTA-0.5 %FCS (*see* **Note 52**), and add to the cell pellet. Resuspend gently by flicking the tube.

6. Incubate for 15 min in a 4 °C cold room (*see* **Note 53**).

7. Dilute to a larger volume with BSS-EDTA-FCS and underlay with FCS-EDTA. Centrifuge the cells to wash away unbound microbeads.

8. Equilibrate a MACS LS column (*see* **Note 54**) by placing it into a cold miniMACS magnet suspended on a MACS multistand and washing with 3 ml BSS-EDTA-0.5 %FCS (*see* **Note 55**).

9. Resuspend the cells in 3 ml BSS-EDTA-0.5 %FCS and apply them to the column. Rinse the column with 3 ml of BSS-EDTA-0.5 %FCS. Add 5 ml of BSS-EDTA-0.5 %FCS to the column and collect the flow through. Run another 5 ml of BSS-EDTA-0.5 %FCS through the column, and pool the flow through containing the unbound CD8$^-$ cells.

10. Remove the column from the magnet, add 5 ml BSS-EDTA-0.5 %FCS, and allow to run to wash the previously bound CD8$^+$ cells from the column. Add a further 5 ml of BSS-EDTA-0.5 % FCS to the column. Use the supplied plunger to force all liquid from the column to ensure all cells are removed. Pool, count, and set aside the recovered CD8$^+$ cells at 4 °C. Discard the column.

11. Count the CD8$^-$ cells and centrifuge them.

12. Add M1/69-FITC (anti-CD24) at 10 µl/10^6 cells and resuspend and incubate at 4 °C for 20 min (*see* **Note 56**).

13. Dilute to a larger volume with BSS-EDTA-FCS and underlay with FCS-EDTA. Centrifuge the cells.

14. Remove the supernatant leaving the FCS-EDTA layer above the cells, and pulse in the centrifuge for 15 s to force any remaining media to wash down the wall of the tube. Remove the remaining media and the FCS-EDTA layer.

15. Prepare anti-FITC microbeads at 1 µl/4×10^6 cells in a final volume of 2.5 µl/10^6 cells BSS-EDTA-0.5%FCS and add to the cell pellet. Resuspend gently by flicking the tube.

16. Repeat **steps 6** and **7**.

17. Equilibrate (*see* **Note 54**) a MACS LD column (*see* **Note 57**) by placing into a cold miniMACS magnet suspended on a MACS multistand and washing with 2 ml BSS-EDTA-0.5%FCS (*see* **Note 58**).

18. Resuspend the cells in 1 ml BSS-EDTA-0.5%FCS and apply them to the column. Rinse the column with 1 ml of BSS-EDTA-0.5%FCS. Repeat the wash once more, and collect the flow through containing the unbound CD24⁻ CD8⁻ fraction of cells. Discard the column (*see* **Note 59**).

19. Count the CD24⁻CD8⁻ cells and centrifuge them.

20. Add M1/70-PE (anti-CD11b) at 10 µl/10^6 cells and resuspend and incubate at 4 °C for 20 min.

21. Dilute up to a larger volume with BSS-EDTA-FCS and underlay with FCS-EDTA. Centrifuge to recover the cells.

22. Remove the supernatant leaving the FCS-EDTA layer above the cells and pulse in the centrifuge for 15 s to force any remaining media to wash down the wall of the tube. Remove the remaining media and the FCS-EDTA layer.

23. Prepare anti-PE microbeads at 1 µl/4×10^6 cells in a final volume of 2.5 µl/10^6 cells BSS-EDTA-0.5%FCS and add to the cells. Gently resuspend by flicking the tube.

24. Repeat **steps 6–8**.

25. Resuspend the cells in 3 ml BSS-EDTA-0.5%FCS and apply them to the column. Rinse the column with 3 ml of BSS-EDTA-0.5%FCS. Add 5 ml of BSS-EDTA-0.5%FCS to the column and collect the flow through. Run another 5 ml of BSS-EDTA-0.5%FCS through the column and discard the flow through containing the unbound CD24⁻ CD8⁻ CD11b⁻ fraction of cells.

26. Remove the column from the magnet and add 5 ml BSS-EDTA-0.5%FCS to wash the previously bound CD24⁻ CD8⁻ CD11b⁺ cells from the column. Add a further 5 ml of BSS-EDTA-0.5%FCS. Use the supplied plunger to force all liquid from the column to ensure all cells are removed. Count the recovered CD24⁻ CD8⁻ CD11b⁺ cells (*see* **Note 60**).

3.7.3 Depletion of NK Cells via Immunomagnetic Beads

1. If depletion of NK cells is deemed necessary, centrifuge cells to a pellet.

2. Add saturating levels of DX5-biotin at 10 µl/10^6 cells, and resuspend and incubate at 4 °C for 30 min. Dilute up to a

larger volume with BSS-EDTA-FCS, underlay with FCS-EDTA, and centrifuge to recover the cells.

3. Remove the supernatant leaving the FCS-EDTA layer above the cells, and pulse in the centrifuge for 15 s to force any remaining media down the wall of the tube. Remove the remaining media and the FCS-EDTA layer.

4. Prepare anti-biotin microbeads (*see* **Note 51**) at 1 $\mu l/4 \times 10^6$ cells in a final volume of 2.5 $\mu l/10^6$ cells BSS-EDTA-0.5%FCS (*see* **Note 52**), and add to the cell pellet. Resuspend gently by flicking the tube.

5. Incubate for 15 min in a 4 °C cold room (*see* **Note 53**).

6. Dilute to a larger volume with BSS-EDTA-FCS and underlay with FCS-EDTA. Centrifuge the cells to wash away unbound microbeads.

7. Equilibrate a MACS LD column (*see* **Notes 54** and **58**) by placing it into a cold miniMACS magnet suspended on a MACS multistand and washing with 3 ml BSS-EDTA-0.5%FCS (*see* **Note 55**).

8. Resuspend the cells in 1 ml BSS-EDTA-0.5%FCS and apply them to the column. Rinse the column with 1 ml of BSS-EDTA-0.5%FCS. Repeat the wash once more and collect the flow through containing the unbound DX5$^-$ DC fraction of cells. Discard the column.

4   Notes

1. Alternatively use recombinant human Flt3L.

2. The B16Flt3L melanoma was derived from male C57/BL6 mice. Injection into mice of other strains, or into female mice, may elicit an immune response and affect the DC recovered.

3. The level of contamination of collagenase with trypsin or trypsin-like proteases can vary between batches, so each should be tested prior to use. Proteases can strip cell surface molecules and alter the surface phenotype of the DC. We test each new batch of collagenase for the presence of these proteases by using them to digest thymocytes for 30 min at 37 °C and then screening for the loss of the trypsin-sensitive cell surface markers CD4 and CD8 by flow cytometry.

4. A pycnometer is used to determine the density of the Nycodenz accurately, by reference to water, using an analytical balance. The pycnometer is a glass flask with a close-fitting ground-glass lid with a capillary tube in it, which allows air bubbles or excess Nycodenz to escape from the vessel. The pycnometer is weighed empty, full of water, and full of Nycodenz and the

specific gravity of the Nycodenz calculated. A correction needs to be made as the density of water will not be 1 g/cm³ at 4 °C.

5. Temperature, pH, and osmolarity all affect the buoyant density of cells, so Nycodenz of higher, or lower, than recommended density will affect the purity and the yield of recovered cells. We calculate the density of Nycodenz at pH 7.2, 308 mOsm/kg, and 4 °C. Temperature is of particular importance during the density cut, so care must be taken to ensure the Nycodenz and the centrifuge to be used for the density cut are at 4 °C.

6. The monoclonal antibodies we have included in the cocktail for depletion of non-DC lineage cells and those we have utilized to identify DC and DC subpopulations are available commercially.

7. Appropriate levels should be determined for each batch of antibody to be included in the cocktail. The rat anti-mouse monoclonal antibodies are individually titrated via flow cytometry using a fluorochrome-conjugated anti-rat Ig secondary reagent in order to determine their working dilution. Antibodies are added to ensure surface saturating quantities of each are contained in 10uL of cocktail. Concentrations determined to result in cell surface saturation of the antigen are considered adequate for efficient depletion and are used in the cocktail.

8. Flt3L should be titrated prior to use by small-scale bone marrow culture. A range of 50–300 ng/ml should be tested. Suboptimal levels will vastly reduce DC yield. Optimal levels can result in DC recoveries of up to 130 % of the starting number of bone marrow cells.

9. All proper precautions should be taken when using sodium azide, particularly when preparing the stock solution. Adequate protective clothing, including safety glasses, gloves, and face mask, should be worn. Sodium azide is extremely toxic if ingested.

10. Do not freeze phycoerythrin (PE), allophycocyanin (APC), and PerCp.Cy5.5 or their conjugates. They are extremely sensitive to freezing and thawing and will lose activity.

11. All proper precautions including protective clothing, safety glasses, gloves, and face mask should be worn during preparation of the stock solution of propidium iodide. Propidium iodide is an irritant and potentially toxic.

12. MSA is used as a carrier protein to minimize loss of Flt3L sticking to tubes, syringes, etc. It may be possible to substitute with endotoxin-free BSA (bovine serum albumin) or even FCS (fetal calf serum), but care must be taken not to elicit an immune response against the carrier protein.

13. Aim to insert the needle and inject as centrally between the shoulder blades as possible. The melanoma will develop rapidly

into a large growth and if not located centrally will make normal movement very difficult for the mouse.

14. Mice should be monitored regularly and any showing signs of distress or illness euthanized.

15. The rate of growth of the melanoma should not adversely affect the mice for at least 10 days after injection. After 10 days, however, mice will begin to show signs of distress and/or illness. It is therefore recommended that the melanoma should not be allowed to develop for longer than 10 days.

16. Provision has been made to cater for the greatly increased size of spleens treated with Flt3L. A proportional increase or decrease of all listed amounts and volumes should also be made to cater for any change in the starting number of organs.

17. Residual fat will reduce cell viability and, combined with undigested connective tissue, will accumulate and cause clumping. It is therefore important to clean the organs as much as possible before commencement of the digestion. We use two 20G needles to perform the cleaning, but any suitable instrument may be used.

18. The organs are cut into very small fragments to increase the surface area available to the enzymes. This ensures adequate digestion and maximizes cell yield.

19. Inadequate digestion will result in a lower recovery and the preferential loss of certain DC populations that are more firmly entrenched in the tissue. A digestion time of 20 min should prove sufficient to digest all but the pulpy tissue from spleen, provided the tissue was cut up adequately prior to the digestion and adequate mixing occurred during the digestion. The digestion may be extended to 25 min if required.

20. EDTA inhibits collagenase and effectively ends the digestion. EDTA also chelates Ca^{2+} and Mg^{2+} ions and will dissociate lymphocyte-DC complexes. EDTA must be added to all media from this point onward to stop the reformation of these multicellular complexes. Failure to do so will cause loss of DC during purification and possible contamination of the recovered DC with lymphocytes.

21. All centrifugation steps are performed at $1000 \times g$ for 7 min at 4 °C unless otherwise stated. Underlaying the sample with FCS, thereby incorporating a zonal centrifugation step, increases the efficiency of separation of cells from smaller particles and soluble material in the supernatant. It therefore eliminates the need for repetitive "washing" of the cells.

22. Nycodenz has a tendency to settle over time, so mix it thoroughly prior to aliquoting and again prior to use to ensure a solution of uniform density.

23. Efficiency of separation will be lost and yields reduced if the density separation is overloaded. Do not load more than four untreated or one Flt3L treated spleen (5×10^8–10^9 cells) per 10 ml of Nycodenz.

24. A discrepancy between the density of the Nycodenz at the top of an aliquot and the bottom, or between different aliquots (most likely due to inadequate mixing), will affect the ability to layer Nycodenz containing cells over the Nycodenz at the bottom of the tube. Ensure Nycodenz has been adequately mixed and is of uniform density before use.

25. Disruption of the interface creates a density gradient rather than a sharp band and increases the efficiency of the density separation.

26. The light-density fraction of cells will be found as a band at the interface zone between the FCS and Nycodenz, while dense cells will have formed a pellet. Cells of intermediate density will be found in the gradient between these zones, so collect all cells down to the 4 ml mark while concentrating on the light-density band at the interface. The 4 ml mark is an arbitrary one, and collecting a few ml either side of this will not alter recovery greatly. It is just important to be consistent.

27. Adequate dilution and mixing of the light-density fraction of cells with EDTA-BSS is essential to recover them as a pellet during centrifugation.

28. This count is used to calculate the appropriate volume of mAb depletion cocktail and immunomagnetic beads required in subsequent steps. It is important that all cells, including any remaining erythrocytes, are included in the count. As a rough guide, the light-density fraction should represent 5–7% of the starting number of cells.

29. Anti-CD45R is included in the cocktail to deplete B cells but will also deplete pDC, which are all CD45R$^+$. In order to include pDC, but continue to deplete B cells, replace anti-CD45R with anti-CD19.

30. For the isolation of spleen DC, we recommend BioMag beads at a 10:1 bead-to-cell ratio. BioMag beads provide optimal economy and reasonably good efficiency.

31. Immunomagnetic beads must be washed prior to use to remove preservative.

32. Centrifuging through a layer of FCS separates cells from unbound (excess) mAb.

33. It is important to carefully remove all the supernatant after washing as any remaining unbound mAb will compete for binding to the beads thus decreasing the efficiency of the depletion.

34. The efficiency of the depletion is greatly increased by maximizing contact between beads and cells in the concentrated slurry.

35. Fitting a wide ring around the top of the tube increases the angle of rotation to 30° from the horizontal and keeps the bead-cell slurry concentrated at the bottom of the tube. The wide ring also serves to slow the rate of rotation.

36. If numbers are critical, more DC can be recovered by washing the beads and attaching the tube to the magnet a second time; however, this will result in reduced purity.

37. We routinely harvest $4–7 \times 10^7$ cells per mouse.

38. RCLM is toxic so it should be added and mixed gently and then washed away quickly. An exposure of 15–30 s is sufficient to lyse erythrocytes in bone marrow.

39. The optional addition of 300 pg/ml GM-CSF to the Flt3L culture medium will improve DC yield.

40. Maximum yield of the DC subpopulations can also vary with culture time, so an optimal culture time should be determined for each batch of Flt3L. Production of pDC peaks at day 8 and cDC production at day 9 of culture.

41. This gentle washing step should be sufficient to remove any slightly adherent DC from the plastic but leave behind any more adherent macrophages.

42. All immunostaining steps should be performed at 4 °C to promote cell viability and to prevent capping of monoclonal antibody from the cell surface.

43. Propidium iodide is used for dead cell exclusion during flow cytometric analysis.

44. Any flow cytometer with sorting capabilities and appropriate lasers and optical setup can be used.

45. Remove autofluorescent cells (mainly macrophages) during fluorescence activated cell sorting or analysis by gating out cells that have low levels of fluorescence in two or more fluorescent channels. Ideally use a combination of PI and an unused channel. During multicolor sorting or analysis, it may be necessary to combine autofluorescence in the PI channel with low fluorescence in a channel that is being used. Choose a parameter where all DC will fluoresce brightly (i.e., CD11c) and gate out cells of low fluorescence.

46. CD45RA, CD45R (B220), and CD317 are alternative markers to siglec H and are often used to separate pDC from cDC.

47. If functional studies are to be undertaken, the cells should be washed to remove propidium iodide post sorting.

48. After Flt-3L treatment, it is advisable to purify the more abundant CD8$^+$ fraction first in order to avoid contamination of the CD11b$^+$ fraction.

49. Too high a concentration of antibody added at this stage tends to increase the nonspecific loss of DC and lower the yield. We routinely use antibodies at a quarter to a half their saturating levels, but would recommend that each user carefully titrate their antibody for optimal performance.

50. It is not important what fluorochrome the anti-CD8 monoclonal antibody is conjugated to as long as an appropriate anti-fluorochrome bead is available.

51. Although directly coupled anti-CD8 beads are available, we find that efficiency of separation is lowered when they are used.

52. This is a lower bead-to-cell ratio than recommended by the manufacturer. We have compensated for this by reducing the volume in order to increase the final concentration and maximize bead-to-cell contact. This has resulted in a more economical and efficient process.

53. Incubation may also be completed in a refrigerator; however, ice should be avoided as temperatures below 4 °C will decrease binding.

54. Do not overload the column. This will result in a lower recovery and reduced purity. An LS column can accommodate up to 5×10^8 cells with a maximum of 10^8 positive cells and an LD column up to 10^9 cells with a maximum of 5×10^8 positive cells. Ensure accurate counts of cells are made and appropriate numbers of columns used.

55. Cell viability is increased at 4 °C, so cooling the magnet before use ensures that the column and cell suspension passing through it are kept cold during the separation. Placing the magnet at −20 °C and the columns at 4 °C for 15 min before use is recommended.

56. A second depletion for CD8-bearing cells using anti-CD24 conjugated to FITC is necessary to improve purity of the CD11b$^+$ cells. This depletion also removes the CD8$^-$CD24$^+$ precursors of the CD8$^+$ DC lineage.

57. LS columns are designed for selection and LD columns for depletion. So when positively selecting a population (CD8$^+$ or CD11b$^+$), we use an LS column, and when trying to select a population we wish to discard (CD24$^+$), we use an LD column.

58. LD columns are more tightly packed and will run much more slowly than LS columns. Allow sufficient time for equilibration of the column, and be aware that the longer exposure of cells to room temperature while they are running through the column may affect purity and/or viability. Consider running LD columns in a cold room.

59. As the bound fraction of cells in this case will be a mix of $CD8^+CD24^+$ and $CD8^-CD24^+$ cells, it is typically discarded. If this fraction is required, it may be recovered using the typical washing procedure used to recover the $CD8^+$ and $CD11b^+$ fractions.

60. Purity of all collected populations may be tested. Fluorochromes on the surface of the cells are not affected by bound beads and can be readily detected using flow cytometry.

Acknowledgements

This work was made possible through Victorian State Government Operational Infrastructure Support and Australian Government NHMRC IRIIS.

References

1. Steinman RM, Kohn ZA (1973) Identification of a novel cell type in peripheral lymphoid organs of mice. I. Morphology, quantitation, tissue distribution. J Exp Med 137:1142–1162

2. Steinman RM (1991) The dendritic cell system and its role in immunogenicity. Annu Rev Immunol 9:271–296

3. Jung S, Unutmaz D, Wong P, Sano G, De los Santos K, Sparwasser T, Wu S, Vuthoori S, Ko K, Zavala F, Pamer EG, Littman DR, Lang RA (2002) *In vivo* depletion of CD11c+ dendritic cells abrogates priming of CD8+ T cells by exogenous cell-associated antigens. Immunity 17:211–220

4. Liu CH, Fan YT, Dias A, Esper L, Corn RA, Bafica A, Machado FS, Aliberti J (2006) Cutting edge: dendritic cells are essential for *in vivo* IL-12 production and development of resistance against Toxoplasma gondii infection in mice. J Immunol 177:31–35

5. Ciavarra RP, Stephens A, Nagy S, Sekellick M, Steel C (2006) Evolution of immunological paradigms in a virus model: are dendritic cells critical for antiviral immunity and viral clearance? J Immunol 177:492–500

6. Watanabe N, Wang YH, Lee HK, Ito T, Cao W, Liu YJ (2005) Hassall's corpuscles instruct dendritic cells to induce CD4+CD25+ regulatory T cells in human thymus. Nature 436:1181–1185

7. Luo X, Tarbell KV, Yang H, Pothoven K, Bailey SL, Dind R, Steinman RM, Suthanthiran M (2007) Dendritic cells with TGF-β1 differentiate naïve CD4+CD25- T cells into islet-protective Foxp3+ regulatory T cells. Proc Natl Acad Sci U S A 104:2821–2826

8. Yamazaki S, Iyoda T, Tarbell K, Olson K, Velinzon K, Inaba K, Steinman RM (2003) Direct expansion of functional CD25+CD4+ regulatory T cells by antigen presenting dendritic cells. J Exp Med 198:235–247

9. Crowley M, Inaba K, Witmer-Pack M, Steinman RM (1989) The cell surface of mouse dendritic cells: FACS analyses of dendritic cells from different tissues including thymus. Cell Immunol 118:108–125

10. Vremec D, Zorbas M, Scollay R, Saunders D, Ardavin C, Wu L, Shortman K (1992) The surface phenotype of dendritic cells purified from mouse thymus and spleen: investigation of the CD8 expression by a subpopulation of dendritic cells. J Exp Med 176:47–58

11. Shortman K, Heath WR (2010) The CD8+ dendritic cell subset. Immunol Rev 234:18–31

12. Shortman K, Caux C (1997) Dendritic cell development: multiple pathways to natures adjuvant. Stem Cells 15:409–419

13. Shortman K, Liu YJ (2002) Mouse and human dendritic cell subtypes. Nat Rev Immunol 2:151–161

14. Vremec D, Shortman K (1997) Dendritic cell subtypes in mouse lymphoid organs: cross-correlation of surface markers, changes on incubation and differences between thymus, spleen and lymph nodes. J Immunol 159:565–573

15. Wilson NS, El-Sukkari D, Belz GT, Smith CM, Steptoe RJ, Heath WR, Shortman K,

Villadangos JA (2003) Most lymphoid organ dendritic cell types are phenotypically and functionally immature. Blood 102:2187–2194

16. Villadangos JA, Schnorrer P (2007) Intrinsic and cooperative antigen-presenting functions of dendritic-cell subsets *in vivo*. Nat Rev Immunol 7:543–555

17. Heath WR, Kurts C, Miller JF, Carbone FR (1998) Cross-tolerance: a pathway for inducing tolerance to peripheral tissue antigens. J Exp Med 187:1549–1553

18. Heath WR (2004) Cross presentation, dendritic cell subsets, and generation of immunity to cellular antigens. Immunol Rev 199:9–26

19. Vremec D, O'Keeffe M, Wilson A, Ferrero I, Koch U, Radtke F, Scott B, Hertzog P, Villadangos J, Shortman K (2011) Factors determining the spontaneous activation of splenic dendritic cells in culture. Innate Immun 17:338–352

20. Vremec D, Pooley J, Hochrein H, Wu L, Shortman K (2000) CD4 and CD8 expression by dendritic cell subtypes in mouse thymus and spleen. J Immunol 164:2978–2986

21. Vremec D (2010) The isolation of mouse dendritic cells from lymphoid tissues and the identification of dendritic cell subtypes by multiparameter flow cytometry. In: Naik SH (ed) Dendritic cell protocols, vol 595, Methods in molecular biology. Humana, New York, pp 205–229

22. O'Keeffe M, Hochrein H, Vremec D, Caminschi I, Miller JL, Anders EM, Wu L, Lahoud MH, Henri S, Scott B, Hertzog P, Tatarczuch L, Shortman K (2002) Mouse plasmacytoid cells: long-lived cells, heterogeneous in surface phenotype and function, that differentiate into CD8+ dendritic cells only after microbial stimulus. J Exp Med 196:1307–1319

23. Naik SH, Shortman K (2007) Steady-state and inflammatory dendritic-cell development. Nat Rev Immunol 7:19–30

24. Asselin-Paturel C, Brizard G, Pin J-J, Briere F, Trinchieri G (2003) Mouse strain differences in plasmacytoid dendritic cell frequency and function revealed by a novel monoclonal antibody. J Immunol 171:6466–6477

25. Naik SH, Proietto AI, Wilson NS, Dakic A, Schnorrer P, Fuchsberger M, Lahoud MH, O'Keeffe M, Shao QX, Chen WF, Villadangos JA, Shortman K, Wu L (2005) Cutting edge: generation of splenic CD8+ and CD8− dendritic cell equivalents in Fms-like tyrosine kinase 3 ligand bone marrow cultures. J Immunol 174:6592–6597

26. Edwards AD, Diebold SS, Slack EM, Tomizawa H, Hemmi H, Kaisho T, Akira S, Reis e Sousa C (2003) Toll-like receptor expression in murine DC subsets: lack of TLR7 expression by CD8α+ DC correlates with unresponsiveness to imidazoquinolines. Eur J Immunol 33:827–833

27. Hochrein H, Shortman K, Vremec D, Scott B, Hertzog P, O'Keeffe M (2001) Differential production of IL-12, IFN-α and IFN-γ by mouse dendritic cell subsets. J Immunol 166:5448–5455

28. Lauterbach H, Bathke B, Gilles S, Traidl-Hoffman C, Luber CA, Fejer G, Freudenberg MA, Davey GM, Vremec D, Kallies A, Wu L, Shortman K, Chaplin P, Suter M, O'Keeffe M, Hochrein H (2010) Mouse CD8alpha+ DCs and human BDCA3+ DCs are major producers of IFN-lambda in response to polyIC. J Exp Med 207:2703–2717

29. Den Haan JM, Lehar SM, Bevan MJ (2000) CD8+ but not CD8− dendritic cells cross-prime cytotoxic T cells *in vivo*. J Exp Med 192:1685–1696

30. Pooley JL, Heath WR, Shortman K (2001) Cutting edge: intravenous soluble antigen is presented to CD4 T cells by CD8− dendritic cells, but cross-presented to CD8 T cells by CD8+ dendritic cells. J Immunol 166:5327–5330

31. Zhang JG, Czabotar PE, Policheni AN, Caminschi I, Wan SS, Kitsoulis S, Tullett KM, Robin AY, Brammananth R, van Delft MF, Lu J, O'Reilly LA, Josefsson EC, Kile BT, Chin WJ, Mintern JD, Olshina MA, Wong W, Baum J, Wright MD, Huang DCS, Mohandas N, Coppel RL, Colman PM, Nicola NA, Shortman K, Lahoud M (2012) The dendritic cell receptor Clec9A binds damaged cells via exposed actin filaments. Immunity 36:646–657

32. Zelenay S, Keller AM, Whitney PG, Schraml BU, Deddouche S, Rogers NC, Schulz O, Sancho D, Reis e Sousa C (2012) The dendritic cell receptor DNGR-1 controls endocytic handling of necrotic cell antigens to favour cross-priming of CTLs in virus-infected mice. J Clin Invest 122:1615–1627

33. Caminschi I, Proietto AI, Ahmet F, Kitsoulis S, Teh JS, Lo JCY, Rizzitelli A, Wu L, Vremec D, van Dommelen SLH, Campbell IK, Maraskovsky E, Braley B, Davey GM, Mottram P, van de Velde N, Jensen K, Lew AM, Wright MD, Heath WR, Shortma K, Lahoud MH (2008) The dendritic cell subtype-restricted C-type lectin Clec9A is a target for vaccine enhancement. Blood 112:3264–3273

34. Sathe P, Pooley J, Vremec D, Mintern J, Jin JO, Wu L, Kwak JY, Villadangos JA, Shortman K (2011) The acquisition of antigen cross-presentation function by newly formed dendritic cells. J Immunol 186:5184–5192

35. Zhan Y, Carrington EM, van Nieuwenhuijze A, Bedoui S, Seah S, Xu Y, Wang N, Mintern JD, Villadangos JA, Wicks IP, Lew AM (2011) GM-CSF increases cross-presentation and CD103 expression by mouse CD8+ spleen dendritic cells. Eur J Immunol 41:2585–2595

36. Lewis KL, Caton ML, Bogunovic M, Greter M, Grajkowska LT, Ng D, Klinakis A, Charo IF, Jung S, Gommerman JL, Ivanov II, Liu K, Merad M, Reizis B (2011) Notch2 receptor signalling controls functional differentiation of dendritic cells in the spleen and intestine. Immunity 35:780–791

37. Dudziak D, Kamphorst AO, Heidkamp GF, Buchholz VR, Trumpfheller C, Yamazaki S, Cheong C, Liu K, Lee HW, Park CG, Steinman RM, Nussenzweig MC (2007) Differential antigen processing by dendritic cell subsets in vivo. Science 315:107–111

38. Proietto AI, O'Keeffe M, Gartlan K, Wright MD, Shortman K, Wu L, Lahoud MH (2004) Differential production of inflammatory chemokines by murine dendritic cell subsets. Immunobiology 209:163–172

39. Onai N, Obata-Onai A, Schmid MA, Manz MG (2007) Flt3 in regulation of type I interferon-producing cell and dendritic cell development. Ann N Y Acad Sci 1106:253–261

40. Schmid MA, Kingston D, Boddupalli S, Manz MG (2010) Instructive cytokine signals in dendritic cell lineage commitment. Immunol Rev 234:32–44

41. McKenna HJ, Stocking KL, Miller RE, Brasel K, De Smedt T, Maraskovsky E, Maliszewski CR, Lynch DH, Smith J, Pulendran B, Roux ER, Teepe M, Lyman SD, Peschon JJ (2000) Mice lacking flt3 ligand have deficient hematopoiesis affecting hematopoietic progenitor cells, dendritic cells and natural killer cells. Blood 95:3489–3497

42. Waskow C, Liu K, Darrasse-Jeze G, Guermonprez P, Ginhoux F, Merad M, Shengelia T, Yao K, Nussenzweig M (2008) The receptor tyrosine kinase Flt3 is required for dendritic ell development in peripheral lymphoid tissues. Nat Immunol 9:676–683

43. Kingston D, Schmid MA, Onai N, Obata-Onai A, Baumjohann D, Manz MG (2009) The concerted action of GM-CSF and Flt3-ligand on in vivo dendritic cell homeostasis. Blood 114:835–843

44. D'Amico A, Wu L (2003) The early progenitors of mouse dendritic cells and plasmacytoid predendritic cells are within the bone marrow hemopoietic precursors expressing Flt3. J Exp Med 198:293–303

45. Maraskovsky E, Brasel K, Teepe M, Roux ER, Lyman SD, Shortman K, McKenna HJ (1996) Dramatic increase in the numbers of functionally mature dendritic cells in Flt3 ligand-treated mice: multiple dendritic cell subpopulations identified. J Exp Med 184:1953–1962

46. Mach N, Gillessen B, Wilson SB, Sheehan C, Mihm M, Dranoff G (2000) Differences in dendritic cells stimulated in vivo by tumors engineered to secrete granulocyte-macrophage colony-stimulating factor or Flt3-ligand. Cancer Res 60:3239–3246

47. Brasel K, De Smedt T, Smith JL, Maliszewski CR (2000) Generation of murine dendritic cells from flt3-ligand-supplemented bone marrow cultures. Blood 96:3029–3039

48. Gilliet M, Boonstra A, Paturel C, Antonenko S, Xu XL, Trinchieri G, O'Garra A, Liu YJ (2002) The development of murine plasmacytoid dendritic cell precursors is differentially regulated by FLT-3-ligand and granulocyte/macrophage colony-stimulating factor. J Exp Med 195:953–958

49. Bedoui S, Prato S, Mintern J, Gebhardt T, Zhan Y, Lew AM, Heath WR, Villadangos JA, Segura E (2009) Characterization of an immediate splenic precursor of CD8+ dendritic cells capable of inducing antiviral T cell responses. J Immunol 182:4200–4207

50. Segura E, Kapp E, Gupta N, Wong J, Lim J, Ji H, Heath WR, Simpson R, Villadangos JA (2010) Differential expression of pathogen-recognition molecules between dendritic cell subsets revealed by plasma membrane proteomic analysis. Mol Immunol 47:1765–1773

Chapter 6

Dendritic Cell Subset Purification from Human Tonsils and Lymph Nodes

Mélanie Durand and Elodie Segura

Abstract

Dendritic cells (DCs) are a rare population of antigen-presenting cells that initiate immune responses in secondary lymphoid organs. In order to better understand the properties of DC in humans, it is essential to analyze DC subsets directly purified from tissues. Here, we describe a protocol allowing the purification of DC subsets from human tonsils and human lymph nodes.

Key words Dendritic cells, DC subsets, Human, Tonsil, Lymph node

1 Introduction

Dendritic cells (DCs) represent a complex population of antigen-presenting cells. Numerous studies in mice and humans have revealed DC heterogeneity and the existence of phenotypically and ontogenically distinct DC subsets. Two main categories of DC can be distinguished: plasmacytoid DC (pDC) and classical DC (cDC). cDC can be further classified according to their migratory capacity. Migratory DCs are first localized in peripheral tissues and can migrate through the lymph to the draining lymph nodes to initiate immune responses. They represent roughly 50 % of draining lymph node DC and are absent from non-draining lymphoid organs (tonsils, spleen, and thymus) [1]. By contrast, resident DCs spend their entire life cycle in lymphoid organs and represent all the DCs found in non-draining lymphoid organs and 50 % of the DCs in draining lymph nodes.

Three resident DC subsets are present in human non-draining lymphoid organs: CD1c/BDCA1⁺ DC, CD141/BDCA3⁺ DC, and pDC [2, 3]. In the tonsils, pDCs are the most abundant DC subset, while CD141⁺ DCs are the rarest. It has been shown that resident DCs from the tonsils are immature, as they do not express co-stimulatory molecules [4].

Elodie Segura and Nobuyuki Onai (eds.), *Dendritic Cell Protocols*, Methods in Molecular Biology, vol. 1423,
DOI 10.1007/978-1-4939-3606-9_6, © Springer Science+Business Media New York 2016

In skin-draining lymph nodes, the three resident DC subsets were identified (CD1c+ DC, CD141+ DC, and pDC) along with three subsets of migratory DC (Langerhans cells (LCs), CD1a+ DC, CD206+ DC) [5]. Phenotypic and functional evidence suggests that the CD206+ DC subset is the lymph node counterpart of CD14+ DC from the dermis [5]. During migration, migratory DCs upregulate MHC II molecules and co-stimulatory molecules [5, 6]. Recent studies describe the existence of CD141+ DC in peripheral tissues [7, 8], suggesting that lymph nodes contain a proportion of previously unnoticed migratory CD141+ DC.

The protocol described here is designed for the purification of the different DC subsets from tonsils and lymph nodes. This protocol consists of three steps: digestion of the organs, DC enrichment, and cell sorting.

2 Materials

2.1 Mechanical and Enzymatic Tissue Digestion

1. Scalpel.
2. 14 and 50 mL tubes.
3. Iscove's Modified Dulbecco's Medium (IMDM).
4. DNase (Roche).
5. Liberase TL Research Grade low Thermolysin concentration (Roche).
6. Multi-Axle-Rotating Mixer.
7. EDTA 0.1 M pH 7.2.
8. 40 μm cell strainer.
9. Digestion medium: IMDM, 0.1 mg/mL of liberase TL, and 0.1 mg/mL of DNase.

2.2 DC Enrichment

1. Iscove's Modified Dulbecco's Medium (IMDM).
2. Lymphoprep (Greiner Bio-One).
3. Staining buffer: PBS, 0.5% human serum, and 2 mM EDTA (see Note 1).
4. Antibodies for T and B lymphocytes, NK cells, and erythrocyte depletion (see Table 1).
5. Magnetic microbeads for T and B lymphocytes, NK cells, erythrocytes, and myelomonocytic cell depletion (see Table 1).
6. LD columns (Miltenyi Biotec).
7. MACS separator magnet (Miltenyi Biotec).

2.3 Cell Sorting of the DC Subsets

1. Staining buffer.
2. Fc receptor blocking reagent (TruStain, BioLegend).

Table 1
Antibodies and microbeads for DC enrichment

Name	Fluorochrome	Clone	Company	Suggested dilution
Mouse anti-CD3	Purified	HIT3a	eBioscience	1/100
Mouse anti-CD19	Purified	HIB19	eBioscience	1/100
Mouse anti-CD235a	Purified	HIR2	eBioscience	1/100
Mouse anti-CD56	Purified	CMSSB	eBioscience	1/100
Anti-mouse IgG1	Microbeads		Miltenyi Biotec	1/10
Anti-mouse IgG2a+b	Microbeads		Miltenyi Biotec	1/20
Antihuman CD15	Microbeads		Miltenyi Biotec	1/20

Table 2
Antibodies for cell sorting of tonsil DC

Name	Fluorochrome	Clone	Company	Suggested dilution
Mouse anti-HLADR	APC efluor780	LN3	eBioscience	1/100
Mouse anti-CD11c	PE Cy7	Bu15	eBioscience	1/200
Mouse anti-CD14	FITC	61D3	eBioscience	1/100
Mouse anti-CD1c	PerCP efluor710	L161	eBioscience	1/100
Mouse anti-CD141	PE	A05-14H2	Miltenyi Biotec	1/200
Mouse anti-CD304	APC	REA380	Miltenyi Biotec	1/20

3. Antibodies for tonsil DC (*see* Table 2) or for lymph node DC (*see* Table 3).

4. DAPI: stock solution at 1 mg/mL.

5. 5 mL polystyrene tubes with a 40 μm cell strainer cap.

6. Cell sorter cytometer.

7. 5 mL polypropylene tubes.

8. Decomplemented fetal bovine serum (FBS).

3 Methods

The use of human samples must be approved by an institutional ethics committee. All steps must be performed in sterile conditions.

Table 3
Antibodies for cell sorting of lymph node DC

Name	Fluorochrome	Clone	Company	Suggested dilution
Mouse anti-HLADR	APC efluor780	LN3	eBioscience	1/100
Mouse anti-CD11c	PE Cy7	Bu15	Biolegend	1/200
Mouse anti-CD14	PE	61D3	eBioscience	1/100
Mouse anti-CD1c	PerCP efluor710	L161	eBioscience	1/100
Mouse anti-CD206	Alexa Fluor 647	15-2	Biolegend	1/50
Mouse anti-CD1a	Alexa Fluor 488	703217	R&D Systems	1/100
Mouse anti-Clec9a	Biotin	8F9	Miltenyi Biotec	1/50

3.1 Purification of DC from Tonsils

3.1.1 Mechanical and Enzymatic Tissue Digestion

To dissociate thoroughly the tissue and be able to isolate DC subsets, we perform first a mechanical dissociation followed by an enzymatic digestion with constant agitation.

1. Use a scalpel to chop the tissue into small pieces until to get a "mash" of tissue.

2. Prepare 40 mL of digestion medium per pair of tonsils (*see* **Note 2**).

3. Place the "mash" of tissue into 40 mL of digestion medium and incubate for 40–45 min at room temperature (RT) (*see* **Note 3**) and with constant agitation (*see* **Note 4**).

4. Add 2 mL of EDTA 0.1 M to the digested cell suspension and incubate for 2 min at RT (*see* **Note 5**).

5. To remove undigested debris, filter the cell suspension into a 40 μm cell strainer.

6. Centrifuge the cell suspension for 10 min at $450 \times g$ at RT.

7. Resuspend the cells into 20 mL of IMDM.

3.1.2 DC Enrichment

Due to the scarcity of DC, it is necessary to enrich the sample for DC by depleting T and B lymphocytes, NK cells, erythrocytes, and myelomonocytic cells (*see* Figs. 1 and 2).

1. Put 17 mL of Lymphoprep into a 50 mL tube, and then slowly layer the 20 mL cell suspension on top of the Lymphoprep solution (*see* **Note 6**).

2. Centrifuge for 15 min at $800 \times g$ at RT with the decelerating brake off (*see* **Note 7**).

3. Collect the cells at the interface with a 5 mL pipette, and wash the cells three times with 50 mL of cold IMDM (*see* **Note 8**).

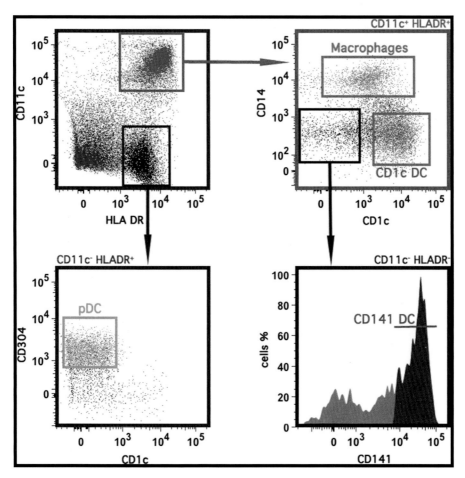

Fig. 1 Gating strategy of the cell sorting of the three tonsil resident DC subsets after DC enrichment. After tissue digestion and DC enrichment, cells were stained for CD11c, HLADR, CD14, CD1c, CD141, and CD304. Dead cells and doublets were excluded by using DAPI staining and SSC-W SSC-A lasers, respectively. Macrophages and cDC are first gated on CD11c and HLA DR (*top left panel*). Macrophages are CD14+, and CD1c DCs are CD1c+CD14−, whereas the double negative population contains CD141 DC (*top right panel*). CD141 DCs are identified using CD141 staining (*bottom right panel*). pDCs are first gated as HLADR+ CD11c− (*top left panel*) and subsequently selected using CD304 staining (*bottom left panel*)

4. Count the cells, centrifuge for 5 min at $450 \times g$ at 4 °C, and resuspend the cells at 200×10^6 cells/mL in the antibodies solution (anti-CD3, anti-CD19, anti-CD56, and anti-CD235a antibodies at 10 μg/mL in staining buffer (*see* Table 1)) (*see* **Note 9**).

5. Incubate for 30 min at 4 °C.

6. Wash the cells with 10–20 mL of cold staining buffer.

7. Resuspend the cells in staining buffer at 400×10^6 cells/mL, and add the microbeads (mouse IgG1 microbeads, mouse IgG2a + b, and CD15 microbeads) (*see* suggested dilution in Table 1).

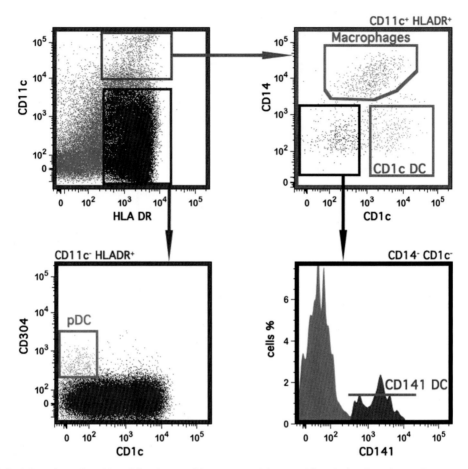

Fig. 2 Staining of tonsil resident DC subsets without pre-enrichment. After tissue digestion, cells were stained for CD11c, HLADR, CD14, CD1c, CD141, and CD304. Macrophages and cDC are contained within CD11c + HLADR+ cells (*top left panel*). Macrophages are CD14+, and CD1c DCs are CD1c + CD14−, whereas the double negative population contains CD141 DC (*top right panel*). CD141 DCs are identified as CD141+ (*bottom right panel*). pDCs are contained within HLADR+ CD11c− cells (*top left panel*) and are identified as CD304+ (*bottom left panel*)

8. Incubate for 20 min at 4 °C.

9. Place the LD columns on the separator magnet and wash them with 3 mL of staining buffer (*see* **Note 10**).

10. Wash the cells with 10 mL of staining buffer.

11. Resuspend the cells at 350×10^6 cells/mL in staining buffer.

12. Distribute 2 mL of cell suspension per LD column.

13. Collect the eluate (negative fraction that is enriched in DC).

14. Wash twice the columns with 1 mL of staining buffer and collect the eluate.

15. Centrifuge the eluate for 5 min at $450 \times g$ at 4 °C (*see* **Note 11**).

16. Remove the column from the separator magnet, add 3 mL of staining buffer at the top of the column, and use the syringe to push the positive fraction out. These cells can be used for unstained and DAPI control tubes for cell sorting.

3.1.3 Cell Sorting of the DC Subsets

1. Resuspend the cells enriched in DC at 10×10^6 cells/100 µL in antibodies solution containing Fc receptor blocking reagent (*see* Table 2 for antibodies and suggested dilutions).

2. Incubate for 30 min at 4 °C.

3. Wash the cells with 10 mL of cold staining buffer.

4. Resuspend the cells in staining buffer containing DAPI (200 ng/mL final dilution) (*see* **Note 12**).

5. Prepare single-stain tubes for the different fluorochromes. Cells from the positive fraction of the DC enrichment can be used to compensate the DAPI (*see* **Note 13**).

6. Prepare collecting tubes: put 1 mL of FBS in 5 mL polypropylene tubes (*see* **Note 14**).

7. Sort the cells using a cell sorter cytometer. CD1c DCs are CD11c+ HLA DR+,CD14- CD1c+ cells. CD141 DCs are CD11c+ HLA DR+ CD14-, CD1c- CD141+ cells. pDCs are CD11c- HLA DR+ CD14- CD1c- CD304+ cells (*see* Fig. 1).

3.2 Purification of DC from Lymph Nodes

To dissociate thoroughly the tissue and be able to isolate DC subsets, we perform first a mechanical dissociation followed by an enzymatic digestion with constant agitation.

3.2.1 Mechanical and Enzymatic Tissue Digestion

1. Use a scalpel to chop the lymph nodes into small pieces until to get a "mash" of tissue.

2. Prepare 5 mL of digestion medium per lymph node sample (*see* **Note 2**).

3. Place the "mash" of tissue into 5 mL of digestion medium, and incubate for 20 min at room temperature (RT) (*see* **Note 3**) and with constant agitation (*see* **Note 4**).

4. Add 0.25 mL of EDTA 0.1 M to the digested cell suspension and incubate for 2 min at RT (*see* **Note 5**).

5. To remove undigested debris, filter the cell suspension into a 40 µm cell strainer.

6. Centrifuge the cell suspension for 10 min at $450 \times g$ at 4 °C.

7. Resuspend the cells into 5 mL of cold IMDM.

3.2.2 DC Enrichment

Due to the scarcity of DC, it is necessary to enrich the sample for DC by depleting T and B lymphocytes, NK cells, erythrocytes, and myelomonocytic cells.

1. Count the cells, centrifuge, and resuspend the cells at 200×10^6 cells/mL in the antibodies solution (anti-CD3, anti-CD19, anti-CD56, and anti-CD235a antibodies) at 10 µg/mL in staining buffer (*see* Table 1) (*see* **Note 9**).

2. Incubate for 30 min at 4 °C.

3. Wash the cells with 5 mL of cold staining buffer.

4. Resuspend the cells in staining buffer at 400×10^6 cells/mL, and add the microbeads (mouse IgG1 microbeads, mouse IgG2a + b, and CD15 microbeads) (*see* suggested dilution in Table 1).

5. Incubate for 20 min at 4 °C.

6. Place the LD columns on the separator magnet and wash them with 3 mL of staining buffer (*see* **Note 10**).

7. Wash the cells with 10 mL of staining buffer.

8. Resuspend the cells at a maximum concentration of 350×10^6 cells/mL in staining buffer. Minimum volume for the cell suspension is 2 mL.

9. Apply the cell suspension on LD columns.

10. Collect the eluate (negative fraction that is enriched in DC).

11. Wash twice the columns with 1 mL of staining buffer and collect the eluate.

12. Centrifuge the eluate for 5 min at $450 \times g$ (*see* **Note 11**).

13. Remove the column from the separator magnet, add 3 mL of staining buffer at the top of the column, and use the syringe to push the positive fraction out. These cells can be used for unstained and DAPI control tubes for cell sorting.

3.2.3 Cell Sorting of the DC Subsets

1. Resuspend the cells at 10×10^6 cells/100 μL in antibodies solution containing Fc receptor blocking reagent (*see* Table 3 for antibodies and suggested dilutions).

2. Incubate for 30 min at 4 °C.

3. Wash the cells with 10 mL of staining buffer.

4. If necessary, incubate the cells 20 min at 4 °C with fluorochrome-conjugated streptavidin and repeat **step 3**.

5. Resuspend the cells in staining buffer with DAPI (200 ng/mL final dilution) (*see* **Note 12**).

6. Prepare single-color stained tubes for the different fluorochrome; you can use the positive fraction of the DC enrichment to compensate the DAPI (*see* **Note 13**).

7. Prepare collecting tubes: put 1 mL of FBS in 5 mL polypropylene tubes (*see* **Note 14**).

8. Sort the cells using a cell sorter cytometer. CD1a DCs are CD11c⁺ HLADR⁺ CD1a⁺ CD14⁻ Epcam⁻ cells. LCs are CD11c⁺ HLADR⁺ CD1a⁺ CD14⁻ Epcam⁺ cells. Macrophages are CD11c⁺ HLADR⁺ CD1a⁺ CD14⁺ cells. CD141 DCs

are CD11c$^+$ HLADR$^+$ CD1a$^-$ CD14$^-$ CD1c$^-$ Clec9a$^+$ cells. CD1c DCs are CD11c$^+$ HLADR$^+$ CD1a$^-$ CD14$^-$ CD1c$^+$ Clec9a$^-$ CD206$^-$ cells. CD206 DCs are CD11c$^+$ HLADR$^+$ CD1a$^-$ CD14$^-$ CD1c$^+$ Clec9a$^-$ CD206$^+$ cells. pDCs are CD11c$^-$ HLADR$^+$ CD304$^+$ cells (*see* Fig. 3 for gating strategy).

4   Notes

1. To minimize DC activation, the staining buffer has to remain at 4 °C during the whole protocol.

2. Liberase is used instead of collagenase because of higher efficiency and lower stripping of cell surface molecules.

3. The digestion is done at RT to minimize DC activation.

4. The digestion has to be performed in constant agitation for improved efficiency.

5. The EDTA stops the enzymatic digestion and allows the dissociation of the cell conjugates (in particular T cell-DC conjugates).

6. Other Ficoll reagents can be used.

7. The use of the function "brake off" is essential for the isolation of low-density cells on Ficoll gradient.

8. At this step you can resuspend the cells in culture medium containing 10 % of FBS, store the cells at 4 °C for 8–12 h, and resume the protocol later. We recommend using Yssel medium for improved DC viability.

9. Include an antihuman CD14 antibody if you wish to remove macrophages.

10. Determine how many LD columns are needed. For optimum enrichment, use one LD column for a maximum of 700×10^6 cells.

11. At this step you can resuspend the cells in culture medium containing 10 % of FBS, store the cells at 4 °C for 8–12 h, and resume the protocol later. We recommend using Yssel medium for improved DC viability.

12. Other viability dyes can be used.

13. Using beads for single-stain compensation controls (for instance, BD CompBeads from BD Biosciences) allows saving cells from the samples.

14. Using polypropylene tubes avoids the attachment of DC onto the tube wall during cell sorting.

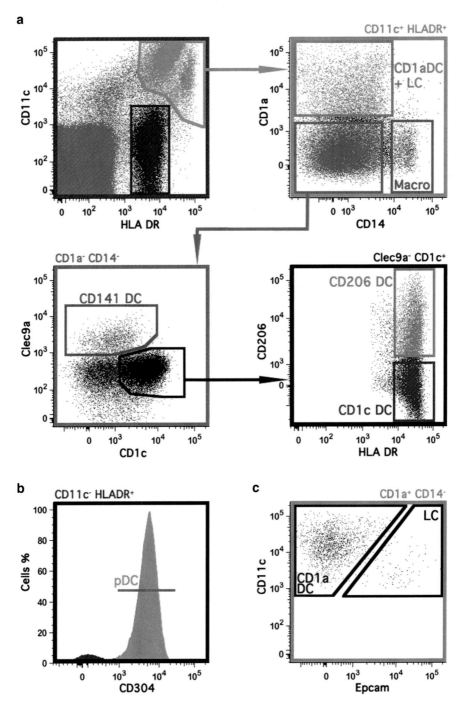

Fig. 3 Gating strategy of the cell sorting of DC subsets from lymph node after DC enrichment. (**a**) After tissue digestion and DC enrichment, cells were stained for CD11c, HLADR, CD1a, Clec9a, CD1c, and CD206. Resident cDC and
migratory DC are gated as CD11c+HLADR+ (*top left panel* **a**). Cells are then separated according to the expression of CD1a and CD14: macrophages are CD1a−CD14+ and CD1a+CD14− contain both Langerhans cells and
CD1a+DC (*top right panel* **a**). CD1a− CD14− are further separated according to the expression of Clec9a and
CD1c (*bottom left panel* **a**). CD141 DCs are identified as Clec9a+CD1c−. CD1c+ cells are finally separated using
CD206 staining allowing the separation of migratory CD206 DC and resident CD1c DC (*bottom right panel* **a**).
(**b**) Additional stainings allow the isolation of pDC. pDCs are first gated as CD11c−HLA DR+, then identified by the
expression of CD304. (**c**) Epcam staining allows the separation of CD1a DC (Epcam−) and Langerhans cells
(Epcam+)

References

1. Villadangos JA, Schnorrer P (2007) Intrinsic and cooperative antigen-presenting functions of dendritic-cell subsets in vivo. Nat Rev Immunol 7(7):543–555

2. McIlroy D, Troadec C, Grassi F, Samri A, Barrou B, Autran B, Debre P, Feuillard J, Hosmalin A (2001) Investigation of human spleen dendritic cell phenotype and distribution reveals evidence of in vivo activation in a subset of organ donors. Blood 97(11):3470–3477

3. Lindstedt M, Lundberg K, Borrebaeck CA (2005) Gene family clustering identifies functionally associated subsets of human in vivo blood and tonsillar dendritic cells. J Immunol 175(8):4839–4846

4. Segura E, Durand M, Amigorena S (2013) Similar antigen cross-presentation capacity and phagocytic functions in all freshly isolated human lymphoid organ-resident dendritic cells. J Exp Med 210(5):1035–1047

5. Segura E, Valladeau-Guilemond J, Donnadieu MH, Sastre-Garau X, Soumelis V, Amigorena S (2012) Characterization of resident and migratory dendritic cells in human lymph nodes. J Exp Med 209(4):653–660

6. Villadangos JA, Heath WR (2005) Life cycle, migration and antigen presenting functions of spleen and lymph node dendritic cells: limitations of the Langerhans cells paradigm. Semin Immunol 17(4):262–272

7. Haniffa M, Shin A, Bigley V, McGovern N, Teo P, See P, Wasan PS, Wang XN, Malinarich F, Malleret B, Larbi A, Tan P, Zhao H, Poidinger M, Pagan S, Cookson S, Dickinson R, Dimmick I, Jarrett RF, Renia L, Tam J, Song C, Connolly J, Chan JK, Gehring A, Bertoletti A, Collin M, Ginhoux F (2012) Human tissues contain CD141(hi) cross-presenting dendritic cells with functional homology to mouse CD103(+) nonlymphoid dendritic cells. Immunity 37(1): 60–73

8. Duluc D, Gannevat J, Anguiano E, Zurawski S, Carley M, Boreham M, Stecher J, Dullaers M, Banchereau J, Oh S (2013) Functional diversity of human vaginal APC subsets in directing T-cell responses. Mucosal Immunol 6(3):626–638

Chapter 7

Isolation and Identification of Conventional Dendritic Cell Subsets from the Intestine of Mice and Men

Charlotte L. Scott, Pamela B. Wright, Simon W.F. Milling, and Allan McI Mowat

Abstract

The identification of conventional dendritic cells (cDCs) in the intestinal mucosa has been hampered by the difficulties associated with isolating cells from the intestine and by the fact that overlapping markers have made it complicated to discriminate them accurately from other intestinal mononuclear phagocytes such as macrophages (MFs). Here we detail the protocols we have developed to isolate live leukocytes from both murine and human small and large intestines and describe reliable strategies which can be used to identify bona fide cDCs in such preparations.

Key words Dendritic cells, Subsets, Isolation, Identification, Intestine

1 Introduction

Given the large surface area of the gut and its continual exposure to a wide variety of agents including dietary constituents, commensal bacteria, and pathogens, the intestine has evolved the largest component of the immune system [1]. The intestinal immune system must be able to discriminate between harmless and harmful antigens. Whereas tolerance must be induced to harmless materials such as commensal, self-, or dietary antigens, active immunity must be generated for the eradication of pathogens. As the sentinels of the immune system, intestinal conventional dendritic cells (cDCs) are absolutely central to this process, continually sampling antigen in their environment and migrating constitutively to the draining mesenteric lymph nodes (MLNs). Here they present the antigen to naïve T cells and initiate tolerance or active immunity as appropriate. When this decision-making process breaks down, aberrant responses to innocuous antigens may lead to a number of pathologies

---

All authors contributed equally

Elodie Segura and Nobuyuki Onai (eds.), *Dendritic Cell Protocols*, Methods in Molecular Biology, vol. 1423,
DOI 10.1007/978-1-4939-3606-9_7, © Springer Science+Business Media New York 2016

including food allergies, celiac disease, and inflammatory bowel diseases such as Crohn's disease and ulcerative colitis.

The gastrointestinal tract consists of the stomach, small intestine, cecum, large intestine (colon), and rectum, with cDCs being found in several compartments of the intestinal immune system. These include the organized lymphoid tissues associated with the intestine such as the Peyer's patches (PPs), microscopic isolated lymphoid follicles (ILFs), and the draining mesenteric lymph nodes (MLNs). PP and ILFs are important for the generation of immune responses against bacteria and viruses, while the MLNs are essential for all aspects of tolerance and immunity in the gut. However, as cDCs in these tissues can usually be isolated and characterized using the same methods that are used for other secondary lymphoid organs, they are not the focus of this chapter. Instead, we concentrate on the large population of cDCs present in the wall of the intestine itself (the mucosa), most of which are found in the lamina propria (LP), the layer of connective tissue immediately below the epithelium. These cDCs have many unusual properties and their isolation and subsequent characterization present several unique challenges to the researcher. As we will discuss, a number of subsets of cDCs are present in the LP of both the small and large intestines. All share the abilities to acquire antigen in LP and migrate to the draining MLNs, where they encounter and present antigen to naïve T cells. As a result, the "migratory" subset of cDCs in MLNs contains analogous populations to those seen in LP and similar phenotypic groups can also be found in PPs ([2–6] and our own unpublished observations). Plasmacytoid DCs (pDCs) are also found in the LP of the small intestine of both mice and humans, but they do not migrate to the draining lymph nodes and are absent from the normal colon [7].

Despite the fact that cDCs were first identified in the LP many years ago [8], it is only recently that progress has been made in their characterization and we still have much to learn about them. Like other cells in the mucosa, cDCs are difficult to isolate in substantial numbers and our understanding of them has been hampered further because many of the markers used to identify them are now known to be nonspecific in nature. As a result, the "DC" populations used in many studies also contain other cells of the mononuclear phagocyte (MP) system, particularly macrophages (MFs), which are more abundant and which share phenotypic markers such as CD11b, CD11c, and MHCII. In this way, intestinal cDCs have often been ascribed functions that are actually attributes of MFs. This is a crucial distinction to make, as cDCs and MFs in the intestine serve complementary yet distinct functions. MFs do not sample antigen and migrate to the draining lymph nodes or prime naïve T cells. Instead they are sessile phagocytes which scavenge bacteria and damaged cells and maintain an anti-inflammatory environment in the mucosa via constitutive production of IL10 [9]. If they do interact with T cells, this is localized to the LP and involves T cells that have migrated there after initial activation in the draining LNs [10]. Therefore, additional markers are needed to separate these cell types [4, 6, 11, 12]. To do this, we exclude MFs on the basis of several pan-MF markers. We find CD64 or F4/80 to be the most useful of these in mice, although CD14 or MerTK can also be used. Using this approach, we have shown that all CD64$^-$/F4/80$^-$ cells in the CD11c$^+$MHCII$^+$ gate are bona fide cDCs, based on their expression of markers such as CD26, CD135 (Flt3), and the cDC-specific gene *Zbtb46*; their development in vivo is dependent on the Flt3L growth factor and they are derived from a committed cDC precursor [6]. Unlike LP MFs, the CD64$^-$/F4/80$^-$ CD11c$^+$MHCII$^+$ cells also express CCR7, migrate via lymphatics to draining lymph nodes and are capable of retinoic acid production [5, 6, 9, 13]. Although often considered to define cells of the monocyte/macrophage lineage, CCR2 expression is not sufficient to discriminate between cDCs and MCs in the murine and human gut, as at least some mature LP DCs express this receptor [6]. It should be noted that cDCs are relatively more numerous in the small intestine, whereas MFs are more frequent in the large intestine [7].

Further complications in our understanding of LP cDCs have come from the advent of multiparameter flow cytometric techniques, which have revealed considerable heterogeneity, with several distinct cDC subsets. cDCs can be separated into four distinct groups defined on the basis of CD103 and CD11b expression, revealing CD103$^+$CD11b$^+$, CD103$^+$CD11b$^-$, CD103$^-$CD11b$^+$, and CD103$^-$CD11b$^-$ subsets. The relative proportions of the subsets vary along the length of the intestine, with the CD103$^+$CD11b$^+$ cDCs predominating in the small intestine and CD103$^+$CD11b$^-$ cDCs being the most abundant population in the colon [7]. Much remains to be understood regarding the in vivo functions of the individual populations. While the CD103$^+$CD11b$^-$ cDCs appear to be equivalent to the lineage of XCR1$^+$CD103$^+$ cDCs that can cross-present exogenous antigen to CD8$^+$ T cells in other murine tissues [14–16], the functions of the other subsets are less well defined. The CD103$^+$CD11b$^+$ cDCs have only been reported in the intestine and they have been shown to drive the differentiation of FoxP3$^+$ regulatory T cells in vitro. However depletion of CD103$^+$CD11b$^+$ cDCs in vivo leads to a defect in Th17 cells in the intestine [2–4, 17, 18], and the numbers of TRegs are only reduced under conditions when both CD103$^+$ cDC populations are absent [19]. The CD103$^-$CD11b$^+$ subset of intestinal cDCs may have a relatively enhanced ability to prime Th17 cells in vitro and this is especially so for a subset within this population which expresses CCR2 [5, 6]. The relationship of these CD103$^-$CD11b$^+$ cDCs to the CD103$^+$CD11b$^+$ subset and to CD11b$^+$ cDCs in other mouse tissues is unclear.

Analogous studies of human intestinal MP cell heterogeneity and phenotype are in their infancy, and very little is known about the ontogeny, development, and functions of human gut cDCs. Our approach has been to follow similar strategies to those described above in mice to allow human cDCs to be characterized accurately. The numbers and viability of cells obtained from human material are highly variable and depend upon whether resected or biopsy samples are being used.

Here we describe how to isolate and characterize cDCs from the mucosa of the small and large intestines of mice and men, including protocols suitable for use with biopsy and resected material from humans.

2 Materials

2.1 Reagents (See Note 1)

1. Ethylenediamine tetraacetic acid (EDTA) solution. Stock at 0.5 M and filter sterilized.

2. Calcium- and magnesium-free (CMF) Hank's buffered salt solution (HBSS) 2%, fetal calf serum (FCS). Store at 4 °C.

3. CMF HBSS 2 mM EDTA solution. Pre-warm and store at 37 °C.

4. CMF HBSS. Pre-warm and store at 37 °C.

5. Complete RPMI (cRPMI): RPMI supplemented with 10% FCS, 2 mM L-glutamine, 100 U/ml penicillin, 100 μg/ml streptomycin, and 50 μM 2-mercaptoethanol. Pre-warm and maintain at 37 °C.

6. FACS buffer: PBS containing 2% FCS and 1 mM EDTA. Store at 4 °C.

7. Enzymatic cocktails specific for tissue to be digested (see Table 1) in complete RPMI. Make up fresh immediately prior to use.

2.2 Equipment

1. 500 ml plastic beakers.

2. Funnels.

3. 9 ml Petri dish.

4. Nitex nylon mesh (50 μm).

5. Tweezers.

6. Scissors.

7. 5 ml syringes.

8. Shaking incubator or water bath at 37 °C.

9. 37 °C incubator.

10. 100 and 40 μM cell strainers.

11. 24-well flat-bottomed tissue-culture plate (human biopsies).

Table 1
Enzyme cocktails for intestinal digestions

Tissue	Species	Vol. (ml)	Enzymes	Supplier	Conc.
Small intestine	Murine	10	Collagenase VIII	Sigma	0.6 mg/ml
Colon	Murine	10	Collagenase V	Sigma	0.425 mg/ml
			Collagenase D	Roche	0.75 mg/ml
			Dispase	Invitrogen	1 mg/ml
			DNase	Roche	30 μg/ml
Intestinal biopsy	Human	2	Collagenase VIII	Sigma	1 mg/ml
			Collagenase D	Roche	1.25 mg/ml
			Dispase	Invitrogen	1 mg/ml
			DNase	Roche	30 μg/ml
Resected intestine	Human	20	Collagenase VIII	Sigma	1 mg/ml
			Collagenase D	Roche	1.25 mg/ml
			Dispase	Invitrogen	1 mg/ml
			DNase	Roche	30 μg/ml

12. 50 ml Falcon tubes (except human biopsies).

13. 6 ml tubes (human biopsies).

14. Analytical balance.

15. Parafilm (human biopsies).

16. 6 ml polypropylene FACS tubes.

17. 96-well round-bottom tissue-culture plates.

18. Cryovials (human biopsies).

19. Hemocytometer.

20. Microscope.

21. Centrifuge.

22. Antibodies (*see* Table 2).

23. Flow cytometer.

3 Methods

3.1 Isolation of Leukocytes from Mouse Small Intestinal Lamina Propria

1. Kill the mouse according to appropriate ethical guidelines and immediately remove the small intestine (SI) from the pyloric outlet of the stomach to the end of ileum and place in a 9 ml Petri dish filled with CMF HBSS 2 % FCS (*see* **Note 2**).

2. Place SI on tissue paper, soak with PBS, and remove fat with tweezers.

3. Carefully remove all Peyer's patches from SI with scissors (*see* **Note 3**).

Table 2
Antibodies and reagents used to identify mouse intestinal DCs by flow cytometry

Antibody	Clone	Isotype	F Fluorochrome
Fc block (anti-CD16/CD32)	2.4G2		N/A
CD45	30-F11	Rat IgG2b	AF700
CD11c	N418	Ham IgG1	PE-Cy7
CD11b	M1/70	Rat IgG2b	APC-Cy7
CD103	M290	Rat IgG2a	PE
MHCII (IA/IE)	M5/114.15.2	Rat IgG2b	eFluor450
CD64	X54-5/7.1	Mouse IgG1	AF647
F4/80	BM8	Rat IgG2a	APC
CD45R (B220)	RA3-6B2	Rat IgG2a	AF488
7-AAD			PerCP/PerCPCy5.5/PECy5 (*see* **Note 29**)

4. Open SI longitudinally using scissors.

5. Wash thoroughly in CMF HBSS 2 % FCS in a 9 ml Petri dish.

6. Cut SI into 0.5 cm segments.

7. Transfer SI segments to a 50 ml Falcon tube containing 10 ml CMF HBSS 2 % FCS and keep on ice (*see* **Note 4**).

8. Manually shake the tubes vigorously and then discard the supernatant (*see* **Note 5** and Fig. 1).

9. Add 10 ml of CMF HBSS 2 mM EDTA to the remaining pieces of SI and transfer to a shaking water bath/incubator at 37 °C for 20 min (*see* **Notes 6** and **7**).

10. Manually shake the tubes vigorously and then discard the supernatant.

11. Wash the pieces of SI by adding 10 ml of warmed CMF HBSS, manually shake the tubes vigorously, and remove the supernatant.

12. Add 10 ml of CMF HBSS 2 mM EDTA and transfer to a shaking water bath/incubator at 37 °C for 20 min.

13. Prepare the appropriate enzyme cocktail (*see* Table 1) in 10 ml of complete RPMI.

14. Manually shake the tubes vigorously and discard the supernatant.

15. Wash by adding 10 ml of warmed CMF HBSS and then manually shake the tubes vigorously and remove the supernatant.

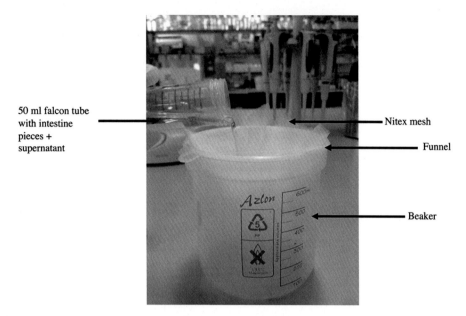

Fig. 1 Removal of supernatants during digestion of murine intestine and resected human material. Supernatants are removed by pouring tube contents through Nitex nylon mesh inserted in a funnel, allowing supernatant waste to drain into a connecting beaker, while tissue is retained on the Nitex nylon mesh

16. Add 10 ml of the prepared enzyme cocktail to each tube.

17. Transfer to a shaking water bath/incubator at 37 °C, for approximately 20 min, or until all tissue pieces are almost completely digested, with manual vigorous shaking performed every 5 min (*see* **Notes 8** and **9**).

18. Add 10 ml of cold FACS buffer and store on ice.

19. Pass the single-cell suspension through a 100 μm cell strainer.

20. Pass the single-cell suspension through a 40 μm cell strainer.

21. Centrifuge at $400 \times g$ for 5 min and resuspend in 5 ml of FACS buffer.

22. Count the cells (*see* **Note 10**) and keep on ice until use.

3.2 Isolation of Leukocytes from Mouse Colonic Lamina Propria

1. Kill the mouse according to appropriate ethical guidelines and immediately remove the colon from the top of cecum to the rectum, (*see* **Note 11**) place on tissue paper, soak in PBS, and remove the cecum and as much fat and feces as possible using tweezers (*see* **Notes 1** and **2**).

2. Open the colon longitudinally with scissors (*see* **Note 3**).

3. Wash thoroughly in CMF HBSS 2 % FCS in a 9 ml Petri dish.

4. Cut the colon into 0.5 cm segments.

5. Transfer the colon to a 50 ml Falcon tube with 10 ml of CMF HBSS 2 % FCS on ice (*see* **Note 4**).

6. Manually shake the tubes vigorously and remove the supernatant (*see* **Note 5** and Fig. 1).

7. Add 10 ml of CMF HBSS 2 mM EDTA and transfer to a shaking water bath/incubator at 37 °C for 15 min (*see* **Notes 6 and 7**).

8. Manually shake the tubes vigorously and remove the supernatant.

9. Wash by adding 10 ml of warmed HBSS and then manually shake the tubes vigorously and remove the supernatant.

10. Add 10 ml of CMF HBSS 2 mM EDTA and transfer to a shaking water bath/incubator at 37 °C for 30 min.

11. Prepare the appropriate enzyme cocktail per Table 1 in 10 ml of cRPMI.

12. Manually shake the tubes vigorously and remove the supernatant.

13. Wash by adding 10 ml of warmed CMF HBSS and then manually shake tubes vigorously and remove the supernatant.

14. Add 10 ml of the enzyme cocktail.

15. Transfer to a shaking water bath/incubator at 37 °C, for approximately 45 min, or until all tissue pieces are almost completely digested, with manual vigorous shaking performed every 5 min (*see* **Notes 8** and **9**).

16. Add 10 ml of cold FACS buffer and store on ice.

17. Pass the single-cell suspension through a 40 μm cell strainer.

18. Centrifuge at $400 \times g$ for 5 min and resuspend in 5 ml of FACS buffer.

19. Count the cells (*see* **Note 12**) and keep on ice until use.

3.3 Isolation of Leukocytes from Human Colonic and Small Intestinal Biopsies

1. Biopsies should be collected into cryovials containing ice-cold PBS.

2. Immediately on arrival in the laboratory, transfer the tissue to a pre-weighed 6 ml tube containing 2 ml of ice-cold CMF HBSS 2% FCS and store on ice (*see* **Note 13**).

3. To allow for the accurate determination of cell number per gram of tissue biopsy, weigh the tissue.

4. Prepare the appropriate enzyme cocktail (*see* Table 1) in 2 ml of cRPMI.

5. Remove the supernatant from the tube containing biopsies by transferring the tube contents into a 24-well plate.

6. Isolate the tissue and return it to the tube using tweezers.

7. Add 2 ml of CMF HBSS 2 mM EDTA solution to the tube (*see* **Note 14**) and transfer to a shaking water bath/incubator at 37 °C for 3 min (*see* **Note 15**).

8. Manually shake the tubes vigorously and discard the supernatant (*see* **Note 5**).

9. Wash by adding 2 ml of warmed CMF HBSS (pre-warmed at 37 °C) and then manually shake the tube vigorously and discard the supernatant.

10. Add 2 ml of the enzyme cocktail and seal the tube by wrapping the lid in parafilm (*see* **Note 16**).

11. Transfer to a shaking water bath/incubator at 37 °C for approximately 30–40 min or until all tissue pieces are almost completely digested, with manual vigorous shaking performed every 5 min (*see* **Notes 17** and **18**).

12. Pass the single-cell suspension through a 40 µm cell strainer into a 50 ml Falcon tube.

13. Add 10 ml of FACS buffer, centrifuge at $400 \times g$ for 5 min, and resuspend in 2 ml of FACS buffer.

14. Count the cells (*see* **Note 19**) and keep on ice until use.

3.4 Isolation of Leukocytes from Resected Human Intestine

1. Resected tissue should be collected and transported in 30 ml of ice-cold PBS. Upon arrival, the tissue should be digested immediately without storing (*see* **Note 20**).

2. Transfer the tissue to a 50 ml Falcon tube containing 20 ml of ice-cold CMF HBSS 2 % FCS and store on ice until use.

3. Transfer the tissue to a 9 ml Petri dish containing CMF HBSS 2 % FCS.

4. Remove fat and muscle layers using scissors and tweezers.

5. Transfer remaining mucosal tissue to a fresh, pre-weighed Petri dish containing 5 ml of CMF HBSS 2 % FCS (*see* **Note 13**).

6. To allow for the accurate determination of cell number per gram of tissue, weigh the tissue.

7. Cut the tissue into 0.5 cm segments (*see* **Note 21**).

8. Place the dissected tissue in a 50 ml Falcon tube containing 10 ml of CMF HBSS 2 % FCS.

9. Manually shake the tubes vigorously and discard the supernatant (*see* **Note 5**).

10. Transfer tissue to a 50 ml Falcon tube containing 20 ml CMF HBSS 2 mM EDTA solution (*see* **Note 14**).

11. Transfer to a shaking water bath/incubator at 37 °C for 15 min.

12. Manually shake the tubes vigorously and discard the supernatant.

13. Wash the tissue by adding 20 ml of pre-warmed CMF HBSS, manually shake the tubes vigorously, and discard the supernatant.

14. Repeat **steps 10–13** two more times.

15. Prepare the appropriate enzyme cocktail (*see* Table 1) in 20 ml of cRPMI.

16. Add 20 ml of the appropriate enzyme cocktail to the remaining tissue.

17. Transfer to a shaking water bath/incubator at 37 °C, for approximately 45–60 min, or until all tissue pieces are almost completely digested, with manual vigorous shaking performed every 5 min (*see* **Notes 22** and **23**).

18. Pass the single-cell suspension through a 40 μm cell strainer.

19. Add 10 ml of FACS buffer, centrifuge at $400 \times g$ for 5 min, and resuspend in 5 ml of FACS buffer.

20. Count the cells (*see* **Note 24**) and keep on ice until use.

3.5 Preparation of Cells for FACS Analysis

1. Add $2–4 \times 10^6$ cells in FACS buffer to individual 6 ml FACS tubes (*see* **Note 25**).

2. Centrifuge at $400 \times g$ for 5 min at 4 °C and remove the supernatant.

3. Block Fc receptors by incubating with Fc block. For mouse: incubate for 20 min at 4 °C in FACS buffer at a dilution 1:200 (*see* Table 2). For human: add 5 μl per 2×10^6 cells for 5 min at 4 °C in 100 μl of FACS buffer (*see* Table 3).

4. Add 1 ml of FACS buffer, centrifuge at $400 \times g$ for 5 min at 4 °C, and remove the supernatant.

5. Stain the cells with primary antibodies at 4 °C for 20–30 min in the dark (*see* Tables 2 and 3; **Notes 26–28**).

Table 3
Antibodies and reagents used to identify human intestinal DCs by flow cytometry

Antibody	Clone	Isotype	Fluorochrome
Fc block (anti-CD16/CD32)	2.4G2		N/A
CD14	M5E2	Mouse IgG2a	AF700
CD45	HI30	Mouse IgG1	BV510
CD64	10,1	Mouse IgG1	AF488
CD103	B-Ly7	Mouse IgG1	PE-CY7
CD11c	B-Ly6	Mouse IgG1	V450
HLA-DR	L243	Mouse IgG2a	APC-CY7
SIRPα	SESA5	Mouse IgG1	Biotin
7-AAD			PerCP/PerCPCy5.5/ PECy5 (*see* **Note 29**)
Streptavidin			BV605

6. Add 1 ml of FACS buffer, centrifuge at $400 \times g$ for 5 min at 4 °C, and remove the supernatant.

7. Repeat **step 6**.

8. Where necessary, stain with 1:200 streptavidin conjugates for 15 min at 4 °C in the dark. Then add 1 ml of FACS buffer, centrifuge at $400 \times g$ for 5 min, and remove the supernatant.

9. Add 12 μl per tube of 7-AAD (7-aminoactinomycin D) to discriminate live and dead cells (*see* **Note 29**).

10. Analyze on a flow cytometer (*see* **Notes 30** and **31**).

3.6 Phenotypic Identification of Conventional DCs in the Intestine

As we have discussed, investigating the biology of intestinal DCs has been hampered by a number of factors, many relating to the use of inappropriate and overlapping phenotypic markers. The main aims of our work have been to establish reliable protocols for obtaining suitable numbers of cells of high viability and then to apply rigorous flow cytometric techniques that can be used to characterize mucosal DCs precisely. Below we describe the gating strategies and staining panels that we have found most useful for analyzing DCs in human and murine intestine.

3.6.1 Identifying DCs in the Murine Intestine

1. Identify live CD45+ cells among a broad FSC/SSC gate, using a live/dead marker such as 7-AAD (*see* **Note 29**) (*see* Fig. 2a).

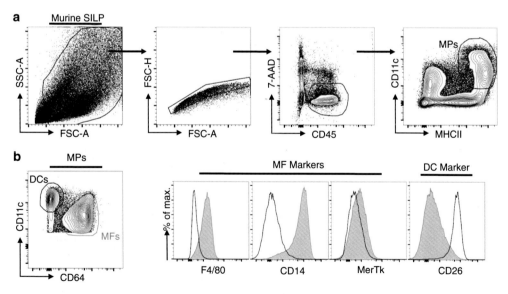

Fig. 2 Identification of cDCs in murine intestine. (**a**) Single live leukocytes are identified among total cells on the basis of FSC and SSC, followed by exclusion of dead cells using 7-AAD⁻ and of non-hematopoietic CD45⁻ cells. Mononuclear phagocytes (MPs) are then identified as CD11c+MHCII+. (**b**) Analysis of CD11c and CD64 expression delineates cDCs and MFs among CD11c+MHCII+ MPs, with CD11c^hi CD64⁻ cells expressing the DC specific marker CD26, whereas the CD11c^int CD64+ cells express the monocyte/MF markers F4/80, CD14, and MerTk. cDCs are shown in *black* and MFs are depicted in *gray*

2. Gate on CD11c⁺MHCII⁺ cells (*see* Fig. **2**a). Although it was this population that was often used in the past as a source of "DCs," it is now clear that it also contains other mononuclear phagocytes (MPs) such as MFs.

3. Exclude MFs on the basis of several pan-MF markers. We find CD64 or F4/80 to be the most useful of these in mice, although CD14 or MerTK can also be used (*see* Fig. **2**b). All CD64⁻/F4/80⁻ cells in the CD11c⁺MHCII⁺ gate are bona fide cDCs, based on their expression of markers such as CD26 (*see* Fig. **2**b) (*see* **Note 32**).

4. Separate cDCs into four distinct groups defined on the basis of CD103 and CD11b expression, revealing CD103⁺CD11b⁺, CD103⁺CD11b⁻, CD103⁻CD11b⁺, and CD103⁻CD11b⁻ subsets (*see* Fig. **3**a) (*see* **Notes 33 and 34**). The subsets can also be characterized further using additional markers (*see* Fig. **3**e) (*see* **Note 35**).

3.6.2 DCs in the Human Intestine

1. Take a broad FSC/SSC gate and identify live CD45⁺ cells (*see* Fig. **4**a).

2. Exclude other MPs such as MFs. In our hands, CD64 and CD14 expression can be used with cDCs then being identified as CD11c⁺ HLA-DR⁺ cells (*see* Fig. **4**b) (*see* **Note 36**).

3. Identify subsets of cDCs analogous to those present in the murine intestine using CD172a (SIRPα) in combination with CD103 (*see* **Note 37**), again revealing four distinct populations (*see* Fig. **4**c) (*see* **Note 38**).

4   Notes

1. Reagents for each experiment can be prepared and incubated the night before they are required, except for enzymes which must be freshly prepared on the day of the experiment.

2. One SI/colon should provide sufficient cells for multiparameter FACS analysis of cDC subsets. If subset purification is required, 6 SIs or 10 colons are needed to obtain approx. 30,000 DCs per subset. If necessary, SIs or colons can be pooled into one 50 ml Falcon tube, but reagents must be scaled up appropriately and no more than three intestines should be pooled per tube. Each volume mentioned in the above protocols is for 1 intestine per tube.

3. PPs must be removed carefully, as they contain cDCs which are different to those in LP. This can be done easily, as PPs are macroscopic. However many 1000s of isolated lymphoid follicles (ILFs) are also found in the intestine, particularly the colon. As ILFs cannot be visualized with the naked eye, it is

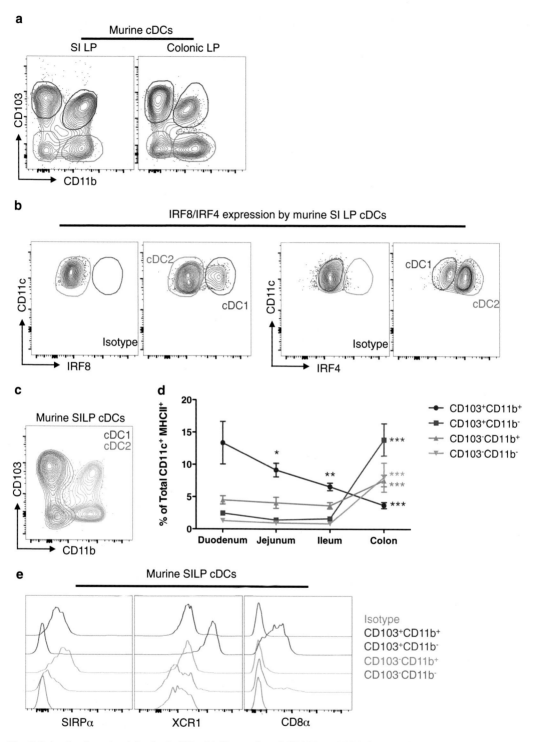

Fig. 3 Subsets of murine intestinal cDCs. (**a**) Expression of CD103 and CD11b on cDCs from mouse colon and small intestine, identified as shown in Fig. 2. (**b**) Expression of IRF8 and IRF4 by total cDCs from mouse small intestine. (**c**) Expression of IRF8 and IRF4 allows identification of cDC1 and cDC2 populations among the subsets of murine small intestinal cDCs identified as in (**a**) above. (**d**) The relative frequencies of cDC subsets in mouse LP vary along the length of the intestine. (**e**) Expression of CD172a (SIRPα), XCR1, and CD8α by the subsets of cDCs in mouse small intestinal LP (see text for details)

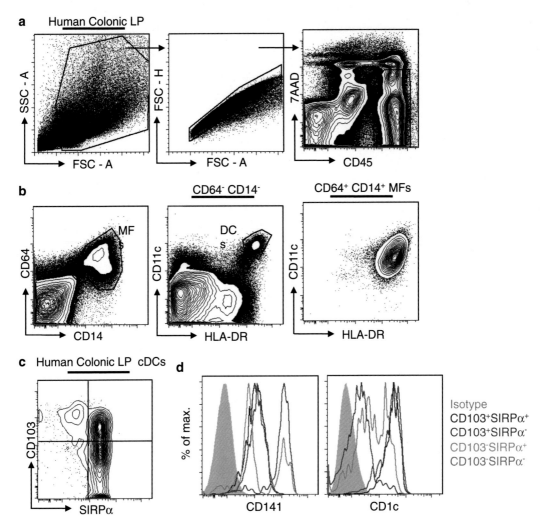

Fig. 4 Identification of cDC subsets in the human intestine. (**a**) Single live leukocytes are identified among total cells on the basis of FSC and SSC, followed by exclusion of dead cells using 7-AAD⁻ and of non-hematopoietic CD45⁻ cells. (**b**) Expression of CD64 and CD14 is then used to identify cDCs and MFs among live CD45⁺ cells, with the CD11c⁺MHCII⁺ cells found in the CD64⁻CD14⁻ population being cDCs. Most CD64⁺CD14⁺ cells are also CD11c⁺MHCII⁺, but these are MFs. (**c**) Expression of CD103 and CD172a (SIRPα) reveals discrete subsets among cDCs in resected human colon. (**d**) Expression of CD1c and CD141 by CD103/CD172a (SIRPα)-defined subsets of cDCs in resected human colon (see text for details)

impossible to remove them and cDCs from these tissues are likely to be present in small numbers in preparations of LP.

4. In our experience, intestinal sections can be kept on ice for up to 6 h before digestion with minimal effect on cell yield and viability.

5. Supernatants are removed by pouring cell suspensions into a beaker through 50 μm Nitex mesh inserted in a funnel (*see* Fig. 1).

6. EDTA is used to remove the surface epithelial layer, allowing access of enzymes to the lamina propria in the next steps.

7. All shaking steps should be performed at 250 rpm to aid digestion.

8. Some tissue should remain in the 50 ml Falcon tube to prevent over-digestion which can compromise the viability of the cells isolated.

9. Try to leave the intestines in the enzyme cocktail for as short a time as possible. Move on to the next step as soon as the tissue has been digested.

10. Typical cell yield from a murine small intestine digestion is $10–15 \times 10^6$ total cells.

11. The cecum is usually not included for the purposes of isolating large intestinal leukocytes, but may be useful to include under certain conditions such as specific infections, which involve the cecum.

12. Typical cell yield from a murine colon digestion is $5–8 \times 10^6$ total cells.

13. It is important to weigh the tubes or the Petri dish containing the appropriate amount of ice-cold CMF HBSS before transfer of tissue to allow accurate measurement of tissue weight and cell numbers.

14. EDTA washing steps are performed to remove the surface epithelial layer from the lamina propria of intestinal tissue.

15. All shaking steps should be performed at 250 rpm to aid digestion.

16. Parafilm prevents leakage of tube contents.

17. Manual vigorous shaking aids tissue digestion and reduces the length of the digestion process, maximizing the isolation of live intestinal leukocytes.

18. Digestion of intestinal biopsies orientated horizontally can aid tissue digestion and prevent surface marker cleavage.

19. Typical yields from a human biopsy range from 0.2 to 0.5×10^6 cells.

20. It is best to process human tissue immediately on arrival, as our experience is that some cDC populations appear to be lost after overnight storage in cRPMI at 4 °C. Alternative storage methods have not been tested for cell recovery.

21. Placing resected human tissue on a piece of blue tissue soaked in CMF HBSS 2 % FCS aids handling for tissue dissection.

22. The fibrous tissue present in resected human material is frequently resistant to digestion. However the digestion should be stopped when the mucosal tissue itself has been fully

digested to avoid compromising cell viability. Fibrous material will be removed from the single-cell suspension by using a 40 μm cell strainer.

23. If resected human tissue is not fully digested after 60 min, the single-cell suspension should be passed through a 40 μm cell strainer and washed in FACS buffer. Any remaining tissue should be transferred to a fresh 50 ml Falcon tube containing 20 ml of fresh enzyme cocktail and incubated again in a shaking water bath/incubator at 37 °C until fully digested. Manual vigorous shaking should be performed every 5–10 min. Once all tissue has been digested, both single-cell suspensions are combined and washed in FACS buffer. The remaining protocol is followed as described.

24. Typical yields from resected human intestinal tissue range from 10 to 20×10^6 cells per gram of tissue.

25. Alternatively, $1-2 \times 10^6$ cells can be added to each well of a 96 well plate.

26. Staining panels included are for basic identification of cDCs in murine and human intestine. CD64 or F4/80 is used to exclude MFs in mouse tissues.

27. We use a broad FSC-A/SSC-A gate to identify cells and a stringent FSC-A/FSC-H gate to identify single cells prior to identifying live CD45$^+$ cells.

28. We do not use Percoll or any other pre-enrichment strategy prior to FACS analysis or sorting, as we find this result in a significant loss of multiple cells types including cDCs. However we have not tested all available options.

29. Detection channel for 7-AAD depends on flow cytometer setup and laser availability. Alternative live/dead dyes can be used, such as the fixable live/dead dyes from eBioscience.

30. For murine analyses, we typically acquire 1×10^6 total cells to be able to visualize sufficient numbers of all cDC subsets.

31. For human analyses, we typically acquire 1×10^6 total cells to be able to visualize sufficient numbers of all cDC subsets.

32. The pDCs found in the small intestine can be identified among live CD45$^+$ cells as CD317$^+$ (PDCA1) SiglecH$^+$ Ly6C$^+$ B220$^+$ CD11b$^-$ CD11c^{lo} MHCII^{lo}.

33. Recent work on other tissues has established a scheme by which cDCs can be divided into two populations (cDC1 and cDC2) on the basis of their expression and dependence on the IRF8 and IRF4 transcription factors, respectively [11]. This overall pattern also applies to the intestine (*see* Fig. 3b), where the CD103$^+$CD11b$^+$ and CD103$^-$CD11b$^+$ subsets express IRF4, whereas the CD103$^+$CD11b$^-$ cDCs express IRF8; the

CD103⁻CD11b⁻ cDCs are heterogeneous for these markers (*see* Fig. 3c).

34. The relative proportions of the subsets varies along the length of the intestine, with the CD103⁺CD11b⁺ cDCs predominating in the small intestine and CD103⁺CD11b⁻ cDCs being the most abundant population in the colon (*see* Fig. 3d) [7].

35. The CD103⁺CD11b⁺ and CD103⁻CD11b⁺ subsets all express CD172a (SIRPα), as do some of the CD103⁻CD11b⁻ cDCs. In contrast, the CD103⁺CD11b⁻ cDCs lack CD172a, but express XCR1 and CD8α. The CD103⁻CD11b⁻ cDCs are heterogeneous for CD172a and CD8α, but lack XCR1 expression (*see* Fig. 3e). The exact nature of this population is unknown. Some may represent precursor cells, while others may be derived from isolated lymphoid follicles (ILFs) and are therefore contaminants of LP preparations [5].

36. pDCs should be identifiable among live CD45⁺ cells in the human SI as CD123⁺ CD303⁺ (BDCA2) CD304⁺ (BDCA4) CD11cˡᵒ, CD14⁻ cells.

37. In our hands, CD11b is not a useful marker in humans, as it does not enable precise separation of myeloid populations.

38. Although it is unknown whether these also express the additional markers that would indicate they are fully homologous to the mouse populations, they do share many of the markers found on the various subsets of cDCs seen in human peripheral blood. Thus, the CD103⁺SIRPα⁻ cDCs homogeneously express CD141, making them analogous to the XCR1⁺ population of cDC1s [11, 20]. Conversely, both SIRPα⁺ cDCs populations in the human LP (CD103⁺SIRPα⁺ and CD103⁻SIRPα⁺) homogeneously express CD1c (*see* Fig. 4d), thought to define the cDC2 population in man [11]. Interestingly, the CD103⁻SIRPα⁻ subset in humans appears to be as heterogeneous as its mouse equivalent, expressing heterogeneous levels of CD141 and intermediate amounts of CD1c (*see* Fig. 4d).

References

1. Mowat AM (2003) Anatomical basis of tolerance and immunity to intestinal antigens. Nat Rev Immunol 3:331–341

2. Scott CL, Murray TFPZ, Beckham KSH et al (2014) Signal Regulatory Protein alpha (SIRPα) regulates the homeostasis of CD103(+) CD11b(+) DCs in the intestinal lamina propria. Eur J Immunol. doi:10.1002/eji.201444859

3. Persson EK, Uronen-Hansson H, Semmrich M et al (2013) IRF4 transcription-factor-dependent CD103(+)CD11b(+) dendritic

cells drive mucosal T helper 17 cell differentiation. Immunity 38:958–969

4. Schlitzer A, McGovern N, Teo P et al (2013) IRF4 transcription factor-dependent CD11b(+) dendritic cells in human and mouse control mucosal IL-17 cytokine responses. Immunity 38:970–983

5. Cerovic V, Houston SA, Scott CL et al (2013) Intestinal CD103(-) dendritic cells migrate in lymph and prime effector T cells. Mucosal Immunol 6:104–113

6. Scott CL, Bain CC, Wright PB et al (2014) CCR2(+)CD103(-) intestinal dendritic cells develop from DC-committed precursors and induce interleukin-17 production by T cells. Mucosal Immunol. doi:10.1038/mi.2014.70

7. Mowat AM, Agace WW (2014) Regional specialization within the intestinal immune system. Nat Rev Immunol 14:667–685

8. Mayrhofer G, Pugh CW, Barclay AN (1983) The distribution, ontogeny and origin in the rat of Ia-positive cells with dendritic morphology and of Ia antigen in epithelia, with special reference to the intestine. Eur J Immunol 13:112–122

9. Cerovic V, Bain CC, Mowat AM et al (2014) Intestinal macrophages and dendritic cells: what's the difference? Trends Immunol 35: 270–277

10. Hadis U, Wahl B, Schulz O et al (2011) Intestinal tolerance requires gut homing and expansion of FoxP3+ regulatory T cells in the lamina propria. Immunity 34:237–246

11. Guilliams M, Ginhoux F, Jakubzick C et al (2014) Dendritic cells, monocytes and macrophages: a unified nomenclature based on ontogeny. Nat Rev Immunol 14:571–578

12. Schraml BU, van Blijswijk J, Zelenay S et al (2013) Genetic tracing via DNGR-1 expression history defines dendritic cells as a hematopoietic lineage. Cell 154:843–858

13. Persson EK, Scott CL, Mowat AM et al (2013) Dendritic cell subsets in the intestinal lamina propria: ontogeny and function. Eur J Immunol 12:3098–3107

14. Bachem A, Hartung E, Güttler S et al (2012) Expression of XCR1 characterizes the Batf3-dependent lineage of dendritic cells Capable of antigen cross-presentation. Front Immunol 3:214

15. Hildner K, Edelson BT, Purtha WE et al (2008) Batf3 deficiency reveals a critical role for CD8alpha + dendritic cells in cytotoxic T cell immunity. Science (New York, NY) 322:1097–1100

16. Cerovic V, Houston SA, Westlund J et al (2014) Lymph borne CD8a+ DCs are uniquely able to cross-prime CD8+ T cells with antigen acquired from intestinal epithelial cells. Mucosal Immunol. doi:10.1038/mi.2014.40

17. Lewis KL, Caton ML, Bogunovic M et al (2011) Notch2 receptor signaling controls functional differentiation of dendritic cells in the spleen and intestine. Immunity 35:780–791

18. Satpathy AT, Briseño CG, Lee JS et al (2013) Notch2-dependent classical dendritic cells orchestrate intestinal immunity to attaching-and-effacing bacterial pathogens. Nat Immunol 14:937–948

19. Welty NE, Staley C, Ghilardi N et al (2013) Intestinal lamina propria dendritic cells maintain T cell homeostasis but do not affect commensalism. J Exp Med 210:2011–2024

20. Bachem A, Güttler S, Hartung E et al (2010) Superior antigen cross-presentation and XCR1 expression define human CD11c + CD141+ cells as homologues of mouse CD8+ dendritic cells. J Exp Med 207:1273–1281

Chapter 8

Isolation of Human Skin Dendritic Cell Subsets

Merry Gunawan, Laura Jardine, and Muzlifah Haniffa

Abstract

Dendritic cells (DCs) are specialized leukocytes with antigen-processing and antigen-presenting functions. DCs can be divided into distinct subsets by anatomical location, phenotype and function. In human, the two most accessible tissues to study leukocytes are peripheral blood and skin. DCs are rare in human peripheral blood (<1 % of mononuclear cells) and have a less mature phenotype than their tissue counterparts (MacDonald et al., Blood. 100:4512–4520, 2002; Haniffa et al., Immunity 37:60–73, 2012). In contrast, the skin covering an average total surface area of 1.8 m^2 has approximately tenfold more DCs than the average 5 L of total blood volume (Wang et al., J Invest Dermatol 134:965–974, 2014). DCs migrate spontaneously from skin explants cultured ex vivo, which provide an easy method of cell isolation (Larsen et al., J Exp Med 172:1483–1493, 1990; Lenz et al., J Clin Invest 92:2587–2596, 1993; Nestle et al., J Immunol 151:6535–6545, 1993). These factors led to the extensive use of skin DCs as the "prototype" migratory DCs in human studies. In this chapter, we detail the protocols to isolate DCs and resident macrophages from human skin. We also provide a multiparameter flow cytometry gating strategy to identify human skin DCs and to distinguish them from macrophages.

Key words Dendritic cells, Skin, Macrophages, Antigen-presenting cells, Mononuclear phagocytes

1 Introduction

Dendritic cells (DCs) and macrophages are antigen-presenting cells (APC) predominantly found in peripheral tissues. Human skin is an accessible epithelial barrier widely used as a model tissue to study primary DCs. The epidermis, the outermost layer of skin, is populated by Langerhans cells with an average of 800 LCs found per mm^2 skin [7]. The dermis is separated from the epidermis by a thin layer of basement membrane (BM) and can be divided into papillary and reticular dermis. Blood vessels and collagen fibers are arranged perpendicular to the BM in the papillary dermis in contrast to the reticular dermis, which supports the horizontal plexus of blood vessels and collagen fibers. DCs and macrophages occupy distinct microanatomical niches within human dermis [3]. DCs form the superficial APC layer and are most abundant in the papillary dermis between 0 and 40 μm beneath the dermo-epidermal

Elodie Segura and Nobuyuki Onai (eds.), *Dendritic Cell Protocols*, Methods in Molecular Biology, vol. 1423,
DOI 10.1007/978-1-4939-3606-9_8, © Springer Science+Business Media New York 2016

junction (DEJ) [3]. Macrophages form the deeper APC layer and are abundant at 40–60 μm beneath the DEJ [3]. Distinct dermal DC and macrophage subsets can be identified by cellular morphology and antigen expression profile: (1) CD1c+ dermal DCs, which coexpress CD1a [5, 8–11]; (2) CD141/BDCA3hi DCs [2]; (3) CD14+ monocyte-derived macrophage-like cell [6, 10, 12]; and (4) CD14+FXIIIA+ dermal macrophages [3, 9, 13]. Whether these subsets occupy distinct microanatomical immune niches is unknown. The murine homologs of human dermal CD1c+ DCs and CD141+ DCs are mouse dermal CD11b+ DCs and CD103+ DCs, respectively [12, 14]. Human dermal CD14+ cells have a transcriptome profile most enriched with dermal macrophages and blood monocytes [12]. Their murine homologs are dermal macrophages [12, 14]. Dermal CD14+FXIIIA+ dermal macrophages are characterized by cytoplasmic melanin granules [13], which led to these cells initially being called melanophages [15, 16]. However, macrophages containing melanin granules have not been demonstrated in the mouse. Figure 1 depicts the composition of human skin DCs and macrophages, the flow cytometry gating strategy to identify them, and the antigens characterizing the different cell populations.

There are two widely used methods to isolate and study human skin DCs and macrophages: (1) enzymatic digestion of skin to obtain single-cell suspension and (2) isolating cells that spontaneously migrate from skin explant cultures, prior to fluorescence-activated cell sorting or magnetic-bead enrichment procedure to purify DCs. Lymphocytes, DCs, and CD14+ cells migrate spontaneously from skin explants but dermal CD14+FXIIIA+ macrophages remain resident [13]. Although spontaneous migration is assumed to recapitulate skin lymphatic migration, evidence in both human and mouse suggests that some leukocytes migrate out without entering lymphatics [12, 17], and this ability must not be equated to lymph node migrating capacity of a cell type. In this chapter, we provide a detailed protocol for identifying and isolating human skin DCs and macrophage subsets.

2 Materials

All instruments and reagents used for DC isolation are sterile and endotoxin-free. Instruments are sterilized by autoclaving and draining in 75% ethanol (EtOH). Commercial culture grade reagents used are sterile and endotoxin-free. All procedures are performed in a class II biological safety cabinet.

2.1 Dermatome-Cut Skin Preparation

1. Skin from mammoplasty or abdominoplasty obtained within 2 h after surgery (see Note 1).

2. Disposable scalpel.

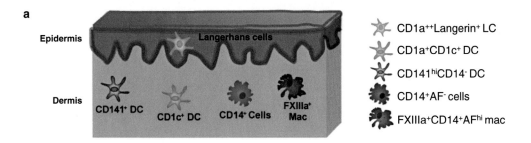

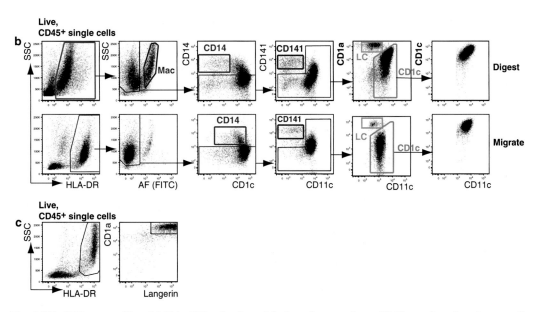

Fig. 1 Skin APC composition. (**a**) Skin APC subsets and their surface markers. (**b**) Flow cytometry of enzymatically digested skin (*upper panel*) and spontaneously migrating cells from skin explant culture ex vivo (*middle panel*). Gating strategy on live, CD45+ single cells used to identify tissue macrophages, CD14+ monocyte-derived macrophage cells, CD141+ DCs and CD1c+ DCs are shown. (**c**) Flow cytometry of enzymatically digested epidermis. Gating strategy on live, CD45+ single cells to identify LCs are shown

3. Teflon/silicone sheets.

4. Dermatome with Pilling-Wecprep blade.

5. Size 8 Goulian guard.

6. One pair of large forceps and two pairs of small forceps.

7. Custom-made block to fix skin (sterilized by draining in biocide, ethanol, and sterile water; air-dry before use) (Fig. 2a). Alternatively, a square wooden or Styrofoam block can be used.

8. Petri dishes 145 mm/20 mm.

9. Sterile phosphate-buffered saline (PBS).

10. Two 19G needles.

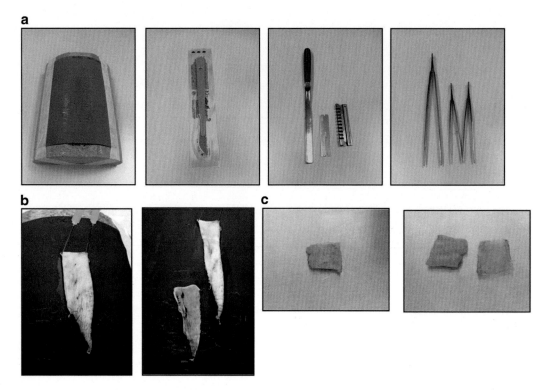

Fig. 2 Epidermis and dermis isolation. (**a**) From *left to right*: custom-made block covered with Teflon sheet secured with masking tape; disposable scalpel; dermatome with Pilling-Wecprep blade and a size 8 Goulian guard; forceps. (**b**) Skin strip prior to cutting with a dermatome (*left*); skin strip after cutting with a dermatome (*right*). (**c**) Whole skin containing both the epidermis and the dermis (*left*); the epidermis peeled off from the dermis following dispase treatment (*right*)

11. Dispase (Gibco): Reconstitute in sterile PBS at 100 U/ml, sterilize over a 0.22 μm filter membrane (Pall) and store in −80 °C.

12. Collagenase type IV (Worthington Biochem): Reconstitute in sterile PBS at 160 mg/ml and store in −80 °C (*see* **Note 3**).

13. PBS.

2.2 Single-Cell Suspension Preparation

1. Pipette aid and serological pipettes.

2. 100 μm cell strainer (BD).

3. Serum-free RPMI.

4. Complete medium: RPMI, 10% heat-inactivated fetal calf serum (FCS), 1% penicillin-streptomycin, 1% glutamine. Store at 4 °C.

5. DNAse I (Roche): Reconstitute in sterile water at 10 mg/ml, and store in −80 °C.

6. Red blood cell (RBC) lysis buffer: 1.5 M NH₄CL, 100 mM KHCO₃, and 1 mM EDTA in sterile water with pH 7.35, sterilize over a 0.22 μm filter membrane (Pall) and store at 4 °C (10× stock). 1× stock is prepared by diluting with sterile water and stored at room temperature, protected from light.

7. PBS 2 % FCS.

2.3 Immunostaining and Cell Sorting

1. Polypropylene FACS tubes.

2. Antibodies against human antigen: CD141 (AD5-14H12; Miltenyi Biotec), CD1c (L161; Biolegend), CD11c (B-ly6), HLA-DR (L243), CD45 (HI30), CD14 (M5E2; Biolegend), and CD1a (HI149). Antibodies were all supplied by BD unless stated otherwise and are denoted as antigen (clone; supplier). Store at 4 °C, protected from light.

3. Collection tubes: polypropylene FACS tubes or 1.5 ml Eppendorf tubes.

4. DAPI stored at room temperature, protected from light.

5. Mouse Ig (Sigma): Reconstitute in sterile PBS at 2.5 mg/ml, and store in −80 °C.

6. Sort buffer: PBS, 0.5 % heat-inactivated FCS, and 2 mM EDTA.

7. 50 μm Partec CellTrics filter.

3 Methods

Large pieces of human surplus skin from mammoplasty and abdominoplasty are first removed of subcutaneous fat if still present. The skin is cut into thin strips prior to dermatome slicing in order to obtain 150–300 μm thin pieces for enzymatic digestion. Dermatome-cut skin can also be separated into the epidermis and dermis by dispase treatment prior to enzymatic digestion or explant culture. Estimate of cell yield by digestion and migration overnight can be found in Table 1.

Table 1
Yield and viability of skin MHC ClassII⁺ cells by the different methods of isolation

	Method	Total cell number (×10³)	% viability	% HLA-DR⁺ cells
Whole skin	Digestion	267.5 (±61.3)	91.3 (±1.6)	22.7 (±3.8)
	Migration	67.5 (±16.2)	85.6 (±11.3)	68.8 (±16.5)

1 cm² of whole skin were either digested or left to migrate overnight (16 h). Average cell numbers and % viability are shown. % HLA-DR⁺ cells were counted as the percentage of CD45⁺HLA-DR⁺ cells from total viable cells. Mean ± SD from three different donors

**3.1 Epidermis
and Dermis Isolation**

1. Cover a flat cutting surface with Teflon or baking silicone and secure with masking tape.

2. Transfer the skin into a large petri dish (150 mm) with a little PBS to prevent the skin from drying out (*see* **Note 4**).

3. Remove subcutaneous fat by holding the skin with large forceps and scraping with a disposable scalpel.

4. Cut the skin into manageable strips using a disposable scalpel, approximately 1.5 cm × 4 cm (*see* **Note 5**).

5. Place the skin strip on Teflon-covered wooden block, using two needles to pin the skin at one end of the block.

6. Flatten and pull the skin away from needles with large forceps and cut the skin by gentle horizontal movement of the hand with slight downward traction using a dermatome (*see* Fig. 2 and **Note 2**).

7. Place dermatome-cut skin pieces into a 50 ml Falcon tube filled with PBS on ice.

8. Repeat process for the remaining skin strips.

9. Collect all the dermatome-cut skin strips and place into a large petri dish filled with PBS with the epidermis side up. The hydrophobic surface of the epidermis will allow the skin strips to float on the PBS.

10. There are several options after this step: to digest (go to **step 12**) or migrate cells (go to **step 13**) from the whole skin without separating the epidermis from the dermis, or to separate the epidermis from the dermis prior to digestion or explant culture for cell migration (go to **step 11**).

11. To separate the epidermis from the dermis (ignore this step if processing the whole skin):

    (a) Float skin pieces in a petri dish, the epidermis facing upwards, in serum-free RPMI containing dispase (1 U/ml). Incubate at 37 °C for 60–90 min (*see* **Note 6**).

    (b) Peel the epidermis from the dermis using two pairs of sterile forceps (*see* Fig. 2). Use one pair of forceps to hold the dermis in place and the other to gently lift and peel off the epidermis from an edge. The translucent dermis has a pink hue in contrast to the hydrophobic epidermis.

    (c) Rinse both layers in fresh complete medium to remove dispase.

    (d) Proceed to skin digestion (go to **step 12**) or skin migration (go to **step 13**).

12. Skin digestion:

    (a) Float skin pieces in a petri dish, the epidermis side facing upward, in complete medium containing collagenase (use collagenase at 1.06 mg/ml for digesting the whole skin or the dermis and at 0.8 mg/ml for digesting the epidermis).

(b) Incubate the skin at 37 °C overnight to digest (*see* **Notes 7** and **8**).

13. Skin migration:

    (a) Float skin pieces in a petri dish, the epidermis side facing upward, in complete medium.

    (b) Incubate the skin at 37 °C for 48–72 h prior to harvesting cells by gentle resuspension and aspiration using a serological pipette.

3.2 Single-Cell Suspension Preparation

1. Top Falcon tubes with 100 μm cell strainers.

2. Gently swirl petri dish containing digested or migrated skin pieces. The digested dermis will form swirls. If the whole skin is digested, the stratum corneum layer remains intact. Gently resuspend and aspirate the supernatant taking care to avoid the residual epidermis using a serological pipette and pass the cell suspension through a 100 μm cell strainer (*see* **Note 9**).

3. Add 25 ml of fresh complete medium to wash the petri dish, and repeat **step 2**.

4. Spin down the cells at 500 g for 5 min.

5. Perform a RBC lysis step if necessary if the cell pellet appears red. Resuspend cells in 1× RBC lysis buffer at a maximum concentration of 10×10^6 cells/ml and incubate at room temperature for 5 min. Stop the lysis reaction with 10 ml of ice-cold medium before spinning cells down.

6. Wash cells twice with PBS 2 % FCS prior to use.

3.3 Immunostaining and Cell Sorting

1. Count cell numbers after PBS washes.

2. Resuspend cells in sort buffer at a maximum concentration of 10×10^6 cells/100 μl in a polypropylene FACS tube (*see* **Note 10**). Supplement with 0.1 mg/ml DNAse to prevent cell clumping.

3. Block with 125 μg/ml of mouse Ig to reduce nonspecific background staining, and incubate at 4 °C for 15 min.

4. Add staining antibodies to the cells, and incubate at 4 °C for 30 min.

5. Spin down and resuspend cells at 10×10^6 cells/ml in sort buffer containing DNAse and keep on ice. Dead cells are excluded by including DAPI in the flow buffer at least 10 min prior to acquisition.

6. Prepare collection tubes filled with 1.5 ml, or 0.5 ml, of complete medium for FACS tubes, or Eppendorf tubes, respectively. Coat the sides of the FACS or Eppendorf collecting tubes with complete medium to prevent cells from sorting stream from drying out.

7. Filter the cells with 50 μm CellTrics filter immediately before sorting to remove any cell clumps (*see* **Note 11**).

8. DC and macrophage subsets of interest can be FACS purified using a suitable instrument, e.g., FACSAria or FACSFusion from BD. For FACS sorting, use a large nozzle (100 μm) and run the sample at a flow rate of 3000 events per second (*see* **Note 12**).

3.4 Identification of Skin DCs and Macrophage Subsets

The same populations of APC subsets can be obtained from digested or migrated cell suspensions (*see* Fig. 1). A flow cytometry gating strategy to identify skin DCs and macrophage subsets using 11–12 parameters is presented here (*see* **Note 13**).

1. Gate on live, CD45$^+$ cells and then single cells using FSC-H versus FSC-A (*see* **Note 14**).

2. Plot SSC versus HLA-DR to distinguish skin APCs which are HLA-DR$^+$SSC^{mid-hi} from lymphocytes (HLA-DR$^-$SSC^{lo}) and mast cells which are in the HLA-DR$^-$SSC^{hi} fraction (*see* Fig. 1b) (*see* **Note 15**).

3. Analyze HLA-DR$^+$ APCs by SSC and a blank channel to detect macrophage autofluorescence (brightest in the FITC (excitation 488/emission 532/30) channel): this allows the separation of SSC^{hi}AF$^+$ cells containing dermal macrophages from SSC^{mid}AF$^-$ fraction containing DCs, CD14$^+$ cells, and LCs. Macrophages can be identified as CD14$^+$ cells within the SSC^{hi}AF$^+$ fraction.

4. Separate AF$^-$ cells into CD14$^+$ cells and CD14$^-$ fraction containing DCs (*see* **Note 16**).

5. CD141$^+$ DCs can be identified as CD14$^-$CD141^{hi}CD11c^{lo} cells. The remaining fraction is then analyzed for the expression of CD11c and CD1a to identify LCs, which are CD11c^{lo}CD1a^{br}. Excluding LCs allows CD1c$^+$ DCs, which coexpress CD1a to be identified (*see* **Note 17**).

4 Notes

1. The skin is collected into a plastic container containing sterile PBS on ice.

2. Small 2–6 mm pieces of the skin obtained from shave (Dermablade™) or punch biopsies are not amenable for Dermablade/split skin graft knife cutting. These samples can be treated with dispase to separate the epidermis from the dermis prior to digestion or explant culture or digested whole after cutting with serrated scissors into small pieces or cultured as a whole skin explant.

3. Collagenase can be replaced by Liberase Blendzyme (Roche).

4. If the skin is large, process one piece at a time, keeping the remaining skin in PBS on ice.

5. Discard mammoplasty skin around the nipple area as it is difficult to process.

6. Check the skin after 1 h in dispase. If the epidermis is difficult to peel, leave for another 30 min.

7. The skin can be cut into smaller pieces using a pair of sterile scissors (serrated scissors are best) to aid digestion if needed in 4 h.

8. Upon complete digestion, the skin will appear more transparent.

9. Handle the epidermis gently to avoid collecting keratinocytes. It may be easier to remove the residual epidermal pieces into a new plate before aspirating the supernatant with serological pipette.

10. Using polypropylene instead of polycarbonate tubes will prevent cells from adhering to the tube wall, which can happen with polycarbonate FACS tubes. This is important when handling adherent macrophages.

11. Separating cell suspensions into a few tubes prior to FACS sorting with smaller volume rather than in one tube will help to prevent cell clumping.

12. Check purity of the isolated population by running a small portion of the sorted cells back into the sorter.

13. The same subsets can be identified by an alternative strategy particularly suited to peripheral blood analysis (detailed in Supplementary Figure 1 of ref. 2).

14. Thresholding on FSC (40,000) maximizes cell event acquisition during flow cytometry analysis. Thresholding is not recommended during cell sorting as it can increase debris in the sorted fractions.

15. Lymphocytes are in the CD45$^+$SSC^{lo}HLA-DR$^-$ fraction. CD3$^+$ lymphocytes are primarily CD4$^+$ memory T cells. Mast cells are CD45$^+$HLA-DR$^-$SSC^{mid-hi} and can be identified by CD117 (c-kit) expression as previously described [3].

16. It is important to identify and gate out CD14$^+$ cells from the HLA-DR$^+$AF$^-$ fraction before identifying CD141$^+$ DCs and CD1c$^+$ DCs. CD14$^+$ cells express high amounts of CD141 and CD11c, which are also expressed by CD141$^+$ DCs and CD1c$^+$ DCs.

17. The identity of LCs can also be confirmed using surface or intracellular Langerin staining (*see* Fig. 1). Notably, LC yield is much higher from epidermal digest following dispase treatment to separate the epidermis from the dermis.

Acknowledgments

We would like to acknowledge funding from the Wellcome Trust (WT088555) and the British Skin Foundation.

References

1. MacDonald KP, Munster DJ, Clark GJ, Dzionek A, Schmitz J, Hart DN (2002) Characterization of human blood dendritic cell subsets. Blood 100:4512–4520

2. Haniffa M, Shin A, Bigley V et al (2012) Human tissues contain CD141(hi) cross-presenting dendritic cells with functional homology to mouse CD103(+) nonlymphoid dendritic cells. Immunity 37:60–73

3. Wang XN, McGovern N, Gunawan M et al (2014) A three-dimensional atlas of human dermal leukocytes, lymphatics, and blood vessels. J Invest Dermatol 134:965–974

4. Larsen CP, Steinman RM, Witmer-Pack M, Hankins DF, Morris PJ, Austyn JM (1990) Migration and maturation of Langerhans cells in skin transplants and explants. J Exp Med 172:1483–1493

5. Lenz A, Heine M, Schuler G, Romani N (1993) Human and murine dermis contain dendritic cells. Isolation by means of a novel method and phenotypical and functional characterization. J Clin Invest 92:2587–2596

6. Nestle FO, Zheng XG, Thompson CB, Turka LA, Nickoloff BJ (1993) Characterization of dermal dendritic cells obtained from normal human skin reveals phenotypic and functionally distinctive subsets. J Immunol 151:6535–6545

7. Collin MP, Hart DN, Jackson GH et al (2006) The fate of human Langerhans cells in hematopoietic stem cell transplantation. J Exp Med 203:27–33

8. Angel CE, George E, Brooks AE, Ostrovsky LL, Brown TL, Dunbar PR (2006) Cutting edge: CD1a+antigen-presenting cells in human dermis respond rapidly to CCR7 ligands. J Immunol 176:5730–5734

9. Zaba LC, Fuentes-Duculan J, Steinman RM, Krueger JG, Lowes MA (2007) Normal human dermis contains distinct populations of CD11c+BDCA-1+ dendritic cells and CD163+FXIIIA+ macrophages. J Clin Invest 117:2517–2525

10. Klechevsky E, Morita R, Liu M et al (2008) Functional specializations of human epidermal langerhans cells and CD14+ dermal dendritic cells. Immunity 29:497–510

11. Segura E, Valladeau-Guilemond J, Donnadieu M-H, Sastre-Garau X, Soumelis V, Amigorena S (2012) Characterization of resident and migratory dendritic cells in human lymph nodes. J Exp Med 209:653–660

12. McGovern N, Schlitzer A, Gunawan M et al (2014) Human dermal CD14(+) cells are a transient population of monocyte-derived macrophages. Immunity 41:465–477

13. Haniffa M, Ginhoux F, Wang XN et al (2009) Differential rates of replacement of human dermal dendritic cells and macrophages during hematopoietic stem cell transplantation. J Exp Med 206:371–385

14. Tamoutounour S, Guilliams M, Montanana Sanchis F et al (2013) Origins and functional specialization of macrophages and of conventional and monocyte-derived dendritic cells in mouse skin. Immunity 39:925–938

15. Cooper KD, Neises GR, Katz SI (1986) Antigen-presenting OKM5+ melanophages appear in human epidermis after ultraviolet radiation. J Invest Dermatol 86:363–370

16. Cerio R, Griffiths CE, Cooper KD, Nickoloff BJ, Headington JT (1989) Characterization of factor XIIIa positive dermal dendritic cells in normal and inflamed skin. Br J Dermatol 121:421–431

17. Ohl L, Mohaupt M, Czeloth N et al (2004) CCR7 governs skin dendritic cell migration under inflammatory and steady-state conditions. Immunity 21:279–288

Chapter 9

Isolation of Mouse Dendritic Cell Subsets and Macrophages from the Skin

Camille Malosse and Sandrine Henri

Abstract

The improvement of dendritic cell subset isolation from tissues and the use of appropriate surface markers allowed to decipher their heterogeneity but also allowed to unravel some specific functions that are valuable for vaccine design as well as for a better understanding of the in situ pathophysiology upon infection. Here, we describe the procedures to extract those cells from the skin and to analyze them by flow cytometry using a combination of appropriate surface markers allowing further transcriptomic analysis and functional assays.

Key words Skin, Epidermis, Dermis, Dispase, Collagenase 4, Percoll, Fc Block, Monoclonal antibodies, Flow cytometry

1 Introduction

Dendritic cells (DCs) are rare cells of the hematopoietic system and are present all over the body where they play a major role as sentinels of the immune system. Indeed, DCs have the unique capacity to detect pathogens, to sample antigens, and to migrate to the lymphoid organs where they present the antigen and subsequently activate naïve T cells. Dendritic cells were firstly described in lymphoid organs in the early 1970s [1], but due to their low numbers and the difficulty to extract them, extensive studies assessing their functions were primarily performed using bone marrow- or monocyte-derived DCs. Indeed, it was shown that large numbers of dendritic cells could be generated from mouse bone marrow cells cultured in vitro in the presence of hematopoietic growth factors such as granulocyte/macrophage colony-stimulating factor (GM-CSF) plus IL-4 *fms*-like tyrosine kinase 3 ligand (Flt3L) [2, 3]. Now we know that GM-CSF-derived DCs correspond mainly to monocyte-derived DC, whereas Flt3L-derived DCs are derived from DC precursors and better reproduce the steady-state resident conventional DCs [4, 5]. Nevertheless, although using appropriate

Elodie Segura and Nobuyuki Onai (eds.), *Dendritic Cell Protocols*, Methods in Molecular Biology, vol. 1423, DOI 10.1007/978-1-4939-3606-9_9, © Springer Science+Business Media New York 2016

hematopoietic growth factors, those in vitro-generated dendritic cells are derived from similar precursors, they would never mimic properly the tissue DC features. Indeed, the tissue barriers, such as the skin, gut, vagina, lung, and oral mucosae where tissue DCs are located, are highly distinct tissues regarding the temperature, the pH, and the commensal flora, and thus as a consequence, DC features would be different from one site to the other as they are most likely regulated according to the tissue where they are residing. Thus, to comprehensively assess tissue DC functions, it is important to be able to extract those cells from the tissue and to discriminate the different subsets.

There was confusion in the field of dendritic cell characterization in tissues as many studies were often mixing conventional tissue dendritic cells derived from DC precursors with monocyte-derived DCs and even tissue macrophages. In this study, we describe how we extract those cells from the mouse skin but also how we combine specific surface markers to stain them and distinguish the different subsets by flow cytometry [6–8].

2 Materials

Keep all the solutions sterile. Do not add sodium azide to the solutions.

1. HEPES-PBS: Add 12.5 mL of 1 M HEPES to 237.5 mL PBS. Pass it through a 0.2 μm filter. This solution is prepared only the day the stock of dispase is prepared.

2. Wash medium: to 500 mL of RPMI 1640, add 4 mL of penicillin-streptomycin, 16 mL of 1 M HEPES, and 10 mL of decomplemented fetal bovine serum (FBS). Store at 4 °C.

3. Dispase II stock solution (from *Bacillus* polymerase grade 2, Roche): 4 mg/mL solution in HEPES-PBS (resuspend 1 g of powder in 250 mL of HEPES-PBS). Aliquot (1–2 mL aliquots) and store at –20 °C.

4. Dispase II working solution: on the day of the experiment, dilute dispase II stock solution in PBS to the required concentration (*see* **Note 1**).

5. DNase stock solution (deoxyribonuclease I from bovine pancreas): 10 mg/mL solution in wash medium (resuspend 100 mg of powder in 10 mL of wash medium). Aliquot (1 mL aliquots) and store at –20 °C.

6. Collagenase 4 stock solution (Worthington): 20 mg/mL solution in wash medium (resuspend 1 g of powder in 50 mL of wash medium). Aliquot (1 mL aliquots) and store at –20 °C.

7. Collagenase 4-DNase working solution: on the day of the experiment, combine 1.5 mL of collagenase 4 stock solution at 20 mg/mL

and 1 mL of DNase stock solution at 10 mg/mL and complete to 20 mL with wash medium. Keep the solution on ice.

8. Percoll 100% stock solution (GE Healthcare): on the day of the experiment, to obtain 100 mL of Percoll 100% stock solution, mix 10 mL of 10× PBS with 90 mL of pure Percoll. Keep the solution on ice.

9. Percoll 40% working solution: to obtain 100 mL of Percoll 40% working solution, add 40 mL of Percoll 100% stock solution to 60 mL of 1× PBS. Keep the solution on ice.

10. Percoll 70% working solution: to obtain 100 mL of Percoll 70% working solution, add 70 mL of Percoll 100% stock solution to 30 mL of 1× PBS. Keep the solution on ice.

11. FACS Buffer: PBS, 2% FBS, 5 mM EDTA. Combine 100 mL of 10× PBS, 10 mL of 0.5 M EDTA (pH 8.0), 20 mL of decomplemented FBS and complete to 1 L with ultrapure water. Pass it through a 0.2 μm filter and store at 4 °C.

12. Fc Block (anti-CD16/CD32, clone 2.4G2).

13. Forceps, surgery scissors.

14. Scalpel.

15. Petri dish.

16. 24-well plates.

17. 15 mL polypropylene (PP) tubes.

18. 1.5 mL Eppendorf tubes.

19. 70 μm cell strainers.

20. Pasteur pipette.

3 Methods

3.1 Tissues Collection

We usually proceed to the cell isolation from mouse ears as they have no fat and no hair (*see* **Note 2**).

1. Sacrifice the mice.

2. Using forceps and small surgery scissors, cut the ears 2 mm from their basis to avoid including hair from the scalp.

3. Place the ears in empty 24-well plates and keep on ice.

3.2 Enzymatic Treatment

1. In a 24-well plate, distribute 1 mL of dispase II working solution per well. You will need four wells per mouse (if collecting both ears).

2. With forceps, separate the internal and external faces of the ears (the internal face will come with the cartilage), and lay it down on 1 mL of dispase II working solution. The "outside" side of each skin layer should be up, whereas the "inside" should be in contact with the solution.

3. Incubate 2 h 30 min at 37 °C or overnight at 4 °C (*see* **Note 1**).

4. In a 24-well plate, distribute 1 mL of collagenase 4-DNase working solution per well. You will need four wells per mouse (if collecting both ears).

5. If you wish to study the dermis and the epidermis separately, take each half ear from the dispase solution, dry it on a paper towel, and using forceps pull off the epidermis from the dermis, before proceeding to **step 6**. If you want to treat the whole skin, then proceed directly to **step 6**.

6. Collect each half ear from the dispase solution, dry it on a paper towel, and using forceps pull off the cartilage layer from the internal face. Then put the half ears on the lid of a petri dish, use a scalpel to chop it into small pieces, and transfer them in 1 mL of collagenase 4-DNase working solution.

7. Incubate 30 min at 37 °C.

8. Add 1 ml of collagenase 4-DNase working solution in each well, mix up and down with a Pasteur pipette or 1 mL tip.

9. Incubate 30 min at 37 °C.

10. Add 1 mL of collagenase 4-DNase working solution in each well, mix up and down with a Pasteur pipette or 1 mL tip.

11. Incubate 30 min at 37 °C.

12. Mix thoroughly and proceed to the filtration in a 15 mL tube using 70 μm cell strainers. Pool the four wells from the same mouse in a single 15 mL tube. Rinse the wells with 1 mL of FACS Buffer.

13. Centrifuge 5 min at $450 \times g$.

14. Resuspend the pellet with 5 mL of FACS Buffer.

15. Centrifuge 5 min at $450 \times g$.

3.3 Percoll Gradient (See Note 3)

1. Add 4 mL of 40% Percoll in 15 mL PP tubes (1 tube per sample).

2. Resuspend the cells in 4 mL of 70% Percoll and add slowly on the bottom of the tube containing the 40% Percoll solution.

3. Centrifuge at $600 \times g$ for 20 min (no brake, no acceleration) at 10 °C.

4. Add 10 mL of FACS Buffer in new 15 mL PP tubes (1 tube per sample).

5. After the Percoll gradient centrifugation, the cells of interest are located at the interface between the two Percoll layers. Collect the cell ring at the interface (around 2 mL) and add the cells to the 15 mL PP tubes containing FACS Buffer.

6. Centrifuge at $450 \times g$ for 5 min.

7. Resuspend the cell pellet in FACS Buffer and proceed to counting and/or staining.

Table 1
Flow cytometry staining strategy

	Conjugate	Name	Clone	Purchased from
Violet laser (405 nm)	BV421	CD11b	M1/70	Pharmingen
	SYTOX Blue	Viability		Life Technologies
Blue laser (488 nm)	FITC	Ly6C	AL-21	Pharmingen
Green-yellow laser (561 nm)	PECy7	CD11c	N418	BioLegend
	PECy5	CD24	M1/69	BioLegend
	PECy5.5	CD45	30-F11	eBioscience
	PE	CCR2	475301	R&D
	Biotin	CD103	M290	Pharmingen
	PECF594	Streptavidin		Pharmingen
Red laser (633 nm)	APCCy7	CD45R	RA3-6B2	BioLegend
	APCCy7	NK1.1	PK136	BioLegend
	APCCy7	CD3	145-2C11	BioLegend
	APCCy7	Ly6G	1A8	BioLegend
	A700	MHCII	M5/114.15.2	BioLegend
	A674	CD64	X54-5/7.1	Pharmingen

3.4 FACS Staining
(See Note 4)

1. Prepare the antibody mix as described in Table 1 (without the SYTOX Blue) (*see* **Note 5**). The antibodies are diluted in FACS Buffer and require to be titrated before use.

2. Cells are distributed in 1.5 mL Eppendorf tubes for staining (with a maximum of one million cells per tube).

3. Centrifuge for 5 min at $400 \times g$.

4. Add Fc Block (0.025 µg/mL).

5. Incubate for 10 min on ice.

6. Centrifuge for 5 min at $400 \times g$.

7. Add 100 µL of antibody mix.

8. Incubate for 30 min on ice.

9. Add 1 mL of FACS Buffer to wash the cells.

10. Centrifuge for 5 min at $400 \times g$.

11. Resuspend the cell pellet in SYTOX Blue diluted (1/1000) in FACS Buffer.

12. Cells are ready to be run on the FACS or cell sorter (*see* **Notes 6** and **7**). If clumps appear, filter one more time on a 70 µm cell strainer.

3.5 Gating Strategy
for FACS Analysis
or Cell Sorting

1. Gate cells according to their size using FSC and SSC, excluding debris.

2. Remove dead cells by gating on SYTOX Blue negative cells.

3. Gate on hematopoietic cells as CD45+ cells.

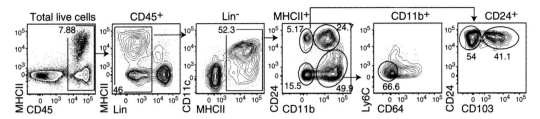

Fig. 1 Conventional DC subset gating strategy. Cells were prepared from the skin and analyzed by flow cytometry. First hematopoietic cells (CD45+) were selected, then NK cells, B cells, and T cells, and neutrophils were excluded (Lin−). A further pre-gating on MHCII+ was then done. Conventional DC subsets can then be discriminated using CD24 versus CD11b. CD24+CD11b− which correspond to CD24/CD207 dermal DC can be further divided into CD103− and CD103+ DC subsets. CD24+CD11b+ correspond to the Langerhans cells (LCs). The CD24−CD11b− is often referred to as DN DC, and finally, within CD24−CD11b+ population, only the Ly6C−CD64− correspond to the conventional CD11b+ dermal DC subset

4. Gate on DC and monocyte-related cells as dump channel negative cells (APCCy7−) within CD45+ cells.

5. DC subsets are gated as follows (*see* Fig. 1): within CD45+ Lin− cells, select MHCII+ CD11c ^low to high^. Then plot them using CD24 versus CD11b (*see* **Note 5**). Four distinct populations should be found: (1) the CD24+CD11b+ cells which correspond to the LCs, (2) the dermal CD24+CD11b− cells which can be further divided as CD103− and CD103+ cells (*see* **Note 8**), and (3) the CD24−CD11b− cells and (4) the CD24−CD11b+ cells which should be further separated using CD64 and Ly6C. The conventional CD24−CD11b+ dermal DC corresponds to the Ly6C−CD64− cells (*see* **Note 9**).

6. Monocyte-related cells are gated as follows (*see* Fig. 2): within CD45+ Lin− cells, select CD24−CD11b+ cells. Then plot them using Ly6C versus CD64. Remove the Ly6C−CD64− cells, which correspond to the conventional CD11b+ dermal DCs. On the remaining cells, plot CD64 versus CCR2, and select CCR2+CD64^low^ and CCR2−CD64+ populations. Within CCR2+CD64^low^ cells, by plotting Ly6C versus MHCII, you can distinguish the monocytes (P1, Ly6C^hi^MHCII−) and monocyte-derived DCs (P2 and P3, Ly6C+MHCII+ and Ly6C−MHCII+, respectively), whereas within CCR2−CD64+ cells, the macrophage populations (P4, Ly6C^low^MHCII− and P5, Ly6C^low^MHCII+) can be distinguished.

4 Notes

1. Dispase treatment allows the separation of the dermis from the epidermis, and it also facilitates the collagenase treatment. Dispase treatment can be performed overnight at 4 °C or 2 h

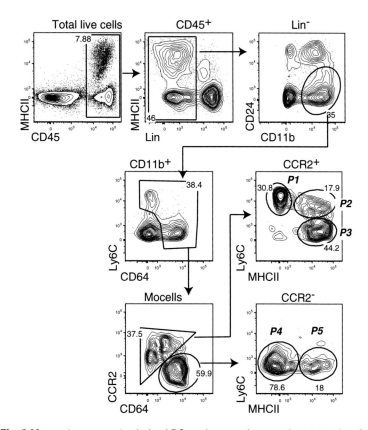

Fig. 2 Monocyte, monocyte-derived DC, and macrophage gating strategies. Cells were prepared from the skin and analyzed by flow cytometry. First hematopoietic cells (CD45+) were selected, then NK cells, B cells, and T cells, and neutrophils were excluded (Lin−). A further pre-gating on total CD11b+ was then done (CD24−CD11b+ population). Conventional CD11b+ DC subset was excluded using Ly6C and CD64. Monocyte-derived cells (Mo cells) were selected based on their expression of Ly6C and CD64. This population can be further divided into CCR2+ and CCR2−. CCR2+ cells include dermal monocytes (P1, Ly6C+MHCII−), dermal monocyte-derived DC (P2, Ly6C+MHCII+), and a further differentiated dermal monocyte-derived DC (P3, Ly6C−MHCII+). CCR2− cells correspond to the dermal macrophages and can be divided in two populations P4 and P5 which are Ly6C−MHCII− and Ly6C−MHCII+, respectively

30 min at 37 °C. If the tissue is treated with dispase for 2 h 30 min at 37 °C, then the dispase II stock solution is diluted ten times and used at a final concentration of 0.4 mg/mL. If the tissue is treated with dispase overnight at 4 °C, then the dispase II stock solution is diluted 20 times and used at a final concentration of 0.2 mg/mL. Further treatment of the skin with collagenase 4 will be the same, but incubating the dispase overnight allows saving precious time of the day of the cell extraction from the skin and of the flow cytometry analysis or sort. Those concentrations are particularly adapted to mouse ear skin. To sepa-

rate dermis and epidermis from flank skin, we recommend overnight incubation with 0.4 mg/mL of dispase II.

2. If your experiments require that you extract cells from flank or back skin of mouse, we recommend to shave and epilate the skin.

3. After the enzymatic treatment and the filtration, the cells can be used directly for flow cytometry staining or any other process. In order to remove most of the dead cells and debris, we perform a Percoll gradient step.

4. The antibody mix provided in Table 1 can be used with the appropriate FACS device with four lasers. In our case, most of the experiments are done using an LSRII and the sort using a FACSAria.

5. In the past we have used CD207 to discriminate DC subsets, but as it requires intracellular staining, we now use CD24 whose expression correlates perfectly with CD207 and whose staining is extracellular. This allows further ex vivo functional assay as well as transcriptomic analysis.

6. For ex vivo functional assay, cells are sorted using a flow cytometer and collected in 5 mL tubes containing 2 mL of 10% FCS EDTA-PBS.

7. For microarray analysis, cells are sorted using a flow cytometer and collected in RNAse-free Eppendorf tubes containing 90 μL of RLT Buffer for further RNA extraction using Qiagen Micro Kit.

8. The CD103$^+$ cells correspond to the XCR1$^+$ dermal cross-presenting DCs [9, 10].

9. Part of the Ly6C$^-$CD64$^-$ cells may produce ALDH, which is involved in Treg induction [11].

Acknowledgment

This work is supported by ANR JCJC Skin DCs.

References

1. Steinman RM, Cohn ZA (1973) Identification of a novel cell type in peripheral lymphoid organs of mice. I. Morphology, quantitation, tissue distribution. J Exp Med 137(5):1142–1162

2. Brasel K, De Smedt T, Smith JL, Maliszewski CR (2000) Generation of murine dendritic cells from flt3-ligand-supplemented bone marrow cultures. Blood 96(9):3029–3039

3. Inaba K, Inaba M, Romani N, Aya H, Deguchi M, Ikehara S, Muramatsu S, Steinman RM (1992) Generation of large numbers of dendritic cells from mouse bone marrow cultures supplemented with granulocyte/macrophage colony-stimulating factor. J Exp Med 176(6):1693–1702

4. Robbins SH, Walzer T, Dembele D, Thibault C, Defays A, Bessou G, Xu H, Vivier E, Sellars M,

Pierre P, Sharp FR, Chan S, Kastner P, Dalod M (2008) Novel insights into the relationships between dendritic cell subsets in human and mouse revealed by genome-wide expression profiling. Genome Biol 9(1):R17

5. Xu Y, Zhan Y, Lew AM, Naik SH, Kershaw MH (2007) Differential development of murine dendritic cells by GM-CSF versus Flt3 ligand has implications for inflammation and trafficking. J Immunol 179(11):7577–7584

6. Malissen B, Tamoutounour S, Henri S (2014) The origins and functions of dendritic cells and macrophages in the skin. Nat Rev Immunol 14(6):417–428

7. Poulin LF, Henri S, de Bovis B, Devilard E, Kissenpfennig A, Malissen B (2007) The dermis contains langerin+dendritic cells that develop and function independently of epidermal Langerhans cells. J Exp Med 204(13):3119–3131

8. Tamoutounour S, Guilliams M, Montanana Sanchis F, Liu H, Terhorst D, Malosse C, Pollet E, Ardouin L, Luche H, Sanchez C, Dalod M, Malissen B, Henri S (2013) Origins and functional specialization of macrophages and of conventional and monocyte-derived dendritic cells in mouse skin. Immunity 39(5):925–938

9. Crozat K, Tamoutounour S, Vu Manh TP, Fossum E, Luche H, Ardouin L, Guilliams M, Azukizawa H, Bogen B, Malissen B, Henri S, Dalod M (2011) Cutting edge: expression of XCR1 defines mouse lymphoid-tissue resident and migratory dendritic cells of the CD8alpha+ type. J Immunol 187(9):4411–4415

10. Henri S, Poulin LF, Tamoutounour S, Ardouin L, Guilliams M, de Bovis B, Devilard E, Viret C, Azukizawa H, Kissenpfennig A, Malissen B (2010) CD207+ CD103+ dermal dendritic cells cross-present keratinocyte-derived antigens irrespective of the presence of Langerhans cells. J Exp Med 207(1):189–206

11. Guilliams M, Crozat K, Henri S, Tamoutounour S, Grenot P, Devilard E, de Bovis B, Alexopoulou L, Dalod M, Malissen B (2010) Skin-draining lymph nodes contain dermis-derived CD103(-) dendritic cells that constitutively produce retinoic acid and induce Foxp3(+) regulatory T cells. Blood 115(10):1958–1968

Chapter 10

Isolation of Conventional Dendritic Cells from Mouse Lungs

Lianne van de Laar, Martin Guilliams, and Simon Tavernier

Abstract

The lungs are in direct contact with the environment. Separated only by a thin layer of mucosa, the lung immune system is being exposed to dangers like pathogens, allergens, or pollutants. The lung dendritic cells form an elaborate network at the basolateral side of the epithelium and continuously sample antigens from the airway lumen. The conventional dendritic cells (cDCs) in the lung can be subdivided into two distinct subsets based on their ontogeny and are described to have distinct immunological functions. High-quality ex vivo isolation of these cells is required for experiments such as functional assays, transfer experiments, or transcriptomics and is crucial to further our knowledge concerning these subpopulations. In this chapter we describe a protocol for the isolation of both CD103+ and CD11b+ cDCs. In our protocol we compare different methods of cell isolation. We propose that the optimal isolation technique is based on the number of cells needed and the type of experiment that will be performed. If low cell numbers are required, simple flow cytometry-assisted cell sorting (FACS) is sufficient. In the case of high cell numbers that will be lysed or fixed upon sorting, positive selection of CD11c+ cells followed by FACS can be utilized. Purification of cDCs through gradient selection and subsequent sorting is found to be optimal for experiments that require large amount of cells for functional assays.

Key words Lung, Dendritic cells, Cell isolation

1 Introduction

In humans, the respiratory tract has an approximate surface of 70 square meters. This large surface is lined by the airway epithelium that is exposed to up to 25 million particles an hour [1]. Among these, allergens (e.g., animal dander, house dust mite feces, or pollen), pathogens (e.g., bacteria, viruses), and air pollutants (e.g., diesel particles, cigarette smoke, or ozone) can potentially disturb lung homeostasis. Intricate cross talk between the mononuclear phagocyte system (MPS) and the airway epithelium underpins the defense against these threats [2].

As a part of the MPS, dendritic cells (DCs) are found at the basolateral side of the airway epithelium, from which they are able to sample particles by extending their processes into the airway

Elodie Segura and Nobuyuki Onai (eds.), *Dendritic Cell Protocols*, Methods in Molecular Biology, vol. 1423,
DOI 10.1007/978-1-4939-3606-9_10, © Springer Science+Business Media New York 2016

lumen. Upon inhalation, DCs recognize, take up, and process antigens, migrate to the mediastinal lymph nodes (mLN), and present these antigens, in an MHC-dependent manner, to cognate T cells. As such, DCs bridge innate and adaptive immunity and are paramount in the induction of long-lasting memory [3, 4].

Two subsets of conventional DCs (cDCs) can be retrieved in human and murine lungs [5]. In mice, these cDC subsets can be distinguished by the expression of the integrins CD103 and CD11b. We and others have found specific, nonredundant functions of these cDC subsets in the immunology of the lung. Whereas CD103+ cDCs are known for effective CTL formation in the context of influenza [6] and uptake and cross-presentation of apoptotic cell-associated antigen [7], CD11b+ cDCs are important in the induction and propagation of Th2 responses in a model of allergic airway inflammation [8]. Note that it was recently demonstrated that CD11b+ DCs can also efficiently cross-present antigen when properly activated [9].

In this protocol, we describe the isolation of both CD103+ and CD11b+ cDCs from the murine lung. Note that we focus here on bona fide conventional DCs and utilize a novel nomenclature system in which the classification is primarily based on ontogeny [10].

2   Materials

2.1   Anesthesia Solution

1. Avertin: To prepare 40 mL, dissolve 1 g of 2,2,2-tribromo-ethanol (Sigma-Aldrich) in 1 mL of 2-methyl-2-butanol (Sigma-Aldrich). Resuspend in 39 mL of PBS and shake overnight at room temperature. Avoid direct exposure to light and store for up to 4 months at 4 °C.

2.2   Buffers

1. PBS.

2. 0.5 M EDTA.

3. HBSS.

4. MACS buffer: 1× PBS supplemented with 0.5–2% BSA and 5 mM EDTA (*see* **Note 1**). Keep at 4 °C.

5. Osmotic lysis buffer: Dissolve 4.145 g ammonium chloride in 450 mL ultra-pure ddH$_2$O. Add 100 μl of 0.5 M EDTA solution. Adjust the pH to 7.1–7.4. Add ultra-pure ddH$_2$O to 500 mL. Keep at 4 °C.

2.3   Media and Supplements

1. RPMI 1640-GlutaMax. Keep at 4 °C.

2. Fetal bovine serum (FBS) (*see* **Note 2**).

3. Gentamicin.

4. 2-mercaptoethanol.

5. R10 medium: RPMI1640-GlutaMax, 10% FBS, 50 μg/mL gentamicin, and 50 μM 2-mercaptoethanol.

2.4 Digestion Medium

1. DNase I stock solution: Dilute 10,000 U DNase I (Roche) in 1 mL PBS, aliquot, and store at −20 °C.

2. Liberase stock solution: Dilute 125 mg (1625 U) Liberase TM (Roche Diagnostics) in 25 mL of RPMI 1640-GlutaMax, aliquot, and store at −20 °C.

3. 2× digestion medium: RPMI 1640-GlutaMax containing 1/500 DNase I stock solution and 1/25 Liberase TM stock solution. Prewarm to 37 °C before use.

2.5 Enrichment

1. OptiPrep (Axis-shield PoC AS).

2. 1 × OptiPrep gradient solution: Add 4 parts PBS to 1 part OptiPrep and mix well. Keep at 4 °C.

3. Anti-CD11c microbeads (Miltenyi Biotec).

4. Anti-FITC microbeads (Miltenyi Biotec).

5. LS columns (Miltenyi Biotec).

6. LD columns (Miltenyi Biotec).

7. Trypan blue.

2.6 Flow Cytometry

1. Antibodies (*see* Table 1).

2. 5 mg/mL DAPI (dilution 1/800).

3. UltraComp eBeads (eBioscience).

2.7 Equipment

1. Cell strainer (100 μm).

2. 15-mL tubes.

3. 5-mL syringe and 25-gauge needle.

Table 1
Antibodies used for flow cytometry

Antibody	Fluorochrome	Clone	Supplier	Dilution
CD3e	FITC	145-2C11	eBioscience	1/500
CD19	FITC	1D3	eBioscience	1/800
CD161	FITC	PK136	eBioscience	1/300
Ly-6G	FITC	1A8	BD-pharmingen	1/1000
CD103	PE	2,00E+07	eBioscience	1/200
CD11c	PE-eFluor610	N418	eBioscience	1/300
CD11b	PE-Cy7	M1/70	BD-pharmingen	1/800
CD64	APC	X54-5/7.1	BD-pharmingen	1/300
MHCII	APC-Cy7	M5/114.15.2	BioLegend	1/1500
CD16/32 (see **Note 3**)	Unlabeled	2.4G2	Produced in-house (3 mg/ml)	1/200

4. Forceps.

5. 37 °C warm water bath.

6. Shaker.

7. Centrifuge.

8. Glass Pasteur pipette.

9. Aspiration device.

10. MACS MultiStand (Miltenyi).

11. Midi MACS (Miltenyi).

12. 1.5-mL Eppendorf tubes.

13. Polystyrene 5-mL tubes (with 100-µM cell strainer cap).

14. Polypropylene 5-mL tubes.

15. Microscope.

16. Bürker-Türk counting plate.

17. Multicolor FACS sorter (BD Aria II/III).

---

3 Methods

3.1 Anesthetization of Mice

1. Inject 0.5 mL of avertin intraperitoneally with a 25-gauge needle. Put the mouse back in the cage and wait for it to stop moving (1–5 min).

2. Once the mouse has stopped moving, check the absence of cerebrospinal reflexes by pinching the paw of the posterior limbs with a forceps. If the mouse shakes avidly, more avertin needs to be injected.

3.2 Dissection

1. Prepare tubes or wells of a 24-well plate containing RPMI 1640 for sample collection. Keep on ice.

2. Spray the animal with 70 % ethanol and immobilize it on a pad by pinning down the paws.

3. Cut into the skin above the abdomen and open up further toward the chin (NB: be sure only to cut the skin). Pull the skin away from the chest.

4. Expose the abdominal cavity by cutting the peritoneal membrane. Lift the sternum and puncture the diaphragm just below the sternum (*see* **Note 4**). Cut the diaphragm from side to side, exposing the thoracical cavity.

5. Use a pair of scissors with a blunt tip to cut the ribs: Upon opening the scissors, move the blunt side of one blade beside the lung, the other blade being on the outside of the chest. By twisting the scissors, the sharp side of the blade moves away from the lungs, reducing the possibility of cutting into the lungs. Cut the ribs open up to the axilla. From this point,

gently move the scissors under the clavicle toward the center of the mouse and cut the upper part of the chest.

6. Pull the ribs away from the lungs and pin down.

7. When lifting up the right lobes with your forceps, the mediastinal lymph nodes (mLN) can be found ventral of the esophagus at the arch of the vena azygos and vena cava superior. Gently pulling the tissue caudal of these veins reveals the lymph nodes. Use forceps to scoop out the mLN and place them in a tube or well containing RPMI 1640 on ice until they can be processed further (*see* **Note 5**).

8. Flushing the lungs: Cut the vena cava at the level of the diaphragm. Make a small incision in the left atrium (*see* **Note 6**). Pinch a 25-gauge needle attached to a 20-mL syringe containing PBS into the right ventricle of the heart. The right ventricle can be found right from the apex of the heart and, in comparison with the left ventricle, has a slightly darker color. Gently push PBS through the lungs. Successful flushing causes the lungs to discolor quickly.

9. After flushing, dissect out the lungs and place them in a tube or well containing RPMI 1640 on ice until further processing.

3.3 Digestion

1. Cut the lungs in new 15-mL tubes, using small scissors. Up to five pairs of lungs can be processed in the same tube, but we recommend cutting two pairs of lungs/tube for optimal cell recovery (*see* Fig. 1). Make sure to cut the lungs thoroughly to ensure a proper digestion of the tissue (*see* **Note 7**).

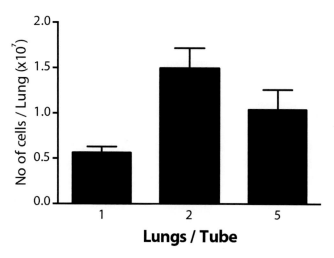

Fig. 1 Comparison cell recovery after processing with 1, 2, or 5 lungs/tube. Dissected lungs were digested according to protocol with 1, 2, or 5 pairs of lungs per tube. Cells were counted with trypan blue exclusion, and the viable cell number recovered per pair of lungs was calculated

2. After cutting, add cold RPMI 1640 (1 mL/pair of lungs). Rinse the scissors and the tube walls to retrieve all cells. Place the tube on ice.

3. When all lungs are cut, add prewarmed (37 °C) 2× digestion medium to each tube (1 mL/pair of lungs). Place the tubes in a warm water bath (37 °C). Incubate 45 min while shaking. Every 15 min, resuspend using a Pasteur pipette.

4. Add 10 mL of cold MACS buffer per tube. Resuspend using the Pasteur pipette and bring onto a 100-μM cell strainer placed on top of a 50-mL tube. Wash the filter with 10 mL of MACS buffer.

5. Spin the tube at $400 \times g$, 7 min at 4 °C.

6. Add osmotic lysis buffer (1 mL/pair of lungs) and incubate 2 min at room temperature.

7. Add at least three times more MACS buffer than osmotic lysis buffer, mix, and spin ($400 \times g$, 7 min at 4 °C).

8. Resuspend in MACS buffer (1 mL/pair of lungs), take a counting sample, and determine the viable cell number (*see* **Note 8**).

9. At this stage, cells from different digestion tubes can be pooled.

3.4 Enrichment

In this protocol, three different enrichment protocols are proposed: (1) positive selection of CD11c⁺ cells by MACS isolation of cells labeled with CD11c⁺ magnetic beads; (2) negative selection of lineage⁻ cells by MACS depletion of cells labeled with FITC-conjugated CD3, CD19, CD161, and Ly-6G antibodies and anti-FITC magnetic beads; and (3) OptiPrep gradient selection. Enrichment can result in a significantly reduced sorting time for experiments in which the researcher requires to sort cDCs from multiple mice (*see* Fig. 2). However, enrichment procedures can also result in loss of cDCs (*see* Table 2) and/or have a negative impact on the viability of the cells when cultured after sorting (*see* Fig. 3). The intended use of the sorted cDCs should be taken into account when selecting a specific enrichment method (*see* **Note 9**).

3.4.1 Enrichment Through Positive Selection with CD11c⁺ Microbeads

1. Resuspend the cells in 400 μl of MACS buffer per 10⁸ cells.

2. Add 100 μl of anti-CD11c microbeads suspension per 10⁸ cells. Mix well and incubate 15 min at 4 °C (*see* **Note 10**).

3. Add 5 mL of MACS buffer and spin the tube at $400 \times g$, 7 min at 4 °C. Remove the supernatant.

4. During the centrifugation step, mount an LS column into the Midi MACS that is attached to the MACS MultiStand. Rinse the column by adding 3 mL of MACS buffer and let it run through. Remove the effluent.

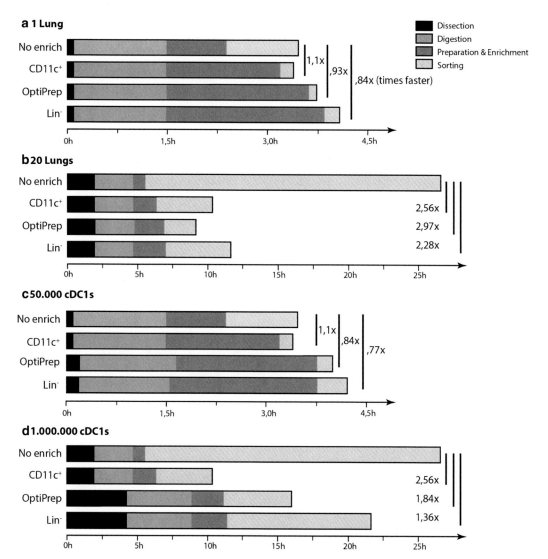

Fig. 2 Time scheme different methods for sorting cells from 1 to 20 mice. Lungs were dissected, digested, prepared for sorting with or without enrichment, and sorted. The time required for each step of the process was determined. The time required to sort cells from 1 (**a**) to 20 mice (**b**) or to obtain 50,000 (**c**) or 1,000,000 cDC1s (**d**) using each preparation method was determined

5. Resuspend the cell pellet in 500 μl of MACS buffer per 10^8 cells. Pipette the cell suspension over a 100-μM cell strainer to remove residual clots and apply onto the column.

6. Wash the column three times with 3 mL of MACS buffer. Allow the liquid to run through completely during each washing step. The resulting effluent contains the CD11c⁻ populations. These can be discarded.

Table 2
Number of cDCs retrieved from one lung

	cDC1	cDC2
No Enrich	41440	59337
CD11c+	40120	58176
OptiPrep	21040	28512
Lin–	24520	34848

Lungs were dissected, digested, and prepared for sorting, either without enrichment or using CD11c⁺ selection, OptiPrep, or lineage⁻ depletion. The number of cDC1 and cDC2 sorted from each sample was determined.

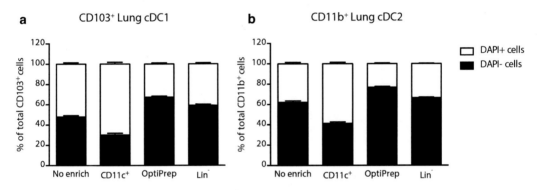

Fig. 3 Survival of cDCs after overnight culture. Lungs were dissected, digested, and prepared for sorting with or without enrichment, and cDC1 and cDC2 were sorted into the same tube. cDCs were then seeded at 30,000 cells/200 μl/well[96] and cultured overnight at 37 °C in R10 medium. After 18 h, the cells were harvested, stained for viability using DAPI, and analyzed by flow cytometry

7. To obtain the CD11c⁺ cells, remove the column from the Midi MACS and put it on a new 15-mL tube. Apply 5 mL of MACS buffer on the column. Take the plunger (supplied with the LS column), position this onto the column, and gently push the CD11c⁺ cells through the column in the 15-mL tube.

8. Spin down the cell suspension at $400 \times g$, 7 min at 4 °C. Resuspend in MACS buffer (0.5 mL/pair of lungs) and determine the viable cell number by counting with trypan blue exclusion (*see* **Note 11**).

9. Proceed to the cell staining and sorting (Subheading 3.5).

3.4.2 Enrichment Through Negative Selection of Lineage⁻ Cells

1. Stain the cells in the antibody mix according to Table 3 (*see* **Note 12**). Use 100 μl of staining mix per 5×10^6 cells. Stain 20 min at 4 °C (*see* **Note 10**).

2. Wash the cells by adding 5 mL of MACS buffer. Spin the tube at $400 \times g$, 7 min at 4 °C. Remove the supernatant.

Table 3
Antibody mix in MACS buffer

Laser	Channel ID	Fluorochrome	Parameter	Clone	Dilution
488 nm	FITC	FITC	CD3e	145-2C11	1/500
	FITC	FITC	CD19	1D3	1/800
	FITC	FITC	CD161	PK136	1/300
	FITC	FITC	Ly-6G	1A8	1/1000
			CD16/32	2.4G2	1/200

3. Resuspend the cells in 90 μl of MACS buffer per 10^7 cells.

4. Add 10 μl of anti-FITC microbeads suspension per 10^7 cells. Mix well and incubate 15 min at 4 °C (*see* **Note 9**).

5. Add 5 mL of MACS buffer (for up to 10^8 cells) and spin the tube at $400 \times g$, 7 min at 4 °C. Remove the supernatant.

6. During the centrifugation step, mount an LD column into the Midi MACS that is attached to the MACS MultiStand. Rinse the column by adding 2 mL of MACS buffer and letting it run through. Throw away the effluent.

7. Place a 15-mL tube on ice under the LD column to collect the flow-through containing the desired cells.

8. Resuspend the cell pellet in 500 μl of MACS buffer per 10^8 cells. Pipette the cell suspension over a 100-μM cell strainer to remove residual clots and apply onto the column.

9. Wash the column twice with 1 mL of MACS buffer. Allow the liquid to run through completely during each washing step. The resulting effluent contains the lineage⁻ populations.

10. Spin down the cell suspension at $400 \times g$, 7 min at 4 °C. Resuspend in MACS buffer (0.5 mL/pair of lungs) and determine the viable cell number by counting with trypan blue exclusion (*see* **Note 11**).

11. Proceed to the cell staining and sorting (Subheading 3.5).

3.4.3 Enrichment Through OptiPrep

1. Resuspend the cells in 5 mL of MACS buffer in a 15 mL tube.

2. Place a Pasteur pipette into the 15-mL tube (*see* Fig. 4a).

3. With a 1-mL tip, carefully pipette 5 mL of the 1× OptiPrep gradient solution into the Pasteur pipette. The solution will settle below the MACS buffer containing the cells (*see* Fig. 4b).

4. Gently move the Pasteur pipette upward until you reach the interface between the OptiPrep gradient solution and MACS buffer (*see* Fig. 4c). The OptiPrep solution still remaining in the Pasteur pipette will flow out during this procedure. Upon reaching the interface, place your finger on top of the Pasteur

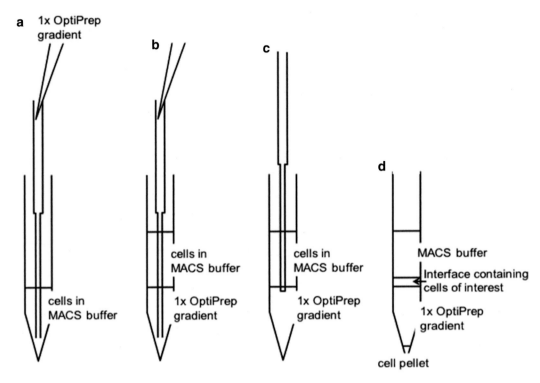

Fig. 4 Enrichment through OptiPrep

pipette to stop the flow and move the pipette upward out of the tube. Make sure not to disturb the interface between the OptiPrep gradient solution and MACS buffer as this will impede cell retrieval.

5. Spin the tube at $600 \times g$, 20 min at 8 °C, with the acceleration as low as possible and without a break at the end of the centrifugation.

6. After spinning, the cDCs are present in a thin layer at the interface of the OptiPrep solution and MACS buffer (*see* Fig. 4d). Collect this layer of cells using a 1-mL tip (approximately 3 mL of fluid should be taken) and put it into a new 15-mL tube containing 10 mL of MACS buffer.

7. Spin at $600 \times g$, 7 min at 4 °C. Resuspend in MACS buffer (1 mL/pair of lungs) and determine the viable cell number by counting with trypan blue exclusion (*see* **Note 11**).

8. Proceed to the cell staining and sorting (Subheading 3.5).

3.5 Cell Staining and Sorting

1. Stain the cells in the antibody mix according to Table 4. Use 100 μl of staining mix per 5×10^6 cells. Stain 30 min at 4 °C (*see* **Note 10**).

2. Wash the cells by adding 5 mL of MACS buffer. Spin the tube at $400 \times g$, 7 min at 4 °C. Remove the supernatant and resuspend in 1 mL of MACS buffer per 10^7 cells.

Table 4
Antibody mix in MACS buffer

Laser	Channel ID	Fluorochrome	Parameter	Clone	Dilution
405 nm	Pacific Blue	DAPI	Viability		1/800
488 nm	FITC	FITC	CD3e	145-2C11	1/500
	FITC	FITC	CD19	1D3	1/800
	FITC	FITC	CD161	PK136	1/300
	FITC	FITC	Ly-6G	1A8	1/1000
	PE	PE	CD103	2,00E+07	1/200
	PE-TR	PE-eFluor610	CD11c	N418	1/300
	PE-Cy7	PE-Cy7	CD11b	M1/70	1/800
633 nm	APC	APC	CD64	X54-5/7.1	1/300
	APC-Cy7	APC-Cy7	MHCII	M5/114.15.2	1/1500
			CD16/32	2.4G2	1/200

3. Single stains are prepared by diluting 1 drop of UltraComp eBeads in 1.5 mL of MACS buffer (*see* **Note 13**). Place 200 µl of this bead suspension into seven 1.5-mL Eppendorf tubes (for six fluorochromes + one unstained sample). Add one antibody in each tube: antibodies conjugated to bright fluorochromes (PE, PE-eFluor610) are diluted 1/2000; other antibodies are diluted 1/400. Stain 15 min on ice. Add 1 mL of PBS and spin down 400×g, 7 min at 4 °C. Remove the supernatant and resuspend the beads in 200 µl of PBS. Proceed to the cell sorter (*see* **Note 14**).

4. Before acquiring your sample, add 0.63 µl of DAPI to the cell suspension tube.

5. Prepare 5-mL polypropylene tubes or 1.5-mL Eppendorf tubes containing 1 or 0.5 mL of R10 medium, respectively, to collect the sorted DCs (*see* **Note 15**).

6. Gate DC subsets (*see* Fig. 5).

4   Notes

1. EDTA minimizes adherence of cells to plastics by chelating Ca²⁺.

2. FBS differs from batch to batch, which can have large effects on in vitro differentiation experiments. Comparison of batches is required to optimize results.

3. Addition of anti-CD16/CD32 minimizes background fluorescence by inhibiting nonspecific binding of antibodies to FcG—receptors present on immune cells.

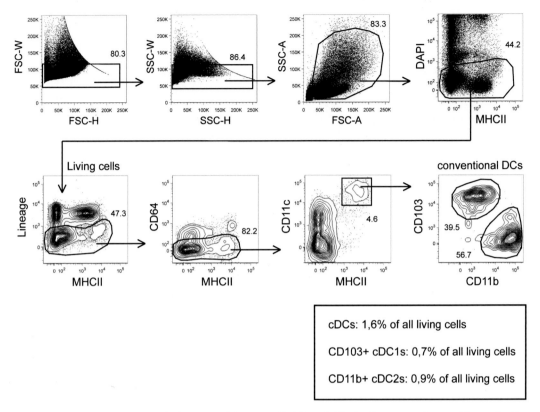

cDCs: 1,6% of all living cells

CD103+ cDC1s: 0,7% of all living cells

CD11b+ cDC2s: 0,9% of all living cells

Fig. 5 Gating strategy to identify CD103$^+$ cDC1s and CD11b$^+$ cDC2s in the mouse lungs. After outgating doublets, debris, dead cells, and lineage-positive cells, plot CD64 against MHCII. To avoid contamination of cDCs with alveolar macrophages (which express high levels of CD11c and are highly autofluorescent) or monocyte-derived cells or interstitial macrophages (which express high levels of MHCII and can express CD11c) gate-out all CD64$^+$ cells, as this marker is highly expressed by alveolar macrophages, monocyte-derived cells, and interstitial macrophages but not by cDCs. Next, gate on MHCII$^+$ CD11c$^+$ cells and divide these cells into Batf3-dependent CD103$^+$ cDC1s and IRF4-dependent CD11b$^+$ cDC2s. Note that as cDCs are a bit autofluorescent, it is practical to keep MHCII as X-axis when outgating dead cells or lineage-positive cells to avoid losing part of the cDCs in the process

4. This causes the lungs to deflate which enables the opening of the chest without causing damage to the lungs.

5. In homeostatic conditions, the mLN is a small organ (1–2 mm) containing $1-2 \times 10^6$ cells, mainly consisting of T and B cells. Both blood-derived (resident) and peripheral organ (e.g., lung, heart, abdominal cavity)-derived (migratory) DC populations reside in the mLN. These two DC populations each contain cDC1 and cDC2. Table 5 gives an overview of expressed surface markers used to identify the different cDC populations and the approximative number of DCs found in the mLN. The isolation of DCs from the mLN through the use of OptiPrep and subsequent sorting has been used previously. In our hands,

Table 5
Overview of cDCs found in the mLN

cDC Subset		Surface markers	No of cells
Resident	cDC1	CD11c^{hi},MHCII$^+$,CD8a$^+$	3555
	cDC2	CD11c^{hi},MHCII$^+$,CD11b$^+$	3281
Migratory	cDC1	CD11c$^+$,MHCII^{hi},CD103$^+$	971
	cDC2	CD11c$^+$,MHCII^{hi},CD11b$^+$	2188

mLNs from ten mice were pooled and digested. Number of cells was determined by manual counting of viable cells in trypan blue. Relative proportions of the different cDC subsets were analyzed by means of FCM. Table 4 depicts surface markers present on the different cDC subsets and number of cells that can be found in homeostatic conditions in the mLN of one mouse

the use of OptiPrep results in a selective loss of resident cDCs, likely due to a difference in density between the migratory and resident populations.

6. This reduces the pressure in the pulmonary circulation and subsequent rupture of capillaries during flushing.

7. If processing multiple pairs of lungs per tube, we recommend cutting them one by one.

8. Typically, digestion of one lung results in $8–15 \times 10^6$ cells.

9. The use of CD11c$^+$ beads will yield a highly purified CD11c$^+$ cell suspension with minimal cDC loss. The downside of this method is the relatively poor survival of the sorted cDCs (probably due to cellular stress elicited by the procedure of positive selection). We advise to use this approach in experiments where the cells are fixed or lysed upon sorting (e.g., for imaging or for RNA or protein-based techniques). Negative selection of lineage$^-$ cells does not have this impact on cell viability. However, this enrichment procedure is slightly less sensitive, leading to a relatively high cDC loss, which may require the dissection and processing of an increased number of lungs when high cDC numbers need to be isolated. The OptiPrep procedure similarly results in enhanced cDC loss, but has the advantage that the cells isolated have a superior survival compared to the use of CD11c$^+$ beads, lineage$^-$ selection, and even sorting without prior enrichment. The high viability of the sorted DCs obtained with OptiPrep renders this density gradient technique optimal for both in vitro experiments and experiments where the sorted DCs are transferred into recipient hosts. We point out that the superior viability of this procedure compared to the no-enrichment procedure could even be more dramatic in these experiments where sorting time increases substantially (e.g., need for high numbers of cells).

10. If the staining volume is high, rolling of the tube is advised to avoid sedimentation of the cells.

11. Typically, one should expect 1–2×10^6 cells after enrichment of one set of lungs.

12. We do not recommend the use of a lineage depletion antibody cocktail for the enrichment of DCs in the lung. This approach results in an increased duration of the preparatory procedure, considerable cDC loss, and additional costs (columns, large set of antibodies).

13. Make sure to shake the beads thoroughly before use as precipitation occurs.

14. Viability dyes do not bind to the UltraComp eBeads. Since we exclude dead cells (defined by a positive staining for DAPI), there is no need to compensate for this fluorochrome.

15. Coating of the collection tubes prior to use is advised.

Acknowledgments

The authors would like to thank current and past members of the Bart Lambrecht laboratory who have made key contributions to the development and optimization of these procedures.

References

1. Osorio F, Lambrecht B, Janssens S (2013) The UPR and lung disease. Semin Immunopathol 35:293–306

2. Hammad H, Lambrecht BN (2011) Dendritic cells and airway epithelial cells at the interface between innate and adaptive immune responses. Allergy 66:579–587

3. Lambrecht BN, Hammad H (2010) The role of dendritic and epithelial cells as master regulators of allergic airway inflammation. Lancet 376:835–843

4. Guilliams M, Lambrecht BN, Hammad H (2013) Division of labor between lung dendritic cells and macrophages in the defense against pulmonary infections. Mucosal Immunol 6(3):464–473

5. Schlitzer A, McGovern N, Teo P et al (2013) IRF4 transcription factor-dependent CD11b+ dendritic cells in human and mouse control mucosal IL-17 cytokine responses. Immunity 38:970–983

6. Kim TS, Gorski SA, Hahn S et al (2014) Distinct dendritic cell subsets dictate the fate decision between effector and memory CD8(+) T cell differentiation by a CD24-dependent mechanism. Immunity 40: 400–413

7. Desch AN, Randolph GJ, Murphy K et al (2011) CD103+ pulmonary dendritic cells preferentially acquire and present apoptotic cell-associated antigen. J Exp Med 208: 1789–1797

8. Plantinga M, Guilliams M, Vanheerswynghels M et al (2013) Conventional and monocyte-derived CD11b(+) dendritic cells initiate and maintain T helper 2 cell-mediated immunity to house dust mite allergen. Immunity 38: 322–335

9. Desch AN, Gibbings SL, Clambey ET et al (2014) Dendritic cell subsets require cis-activation for cytotoxic CD8 T-cell induction. Nat Commun 5:4674

10. Guilliams M, Ginhoux F, Jakubzick C et al (2014) Dendritic cells, monocytes and macrophages: a unified nomenclature based on ontogeny. Nat Rev Immunol 14:571–578

Chapter 11

Purification of Human Dendritic Cell Subsets from Peripheral Blood

Solana Alculumbre and Lucia Pattarini

Abstract

Blood represents the most accessible source of human dendritic cells (DCs). We present here a method to isolate three DC subtypes, as identified until now, from peripheral blood: plasmacytoid dendritic cells (pDCs), CD141+ myeloid DCs, and CD1c+ myeloid DCs. The method is based on the sequential depletion of non-DCs. First, depletion of granulocytes, erythrocytes, and platelets is obtained by blood centrifugation over a Ficoll gradient. Then, antibodies recognizing non-DCs, combined with magnetic beads, allow enrichment of DCs from peripheral blood mononuclear cells (PBMCs). Finally, enriched DCs are purified and separated into the different subtypes by immunolabeling and fluorescence-activated cell sorting (FACS) using DC-specific surface markers.

DC studies might contribute to the comprehension of human immune processes in physiological and pathological conditions. Human blood DCs targeting might be a useful tool to ameliorate inflammatory diseases and improve vaccination strategies.

Key words Dendritic cell, Human, Blood, pDC, Subtypes

1 Introduction

Identification of dendritic cells in blood, the easiest tissue to obtain from human subjects, paved the way to dendritic cell research in humans. Originally, human DCs were enriched to 20–60 % purity from peripheral blood mononuclear cells (PBMCs) using a protocol based on cell aggregation and differential adhesion, adapted from the one used by Steinman to identify DCs in mouse lymphoid organs [1, 2].

Several years of DC research, in combination with technical advances in magnetic immune selection and fluorescence-activated cell sorting, improved human DC enrichment: the method we describe in this chapter allows isolation of DCs from blood to 99 % purity in about 7 hours.

Currently, the combination of phenotypic markers, transcriptomic analysis, and functional characterization allows classification

Elodie Segura and Nobuyuki Onai (eds.), *Dendritic Cell Protocols*, Methods in Molecular Biology, vol. 1423,
DOI 10.1007/978-1-4939-3606-9_11, © Springer Science+Business Media New York 2016

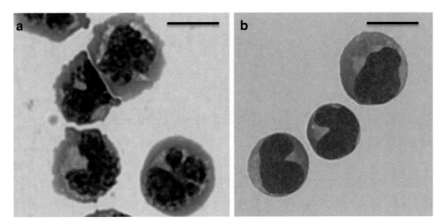

Fig. 1 Morphology of human pDCs and DCs. Giemsa staining of purified and sorted (**a**) myeloid DCs (Lin⁻, CD4⁺, CD11c⁺) and (**b**) blood pDCs (Lin⁻, CD4⁺ and CD11c⁻). Bar is 10 μm

of three subtypes of DCs in human blood: the two myeloid subtypes CD141⁺ DCs and CD1c⁺ DCs and plasmacytoid dendritic cells (pDCs) (*see* Fig. 1) [3]. Our protocol was designed to purify the three DC subtypes at the same time.

DC's main feature is the initiation of adaptive immune response through antigen capture, processing, and presentation to T cells. Spontaneous human DC deficiency recently showed that in vivo DCs are important against mycobacteria infection [4]. To fully exert their function, DCs population evolves from an immature to a mature state. In the immature state, DCs patrol the environment and pick up antigens: this stage is characterized by phagocytic and migratory capacities. In the mature state, DCs express a broad range of communication molecules, such as costimulatory factors (among them CD40, CD80, CD86), as well as high levels of major histocompatibility complex class I and II (MHC I and II) and cytokines. All these factors enable the DC to efficiently present antigens and induce different types of T cell responses. Blood myeloid DCs—about 1 % of PBMCs—appear to be in transit toward other tissues, even if the formal demonstration in humans is still lacking [5]. Myeloid DCs in blood are mainly in the immature state and our DC purification method has been designed to preserve this steady state and minimize DC activation. Therefore DCs isolated following this protocol constitute a useful in vitro model to study human DC activation and plasticity [6, 7]. Endotoxins present in several reagents such as serum or saline buffers may activate DCs and must be controlled during the procedure.

Myeloid DCs are mainly CD1c⁺ (also called BDCA1) cells, while CD141⁺ cells (BDCA3) represent a small percentage [8, 9]. Phenotypic and functional differences between these two human subsets are subject of debate and of intense investigation [10–13].

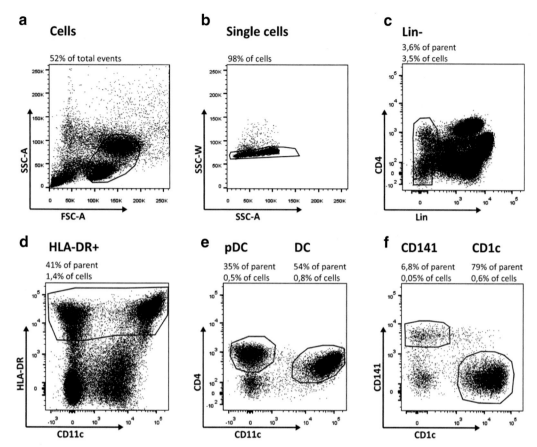

Fig. 2 Identification of dendritic cell subsets in human PMBCs by flow cytometry. Sequential gating strategy applied to total PBMCs to separate DC subtypes. (**a**) Cells are selected based on size and morphology by FSC and SSC. (**b**) Single cells to exclude doublets by SSC-A and SSC-W. (**c**) Lineage⁻ (lineage markers include CD3, CD14, CD16, and CD19). (**d**) HLA-DR⁺ selection of DCs and pDCs. (**e**) Separation between pDC: CD4⁺ CD11c⁻ and mDCs: CD4⁺CD11c⁺. (**f**) Discrimination of myeloid DC by CD1c⁺CD141⁻ and CD1c⁻CD141⁺. Numbers above each panel represent the percentage of each gate in respect to their parent gate and to total cells

Our protocol includes anti-CD141 and CD1c antibodies in the sorting panel to enable separation of these two subtypes.

Both myeloid DC subtypes in the blood are characterized by the expression of CD4, CD11c, and HLADR (*see* Fig. 2). Unfortunately, these are not DC-exclusive markers: CD4 is expressed at similar levels by some T lymphocytes, while the large majority of human blood monocytes express CD11c [14]. Scarcity of exclusive markers requires a depletion approach for blood DC purification: i.e., the removal of non-DCs from PMBCs, in particular cells expressing markers common to DCs such as CD4, CD11c, or HLADR. Our protocol exploits some lineage markers, whose expression is low or absent in DCs, to exclude contaminant cells. These markers include CD3, CD19, CD56, CD14, and CD16 (collectively defined lineage markers), used to deplete T cells, B cells, natural killer (NK), and monocytes, respectively [15, 16]. A DC lineage-specific marker has been recently reported by two independent studies [17, 18]. The zinc finger transcription factor Zbtb46 is specifically expressed in mouse and human myeloid DCs and their precursors, but not in monocytes or pDCs; nonetheless, at the moment, the use of this intracellular marker is not compatible with the purification of viable DCs, while it is really useful to control the identity of purified cells.

PDCs represent 0.3–0.5 % of human PMBCs, they show round plasma cell-like morphology and ability to differentiate into DCs [19, 20]. Cells circulating in blood are thought to be in a nonactivated state. PDCs play an important role in the protection from viral infections [21]. PDCs sense viruses or viral nucleic acids through the endosomal TLR7 and TLR79 and efficiently secrete important amounts of type I interferons (IFNα/β), 1000-fold higher than other cell types [22, 23]. Upon TLR triggering they also upregulate costimulatory molecules and secrete cytokines, such as TNF-α and IL-6 [24]. Similarly to myeloid DCs, activated pDCs have antigen presentation ability, and polarize naïve CD4 T cells [25].

While both myeloid DCs and pDCs express CD4, pDCs are negative for the integrin CD11c (*see* Fig. 2) and express lower levels of the HLADR compared to myeloid DCs. Additionally, they present some specific markers such as CD303 (BDCA2), CD304 (BDAC4), and CD123 (IL3RA), even if not all of them are pDC-restricted [24]. The use of single positive markers to purify pDCs should be avoided: as an example, isolation using anti BDCA2 antibody impairs production of IFNα [26]. Additionally, some of those markers, as BDCA2, may be downregulated after activation [27]. Our protocol is designed to purify pDCS as Lin$^-$, CD4$^+$, CD11c$^-$ cells: this approach leaves the cells fully responsive for further functional studies [28].

2   Materials

2.1   Total PBMCs Purification from Blood

1. Dulbecco's Phosphate Buffered Saline (DPBS) pH 7.2. Salt solution containing 200 mg/mL KCl, 8000 mg/mL NaCl, 200 mg/mL KH_2PO_4, 2160 mg/mL Na_2HPO_4 $\cdot 7H_2O$ (No Ca^{2+}, No Mg^{2+}). Adjust pH to 7.0–7.2. Keep the sterile buffer at 4 °C (*see* **Note 1**).

2. Ficoll-Paque PLUS ($\rho = 1.077 \pm 0.001$ g/mL), sterile, with a low (<0.12 EU/mL) endotoxin activity (GE Healthcare Life Science, Pittsburgh, PA, USA). Store at room temperature until open, then store at 4 °C.

3. 50 mL conical tubes (BD Falcon San Jose, CA, USA, or TPP Trasadingen, Switzerland).

4. FBS: Research grade Fetal Bovine Serum, triple 0.1 μm pore size filtered and decomplemented (*see* **Notes 2** and **3**). Aliquot and store at –20 °C.

5. NEAA: Nonessential amino acid solution. Solution containing 750 mg/mL glycine, 890 mg/mL L-alanine, 1320 mg/mL L-asparagine, 1330 mg/mL L-aspartic acid, 1470 mg/mL L-glutamic acid, 1150 mg/mL L-proline, 1050 mg/mL L-serine. Store at 4 °C.

6. Sodium pyruvate: 100 mM sodium pyruvate solution in water. Store at 4 °C.

7. Penicillin/ streptomycin: Solution containing 10^4 U/mL of penicillin and 10^4 μg/mL streptomycin. Aliquot and store at –20 °C.

8. Complete RPMI: RPMI 1640, 446 mg/mL L-alanyl-glutamine, 5 mg/mL phenol red, 10 % FBS, 1 % NEAA, 1 % sodium pyruvate, 1 % penicillin/streptomycin. Store at 4 °C.

2.2 Total DC Enrichment from PBMCs

1. HS: Human serum from male AB group. Tested for virus and sterility. Double 0.2 μm pore size filtered. Store at –20 °C before use. Decomplement at 57 °C for 1 h and store then at 4 °C.

2. EDTA: 0.5 M EDTA adjusted to pH 8.0 with NaOH, 0.2 μm filtered.

3. PBS HS EDTA: DPBS, 1 % HS, 2 mM EDTA. Add 5 mL of 0.2 μm filtered HS and 2 mL of EDTA to 500 mL of DPBS. Store at 4 °C, discard after 1 month.

4. Human Pan-DC Pre-Enrichment Kit (STEMCELL Technologies, Vancouver, Canada).

5. Easy 50 EasySep™ Magnet (STEMCELL Technologies, Vancouver, Canada). We recommend keeping the magnet at 4 °C.

2.3 Immunofluo-rescence Staining and FACS

1. Fluorochrome-conjugated monoclonal mouse anti-human antibodies (*see* Table 1). All antibodies are kept at 4 °C and protected from light.

2. Polypropylene collection tubes.

3. Cell sorter: FACS Aria III (BD Bioscience) or any flow cytometer with the appropriate lasers and setup (*see* Table 2).

3 Methods

The purification method presented here has been optimized to maximize DC recovery, purity, and viability, as well as to minimize cell modification and experimental procedure time and cost. Nonetheless, DC purification remains an expensive procedure.

Table 1
Monoclonal antibodies used for the specific recognition of human DC subsets from blood

Specificity	CD3	CD14	CD19	CD16	CD4	CD4	CD4	HLA-DR	CD11c	CD11c	CD1c	CD141
Clone	HIT3a	TÜK4	LT19	NKP15	13B8.2	RPA-T4	VIT4	LN3	Bu15	MJ4-27G12	L161	AD5-14H12
Fluorescent conjugate	FITC	FITC	FITC	FITC	PE Cy5	APC Cy7	VioGreen	APC-eFluor 780	PE Cy7	PE Vio770	PerCP-eFluor 710	PE

Table 2
Optical requirements for detection of fluorochromes used for the separation of human dendritic cell subsets from blood

Laser	Fluorochrome	Peak emission	Bandpass filter	Longpass dichroic mirror
Blue 488 nm				
	PerCP-eFluor 710	710	675–715	655LP
	FITC	518	515–545	502LP
Yellow-Green 561 nm				
	PE Cy7/PE Vio770	785/775	750–810	750LP
	PE Cy5	667	685–735	630LP
	PE	575	575–589	
Red 633–640 nm				
	APC-eFluor 780/ APC Vio770/ APC Cy7	780/775/785	750–810	735LP
	APC	668	650–670	
Violet 406 nm				
	VioGreen	520	500–575	502LP
	VioBlue	455	420–470	

This configuration corresponds to a BD ARIA III

Three sequential steps compose our method:

1. Total PBMCs purification from blood. Centrifugation over a Ficoll gradient followed by several washes results in exclusion of granulocytes, erythrocytes, and platelets, as well as serum. We optimized the protocol of Ficoll to maximize DC recovery and viability. This step takes about two hours, and PBMCs obtained here can be kept overnight for the following steps.

2. Total DC enrichment from PBMCs. This is the crucial step in DC purification and is based on the immunodepletion of non-DCs. To this purpose, cells are incubated with a cocktail of antibodies recognizing surface markers expressed by other PMBCs populations but absent on DCs. Coupling the antibodies to magnetic particles depletes non-DCs. In Fig. 4 we show the DC proportions after the step of enrichment. This step takes 2 hours, and enriched DCs can be sorted the day after.

3. Immunofluorescent staining and DC subtypes sorting. This final step is required to obtain highly (99 %) pure cells and to separate DCs in the three DC subtypes. Enriched DCs are

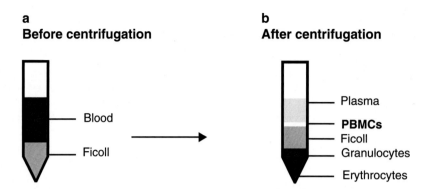

Fig. 3 Gradient centrifugation of human blood. Blood centrifugation on a Ficoll gradient allows the separation of PBMCs from erythrocytes and granulocytes. (**a**) Before centrifugation, the blood sample is located over the Ficoll. (**b**) After centrifugation, a layer of PBMCs is visible between the plasma and the Ficoll. Due to their higher density, both granulocytes and erythrocytes locate under the Ficoll

stained using fluorescent antibodies that exclude remaining contaminant cells and allow a positive discrimination of the three DC subsets based on CD11c, HLA-DR, CD4, CD141, and CD1c. FACS of these enriched DCs results in three separated population of DCs.

The whole purification protocol should be performed within 36 h after blood sampling to obtain optimal DC recovery and survival.

3.1 Total PBMCs Purification from Blood

Work under a laminar flux during the whole process.

1. Prepare the Ficoll-Paque, PBS, and centrifuge at room temperature (18–20 °C) (*see* **Note 4**).

2. Collect the enriched leukocyte blood sample from the buffy coat and dilute the sample twofold with room temperature PBS (*see* **Note 5**).

3. Prepare 15 mL of room temperature Ficoll-Paque in 50 mL conical tubes. Prepare one tube for every 25 mL of final diluted blood sample (*see* **Note 6**).

4. Carefully, and using the pipetboy at slow speed, layer 25 mL of the diluted blood over the Ficoll-Paque. To maintain the gradient, the blood should not mix with the Ficoll-Paque. It is recommended to tilt the tube to approach the liquid to the top of it, which will help to layer the sample (*see* **Note 7**) (*see* Fig. 3).

5. Place the tubes in the centrifuge with extreme caution to the balancing. Centrifuge at $1400 \times g$ for 20 min at 18 °C without brake or acceleration (*see* **Note 8**).

6. Transfer the mononuclear cell layer to a new 50 mL tube (*see* Fig. 3) (*see* **Note 9**). You can pool the content of two tubes into one (*see* **Note 10**).

7. Fill up the tube with cold PBS to 50 mL to wash the remaining Ficoll as well as platelets (*see* **Note 11**).

8. Centrifuge at $500 \times g$ for 10 min, with brake and acceleration. Carefully remove the supernatants. Do not pour (*see* **Note 12**).

9. Resuspend the cell pellet in 50 mL of cold PBS and centrifuge at $200 \times g$ for 10 min at 4 °C. Carefully remove the supernatant. This step improves removal of platelets as well as minimizes Ficoll-Paque toxicity.

10. Repeat the **step 9** and count the cells.

11. After the last wash, resuspend the cell pellet in the appropriate amount of PBS HS EDTA to continue with the protocol of DC enrichment. If the enrichment is to be performed the next day, then resuspend the cell pellet in complete RPMI and keep the cells overnight at 4 °C in a tube. If not, proceed to the total DC enrichment.

3.2 Total DC Enrichment from PBMCs

Different methods can be applied for total DC enrichment before sorting of the different DC subsets. Here we expose in details one of those methods and we comment on other available possibilities (*see* **Note 13**). At the end of the protocol, the sample is highly enriched in the populations of interest and is ready for flow cytometry separation of the different DC subtypes (*see* Fig. 4).

We adapted our protocol from the one provided by the manufacturer. It corresponds to a sample starting size of 10^9 PBMCs, which corresponds to an average preparation from a buffy coat. The amounts should be adjusted proportionally to the amount of cells (*see* **Note 14**). If starting from more than 1.5×10^9 PBMC, it is recommended to split the sample in two.

1. Resuspend the PBMCs with 20 mL of PBS HS EDTA in a 50 mL falcon tube (*see* **Note 15**).

2. Add 600 μL of anti-human CD32 (Fcg RII) blocker (*see* **Note 16**) and 1 mL of Human Pan-DC Pre-Enrichment Cocktail. Mix well and incubate 30 min at 4 °C (*see* **Notes 17 and 18**).

3. Vortex the magnetic particles for 30 seconds or until a uniform suspension without aggregates is observed (*see* **Note 19**).

4. Add 2 mL of magnetic particles to the cells. Mix well and incubate 10 min at 4 °C.

5. Add 10 mL of PBS HS EDTA to the cell suspension and mix well by pipetting up and down. Close the tube loosely and place it in the Easy 50 EasySep Magnet. Incubate during 10 min at 4 °C (*see* **Notes 20 and 21**).

6. Carefully pipette the cell suspension into a new 50 mL tube. Do not pour. Avoid touching the sides and the bottom of the tube. The non-targeted, magnetically labeled cells will remain loosely attached to the wall of the original tube.

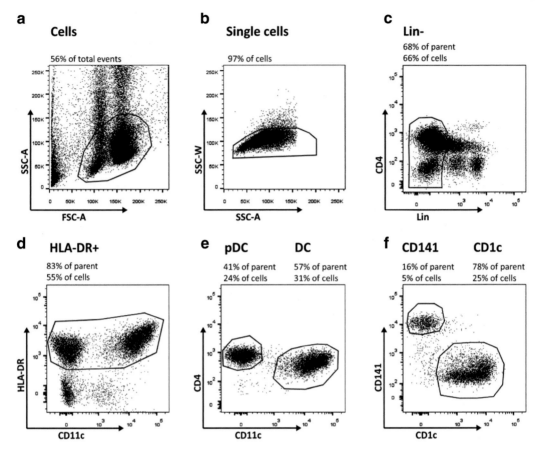

Fig. 4 DC enrichment from human PBMCs. Gating strategy as in Fig. 2 shows the increment on DC percentages after the enrichment step. This gating can be used for cell separation during FACS. (**a**) Cells by FSC and SSC, (**b**) single cells by SSC-A and SSC-W, (**c**) lineage⁻ (Lin), including CD3, CD14, CD16, and CD19. Careful gating of Lin⁻ cells is important to avoid contaminant cells if the separation between Lin⁻ and Lin⁺ is not clear, we recommend to restrict this gate; (**d**) HLA-DR⁺, (**e**) pDC: CD4⁺CD11c⁻ and myeloid DCs: CD4⁺CD11c⁺, (**f**) from the myeloid DCs gate: CD1c⁺CD141⁻ and CD1c⁻CD141⁺

7. Place the new tube containing the enriched fraction again into the magnet and incubate 10 min at 4 °C.

8. Proceed as in **step 6**.

9. Count the cells and continue with the staining for FACS (*see* **Note 22**). If the FACS is to be performed the next day, centrifuge the cells at $200 \times g$ for 10 min at 4 °C, resuspend the cell pellet in complete RPMI, and keep the cells overnight at 4 °C.

3.3 Immunofluorescence Staining and FACS

1. Prepare a mix of pre-titrated fluorescent-conjugated monoclonal antibodies at the appropriate concentration in PBS HS EDTA instantly before using it (*see* **Notes 23–25**).

2. Centrifuge the cells at $200 \times g$ for 10 min at 4 °C (*see* **Note 26**). Remove the supernatant completely.

3. Add 100 μL of antibody mix per 10×10^6 cells and carefully resuspend the cell pellet by pipetting (*see* **Note 27**). Incubate at 4 °C for 15–20 min in the dark.

4. Add a large volume of PBS HS EDTA and centrifuge to wash the cells.

5. Remove the supernatant and resuspend the cell pellet in the appropriate volume of PBS HS EDTA to reach the required concentration for FACS.

6. Filter the cells with a 0.2 μm pore size filter and transfer them to a FACS tube. Keep them ice cold in the dark until sorting (*see* **Note 28**).

7. Prepare collection tubes according to the sorter used. Add at least 1 mL of complete RMPI to polypropylene collection tubes, coating the walls of the tube with medium (*see* **Note 29**).

8. Sort DCs using a cell sorter. Select the DCs by gating high forward and middle side scatter, excluding debris and possible dead cells with lower forward and variable side scatter (*see* **Note 30**). Exclude doublets. Select the Lin– population based on the FITC channel (*see* Fig. 4). Gate on the HLA-DR+ cells and subsequently identify the pDC population as CD4$^+$ and CD11c$^-$ and the total mDC as CD4$^+$ and CD11c$^+$. Finally, from the myeloid DC gate, differentiate CD1c and CD141 DCs (*see* Fig. 4). Cell separation should be performed under low pressure to prevent activation and increase the viability (*see* **Note 31**).

4   Notes

1. Sterility and low presence of endotoxins in PBS are crucial to avoid contaminations and DC activation during the isolation procedure. One possibility is to use commercial sterile PBS.

2. Decomplement serum by incubation at 57 °C for 1 hour.

3. Different serum lots should be tested for endotoxin activity prior to use.

4. 18–20 °C is the optimal temperature for Ficoll separation.

5. Depending on blood banks, human blood is available as whole blood, buffy coat, and platelet apheresis blood. For optimal DC recovery and viability, blood should be handled as soon as possible, maximally within 24 h after sampling. This might not be compatible with testing for infections; therefore every sample should be handled as potentially infected. Careful records of blood samples are crucial for safe and reproducible research.

6. Ficoll-Pacque is toxic for DCs; therefore, all the steps in this section should be performed as quickly as possible.

7. For optimal PBMCs recovery it is crucial to keep a good interface between blood and Ficoll.

8. The absence of acceleration and brake is strictly required for PMBCs separation over Ficoll-Paque gradient.

9. To facilitate the visualization of the cellular white ring you can aspirate the upper plasma layer from the gradient, without perturbing the mononuclear cell layer.

10. Avoid pipetting red blood cells; if that is the case, throw away the pipette content.

11. Keep samples on ice between each step, in particular when processing several tubes at the same time.

12. The cell pellet is loosely attached to the tube, and pouring could result in cell loss.

13. Alternatively to the StemCell kit, there is another commercial kit from Miltenyi Biotec for the enrichment of total blood DC. The results obtained are similar between both kits. In addition to these commercial kits, it is also possible to perform the DC enrichment by preparing a monoclonal mouse antibodies cocktail specific for the nontargeted cells followed by the addition of Dynabeads. The rationale for homemade DC enrichment is to deplete non-DCs using a homemade mix of antibodies, and not a commercial kit. The main advantage of this protocol is the possibility to optimize the mix of antibodies, both by changing the quantities and adding or removing antibodies. The main disadvantages are the length of the protocol and the final efficiency.

14. Since DCs correspond to 1 % of total PMBCs, and pDCs to 0.5 %, and the efficiency of our protocol is about 30 %, starting from low numbers of PMBCs (less than 300×10^6) could result in very poor DC recovery.

15. When incubating PBMCs or DCs with antibodies or beads, always use PBS HS EDTA. The HS impairs unspecific binding while improving DC viability. It can be substituted by 5 % final bovine serum albumin, sterile and filtered. EDTA, by chelating divalent ions, minimizes cell aggregation.

16. This step is optional and it is meant to reduce the unspecific binding of the antibodies to the cell surface Fcg RII.

17. The manufacturer protocol proposes room temperature incubation, nevertheless, we recommend to perform all incubation steps at 4 °C to increase cell viability and reduce unspecific binding of the antibodies.

18. To obtain an acceptable balance among DC enrichment and cell recovery during the FACS, it is recommended to titrate the Enrichment Cocktail. Higher enrichment will facilitate the population gating and will shorten the time of the sorting but may reduce the cell recovery. This is due to a proportionally

higher amount of cells of interest that are lost during the staining and particularly during the sorting. An acceptable enrichment should be higher than tenfold.

19. For a correct maintenance of the particles, ensure the tubes are kept in a vertical position; this will avoid a dehydration of the particles and formation of aggregates on the walls of the tubes.

20. Alternatively, it is possible to place the tube in the magnet without the cap and keep it under the laminar flux at room temperature during 5 min.

21. Since the protocol we present here is meant for 10^9 PBMCs, the appropriate magnet for this volume is the Easy 50. When starting from lower amount of PBMCs the use of smaller magnets may be recommended. The manufacturer also sells the Purple EasySep magnet for up to 10^6 PBMCs and the "The Big Easy" silver EasySep magnet for up to 4×10^6 PBMCs.

22. If necessary, to increase DC enrichment and reduce the amount of cells to be subsequently sorted, it is possible to add one round of magnetic particles/magnet depletion. In this case, add 400 μL of magnetic particles and incubate 10 min at 4 °C. Place the tube into the Magnet and proceed as in **step 6**.

23. The selection of antibodies will depend on the populations that will be sorted as well as the markers chosen to identify them. The election of fluorochromes should be based on the lasers and filters available in the flow cytometer (*see* Table 2).

24. We propose a minimal staining based on Lin-, CD4, CD11c, CD1c, and CD141. Nevertheless, different alternatives are possible. The CD4 staining may be replaced by HLA-DR and specific markers for pDC as BDCA2 or CD123. We recommend foreseeing the staining of surface antigens that might be required during the culture conditions.

25. It should be taken into account that the cells remain stained with the fluorochromes used for the separation. When planning a subsequent staining (ex vivo or after culture), the new antibodies should not be coupled to fluorochromes already used to identify the DC subset. Additionally, the fluorochromes used for the isolation should be also considered in the compensations of new staining.

26. All centrifugation steps from now on are performed at $200 \times g$ for 10 min at 4 °C unless otherwise stated.

27. It is recommended to perform the staining in a minimal volume of 500 μL.

28. Immediate sorting improves DC recovery and viability.

29. An alternative to improve DC viability is to use pure HS to coat the tubes, or to coat them by incubating them overnight at 37 °C filled with complete RPMI or HS.

30. Usually, the viability after the enrichment protocol is very high and there is no need to stain the cells to distinguish live/dead populations. Nevertheless, if any doubts about the cell viability are present, it is possible to add DAPI during the immunostaining and exclude the dead cells as DAPI$^+$.

31. We recommend collecting the sorted cells in tubes containing 1–2 mL complete RPMI, as well as keeping them on ice until their use. This will increase their viability.

Acknowledgments

We are grateful to Vassili Soumelis for his support and suggestions.

We thank Zofia Maciorowski and the Cytometry platform of Curie Institute for help in cell sorting. We thank Cristina Ghirelli and Raphael Zollinger for DC protocol tests and improvement, Carolina Martinez, and Coline Trichot for critical reading and suggestions. S.A. and L.P. were supported by ANR-10-IDEX-0001-02 PSL* and ANR-11-LABX-0043 grants.

References

1. Van Voorhis WC, Hair LS, Steinman RM, Kaplan G (1982) Human dendritic cells. Enrichment and characterization from peripheral blood. J Exp Med 155(4):1172–1187

2. Steinman RM, Cohn ZA (1973) Identification of a novel cell type in peripheral lymphoid organs of mice. I. Morphology, quantitation, tissue distribution. J Exp Med 137(5):1142–1162

3. Guilliams M, Ginhoux F, Jakubzick C, Naik SH, Onai N, Schraml BU, Segura E, Tussiwand R, Yona S (2014) Dendritic cells, monocytes and macrophages: a unified nomenclature based on ontogeny. Nat Rev Immunol 14(8):571–578

4. Collin M, Bigley V, Haniffa M, Hambleton S (2011) Human dendritic cell deficiency: the missing ID? Nat Rev Immunol 11(9):575–583

5. Randolph GJ, Ochando J, Partida-Sanchez S (2008) Migration of dendritic cell subsets and their precursors. Annu Rev Immunol 26:293–316

6. Martinez-Cingolani C, Grandclaudon M, Jeanmougin M, Jouve M, Zollinger R, Soumelis V (2014) Human blood BDCA-1 dendritic cells differentiate into Langerhans-like cells with thymic stromal lymphopoietin and TGF-beta. Blood 124(15):2411–2420

7. Volpe E, Pattarini L, Martinez-Cingolani C, Meller S, Donnadieu MH, Bogiatzi SI, Fernandez MI, Touzot M, Bichet JC, Reyal F, Paronetto MP, Chiricozzi A, Chimenti S, Nasorri F, Cavani A, Kislat A, Homey B, Soumelis V (2014) Thymic stromal lymphopoietin links keratinocytes and dendritic cell-derived IL-23 in patients with psoriasis. J Allergy Clin Immunol 134(2):373–381

8. Dzionek A, Fuchs A, Schmidt P, Cremer S, Zysk M, Miltenyi S, Buck DW, Schmitz J (2000) BDCA-2, BDCA-3, and BDCA-4: three markers for distinct subsets of dendritic cells in human peripheral blood. J Immunol 165(11):6037–6046

9. Ito T, Inaba M, Inaba K, Toki J, Sogo S, Iguchi T, Adachi Y, Yamaguchi K, Amakawa R, Valladeau J, Saeland S, Fukuhara S, Ikehara S (1999) A CD1a+/CD11c+subset of human blood dendritic cells is a direct precursor of Langerhans cells. J Immunol 163(3):1409–1419

10. Bachem A, Guttler S, Hartung E, Ebstein F, Schaefer M, Tannert A, Salama A, Movassaghi K, Opitz C, Mages HW, Henn V, Kloetzel PM, Gurka S, Kroczek RA (2010) Superior antigen cross-presentation and XCR1 expression define human CD11c+CD141+ cells as homologues of mouse CD8+ dendritic cells. J Exp Med 207(6):1273–1281

11. Jongbloed SL, Kassianos AJ, McDonald KJ, Clark GJ, Ju X, Angel CE, Chen CJ, Dunbar PR, Wadley RB, Jeet V, Vulink AJ, Hart DN, Radford KJ (2010) Human CD141+ (BDCA-3)+dendritic cells (DCs) represent a unique myeloid DC subset that cross-presents necrotic cell antigens. J Exp Med 207(6):1247–1260

12. Merad M, Sathe P, Helft J, Miller J, Mortha A (2013) The dendritic cell lineage: ontogeny and function of dendritic cells and their subsets in the steady state and the inflamed setting. Annu Rev Immunol 31:563–604

13. Segura E, Amigorena S (2014) Cross-presentation by human dendritic cell subsets. Immunol Lett 158(1-2):73–78

14. Hogg N, Takacs L, Palmer DG, Selvendran Y, Allen C (1986) The p150,95 molecule is a marker of human mononuclear phagocytes: comparison with expression of class II molecules. Eur J Immunol 16(3):240–248

15. MacDonald KP, Munster DJ, Clark GJ, Dzionek A, Schmitz J, Hart DN (2002) Characterization of human blood dendritic cell subsets. Blood 100(13):4512–4520

16. Ziegler-Heitbrock L, Ancuta P, Crowe S, Dalod M, Grau V, Hart DN, Leenen PJ, Liu YJ, MacPherson G, Randolph GJ, Scherberich J, Schmitz J, Shortman K, Sozzani S, Strobl H, Zembala M, Austyn JM, Lutz MB (2010) Nomenclature of monocytes and dendritic cells in blood. Blood 116(16):e74–80

17. Meredith MM, Liu K, Darrasse-Jeze G, Kamphorst AO, Schreiber HA, Guermonprez P, Idoyaga J, Cheong C, Yao KH, Niec RE, Nussenzweig MC (2012) Expression of the zinc finger transcription factor zDC (Zbtb46, Btbd4) defines the classical dendritic cell lineage. J Exp Med 209(6):1153–1165

18. Satpathy AT, Kc W, Albring JC, Edelson BT, Kretzer NM, Bhattacharya D, Murphy TL, Murphy KM (2012) Zbtb46 expression distinguishes classical dendritic cells and their committed progenitors from other immune lineages. J Exp Med 209(6):1135–1152

19. Grouard G, Rissoan MC, Filgueira L, Durand I, Banchereau J, Liu YJ (1997) The enigmatic plasmacytoid T cells develop into dendritic cells with interleukin (IL)-3 and CD40-ligand. J Exp Med 185(6):1101–1111

20. Liu YJ (2005) IPC: professional type 1 interferon-producing cells and plasmacytoid dendritic cell precursors. Annu Rev Immunol 23:275–306

21. Gilliet M, Cao W, Liu YJ (2008) Plasmacytoid dendritic cells: sensing nucleic acids in viral infection and autoimmune diseases. Nat Rev Immunol 8(8):594–606

22. Cella M, Jarrossay D, Facchetti F, Alebardi O, Nakajima H, Lanzavecchia A, Colonna M (1999) Plasmacytoid monocytes migrate to inflamed lymph nodes and produce large amounts of type I interferon. Nat Med 5(8):919–923

23. Siegal FP, Kadowaki N, Shodell M, Fitzgerald-Bocarsly PA, Shah K, Ho S, Antonenko S, Liu YJ (1999) The nature of the principal type 1 interferon-producing cells in human blood. Science 284(5421):1835–1837

24. Reizis B, Bunin A, Ghosh HS, Lewis KL, Sisirak V (2011) Plasmacytoid dendritic cells: recent progress and open questions. Annu Rev Immunol 29:163–183

25. Rissoan MC, Soumelis V, Kadowaki N, Grouard G, Briere F, de Waal MR, Liu YJ (1999) Reciprocal control of T helper cell and dendritic cell differentiation. Science 283(5405):1183–1186

26. Dzionek A, Sohma Y, Nagafune J, Cella M, Colonna M, Facchetti F, Gunther G, Johnston I, Lanzavecchia A, Nagasaka T, Okada T, Vermi W, Winkels G, Yamamoto T, Zysk M, Yamaguchi Y, Schmitz J (2001) BDCA-2, a novel plasmacytoid dendritic cell-specific type II C-type lectin, mediates antigen capture and is a potent inhibitor of interferon alpha/beta induction. J Exp Med 194(12):1823–1834

27. Wu P, Wu J, Liu S, Han X, Lu J, Shi Y, Wang J, Lu L, Cao X (2008) TLR9/TLR7-triggered downregulation of BDCA2 expression on human plasmacytoid dendritic cells from healthy individuals and lupus patients. Clin Immunol 129(1):40–48

28. Ghirelli C, Zollinger R, Soumelis V (2010) Systematic cytokine receptor profiling reveals GM-CSF as a novel TLR-independent activator of human plasmacytoid predendritic cells. Blood 115(24):5037–5040

Chapter 12

Protocols for the Identification and Isolation of Antigen-Presenting Cells in Human and Mouse Tissues

Naomi McGovern, Andreas Schlitzer, Baptiste Janela, and Florent Ginhoux

Abstract

The efficient processing of both mouse and human tissues is a valuable technique for characterizing tissue-associated immune cells. Here, we describe the techniques used and optimised within our laboratory for the enrichment and identification of antigen-presenting cells across a number of mouse and human tissues.

Key words Human tissue processing, Antigen-presenting cells, Flow cytometry, Dendritic cells

---

1 Introduction

Both mouse and human nonlymphoid tissues contain a high concentration of antigen-presenting cells (APCs), which are specialized in regulating and directing immune responses, tissue repair, and maintaining healthy tissue homeostasis [1]. The human APC network is comprised of a heterogeneous population of mononuclear phagocytes, including classical dendritic cells (cDCs), namely, CD1c+ DCs and CD141+ DCs (also known as cDC2 and cDC1 respectively) [2], plasmacytoid dendritic cells (pDCs), monocyte-derived cells (termed CD14+ cells here), and tissue-resident macrophages [3]. Meanwhile, in the mouse, several APC subsets can be identified including pDCs, CD8α+/CD103+ cDCs (also named cDC1), and CD11b+ cDCs (also named cDC2) in both lymphoid and nonlymphoid organs [2]. In addition to DCs, the mouse APC population also includes Ly6c+ and Ly6clow monocytes and tissue-resident macrophages, which are particularly important in mucosal tissues [1].

As APCs make attractive therapeutic targets [4], developing and improving methods for their detection, isolation, enrichment, and characterization is important for the advancement of this area of research. Through a number of recent developments in the

Elodie Segura and Nobuyuki Onai (eds.), *Dendritic Cell Protocols*, Methods in Molecular Biology, vol. 1423,
DOI 10.1007/978-1-4939-3606-9_12, © Springer Science+Business Media New York 2016

techniques used for APC enrichment and isolation, the human APC network has now been well defined across a number of human lymphoid and nonlymphoid tissues, including the skin, liver, lung, and gastrointestinal tract. Here, we describe the techniques frequently used in our laboratory to isolate and characterize these cells, from both human and mouse nonlymphoid tissues.

2 Materials

2.1 Human Tissue Processing

1. Medium: RPMI 1640, 10 U/mL penicillin/streptomycin, 50 µg/mL gentamycin.

2. RF10: medium supplemented with 10 % FBS.

3. Phosphate buffered saline (PBS).

4. Collagenase: Worthington's Collagenase Type IV.

5. Dermatome knife.

6. Surgical scissors.

7. Dispase (Gibco).

8. 70 µm cell strainers.

9. Petri dishes.

10. Incubator at 37 °C, 5 % CO_2.

11. Recombinant human GM-CSF.

12. Blocking buffer: PBS, 1 % mouse serum, 1 % rat serum, 1 % human serum albumin, 2 mM EDTA.

13. Bovine serum albumin (BSA).

14. FACS buffer: PBS, 5 g/L BSA, 2 mM EDTA.

15. DAPI.

16. Antibodies for skin samples: CD45 (HI30), HLA-DR (LN3), CD14 (61D3), CD1a (H149), CD1c (L161), CD11c (B-ly6), CD141 (AD5-14H12). Use at concentrations recommended by manufacturers.

17. Antibodies for soft tissue samples: CD3 (SK7), CD19 (HIB-19), CD20 (2H7), CD56 (MEM188), CD123 (7G3), HLA-DR (LN3), CD14 (61D3), CD1c (L161), CD11c (B-ly6), CD141 (AD5-14H12). Use at concentrations recommended by manufacturers.

18. Flow cytometer and FlowJo software (Treestar).

19. 6-well plates.

20. 30 % sucrose solution in PBS.

21. Microscopy fixative (PBS, 2 % paraformaldehyde, 30 % sucrose).

22. Microscopy wash buffer (PBS, 0.2 % BSA, 0.1 % TritionX-100).

23. Microscopy blocking buffer (PBS, 0.5 % BSA, 0.3 % TritionX-100).

24. Antibodies for microscopy (*see* **Note 1**).

25. Confocal microscope.

2.2 Mouse Tissue Processing

1. Ketamine/xylazine, for anesthesia.

2. PBS.

3. Surgical scissors and forceps.

4. Digestion mix I: RPMI 1640, 10% FBS, 0.2 mg/mL collagenase IV (Roche), 20000 U/mL DNAse I (Roche).

5. Digestion mix II: PBS, 2 mM dithiothreitol (DTT), 5 mM EDTA. Keep at room temperature for efficient digestion.

6. Digestion mix III: RPMI 1640, 2.4 U/mL dispase II (Gibco).

7. Petri dishes.

8. 15 and 50 mL tubes.

9. Tube rotator.

10. Red blood cell lysis buffer.

11. Blocking buffer: PBS, 1% mouse serum, 1% rat serum, 2 mM EDTA.

12. 3 mL syringes and plunger.

13. 19 G needle.

14. 70 μm tube top cell strainers.

15. 41 μm nylon mesh.

16. Antibodies: CD11c (HL3), CD103 (E27), Epcam (EBA-1), CD14 (Sa14-2), CD11b (M1/70), CD24 (M1/69), CD45 (30-F11), CD64 (X54-5/7.1), MHC-II (M5/114.15.2), CCR2 (475301), and Ly6C (HK1.4). Use at concentrations recommended by manufacturers.

17. FACS buffer: PBS, 5 g/L BSA, 2 mM EDTA.

18. DAPI.

19. Flow cytometer and FlowJo software (Treestar).

3 Methods

3.1 Human Tissue Processing

Carry out all procedures at room temperature, unless otherwise specified. Washing steps are performed by centrifugating at $400 \times g$ for 5 min at 4 °C.

3.1.1 Isolation of Antigen-Presenting Cells from Human Skin

For all skin incubation steps, place the skin with epidermal side up.

1. Upon the removal of skin samples from patients, immediately place into ice-cold PBS and process as soon as possible for

enhanced cell yield and viability, i.e., within 4 h of removal from patient.

2. Remove all of the subcutaneous fat.

3. With a dermatome knife, cut the skin into sections (approximately 200 μm thick).

4. Optional—Separate the epidermal layer from the dermis by incubating for 1–2 h with 1 mg/mL dispase at 37 °C in medium without serum (*see* **Notes 2** and **3**).

5. Option 1—Obtaining skin cells via digestion: Cut the large dermal sheets into 1 cm square pieces. Incubate with 0.8 mg/mL collagenase in RF10 in a sterile bacterial Petri dish (to reduce adherence of leukocytes) in the incubator. When the epidermal and dermal layers are separated, the digestion takes approximately 6–8 h. If digesting the whole skin (i.e., the epidermal layer is still attached), incubate for 18 h to maximize cell yield (*see* **Notes 4, 5,** and **6**).

6. Option 2—Obtaining skin cells via ex vivo migration (*see* **Note 7**): Cut the skin into approximately 4×4 cm square pieces. Incubate the skin pieces in RF10 with GM-CSF (500 U/mL) (*see* **Note 8**) in a Petri dish for 48–72 h in the incubator.

7. Pass the supernatant from skin cultures through 70 μm sterile cell strainers.

8. Wash the cells with PBS.

3.1.2 Enrichment of Antigen-Presenting Cells from Human Soft Tissues Including the Lung, Spleen, and Liver

1. Using surgical scissors, mechanically shred soft tissue.

2. Place tissue pieces in RF10 containing 0.8 mg/mL collagenase for 10–40 min (*see* **Note 9**).

3. Every 10 min during digestion, re-disperse the sample by pipetting up and down.

4. Pass digested sample through 70 μm sterile cell strainers.

5. Wash the cells with PBS twice.

3.1.3 Flow Cytometry of Antigen-Presenting Cells

1. Resuspend the cell pellet in blocking buffer and incubate for 15 min at 4 °C.

2. Add the required antibodies and incubate for 30 min at 4 °C.

3. Wash the cells twice in FACS buffer.

4. Resuspend the pellet at a cell: volume ratio of 2×10^6 cells in 300 μL of FACS Buffer.

5. Add DAPI for dead cell exclusion.

6. Acquire the samples on a flow cytometer.

7. *See* Fig. 1 and **Note 10** for a description on how to identify antigen-presenting cells from human skin and soft tissues.

a

Skin, commonly used gating strategy

Digested skin

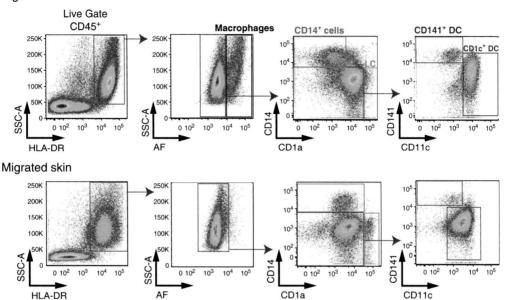

Migrated skin

b

Soft tissue, commonly used gating strategy

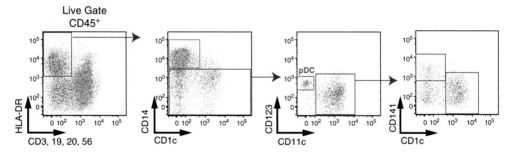

Fig. 1 Gating strategy commonly used to identify human tissue antigen-presenting cells. (**a**) By successive gating on DAPI⁻Live cells, CD45⁺ cells, and HLA-DR⁺ cells, antigen-presenting cell subsets can be identified, including autofluorescent (AF) tissue-resident macrophages (*purple gate*), CD14⁺ cells (*green gate*), CD141⁺ DCs (*blue gate*), CD1c⁺ DCs (*red gate*) in the skin. (**b**) By successive gating on DAPI⁻Live cells and CD45⁺ HLA-DR⁺ cells and exclusion of B cells, NK cells, and T cells (in lineage), soft tissue antigen-presenting cell subsets can be identified, including CD14⁺ cells (*green gate*), plasmacytoid dendritic cells (pDCs), CD141⁺ DCs (*blue gate*), CD1c⁺ DCs (*red gate*) in the skin

3.1.4 Whole Mount Microscopy of Human Dermis

1. Cut with a dermatome knife whole-skin sheets (200 μm thick).

2. Remove the epidermal layer of skin by culturing the skin for 1 h with 1 mg/mL Dispase at 37 °C in media without serum.

3. Spread out skin pieces in a 6-well plate with epidermal side up, and leave in fixative for 24–48 h.

4. After fixation, wash skin pieces in 30 % sucrose for 2 h and store in PBS at 4 °C.

5. Prior to staining for microscopy, cut skin into small pieces, approximately 0.5 cm × 0.5 cm.

6. Block and permeabilize dermal sheets by incubating at 4 °C overnight in microscopy blocking buffer.

7. Add primary antibodies (*see* **Note 1**) at the necessary concentration in microscopy blocking buffer overnight.

8. Wash skin for 1 h with microscopy washing buffer, changing wash buffer every 15 min.

9. Incubate skin pieces in microscopy blocking buffer overnight.

10. Add secondary antibodies overnight (*see* **Note 1**).

11. Image the immunofluorescence-stained specimens using a confocal microscope. Three-dimensional reconstruction of the dermis and leukocyte enumeration can be performed using software such as Imaris software (Bitplane).

3.2 Mouse Tissue Processing

Carry out all procedures at room temperature, unless otherwise specified. Washing steps are performed by centrifugating at $400 \times g$ for 5 min at 4 °C.

3.2.1 Murine Lung, Liver, and Kidney Tissue Processing

1. Optional—Perfusion of lung and liver.
   (a) Anesthetize mice.
   (b) Expose the heart.
   (c) Slowly inject ice-cold PBS into the left ventricle of the heart to perfuse lungs or into the right ventricle to perfuse the liver, until the organs turn white.
   (d) Remove the organs.

2. Remove the lung, liver, or kidney and immediately place them into 3–5 mL of digestion mix I. **Important**: Remove kidney membrane prior to digestion to ensure efficient digestion.

3. Using surgical scissors, cut organs into confetti-sized pieces.

4. Digest the pieces by incubating at 37 °C for 45 min.

5. After digestion, homogenize tissue pieces using a 19 G needle attached to a 3 mL syringe, by syringing the tissue piece solution up and down at least five times.

6. Strain resulting dispersed tissue through 70 μm cell strainer.

7. Wash cells with PBS.

8. Lyse red blood cells.

9. Wash cells with PBS.

10. Resuspend the pellet at a cell: volume ratio of 2×10^6 cells in 300 μL of FACS Buffer.

3.2.2 Enrichment of the Lamina Propria and Intra-Epithelial Antigen-Presenting Cells from Murine Small Intestine and Colon

1. Remove the small intestine and colon and immediately place them on ice. **Important**: Remove Peyer's patches from the small intestine prior to tissue processing to avoid contamination with lymphoid organ cells.

2. Using surgical scissors, cut open the small intestine and colon longitudinally.

3. Wash the tissue by shaking vigorously in ice-cold PBS until fecal matter and mucus is washed away and the tissue starts to appear white.

4. Cut the tissue into 3–5 mm pieces and resuspend in 15 mL of digestion mix II and incubate for 30 min on a tube rotator (300 rpm) at 37 °C.

5. *Option A: Isolation of intra-epithelial APCs.* Collect the digested solution, strain through a 70 μm cell strainer, and wash the cells with PBS.

6. *Option B: Isolation of lamina propria APCs.* Wash tissue pieces in ice-cold PBS. Transfer tissue pieces into 5 mL of digestion mix I and incubate for 30 min at 37 °C. Force tissue pieces through a 70 μm cell strainer using a syringe plunger. Wash the cells twice with PBS.

7. Resuspend the pellet at a cell: volume ratio of 2×10^6 cells in 300 μL of FACS Buffer.

3.2.3 Enrichment of Antigen-Presenting Cells from Mouse Ear Skin

1. Cut the mouse ears and split ear skin into dorsal and ventral surfaces with curved forceps.

2. Place the pieces floating in a Petri dish containing 2 mL of Digestion mix III and incubate for 1 h at 37 °C.

3. Detach dermal and epidermal sheets with curved forceps and place in separate Petri dishes containing 2 mL of Digestion mix I.

4. Cut in small pieces the dermal and epidermal fractions and incubate for 1 h at 37 °C.

5. Disperse epidermal and dermal cells by pipetting up and down ten times using a syringe and needle (19 G).

6. Filter the suspensions through a nylon mesh to remove clumps, and, to reduce cell loss, wash the well with PBS and add the PBS to the suspensions.

7. Wash the cells.

8. Wash again the cells with PBS.

9. Resuspend the pellet at a cell: Volume ratio of 2×10^6 cells in 300 μL of FACS buffer.

3.2.4 Flow Cytometry Analysis of Murine Antigen-Presenting Cells

1. Centrifuge the cells.

2. Resuspend the cell pellet in blocking buffer and incubate for 15 min at 4 °C.

3. Add the required antibodies and incubate for 30 min at 4 °C.

4. Wash the cells twice in FACS buffer.

5. Resuspend the pellet at a cell: volume ratio of 2×10^6 cells in 300 μL of FACS buffer.

6. Add DAPI for dead cell exclusion.

7. Acquire the samples on a flow cytometer.

8. *See* Fig. 2 and **Notes 11** and **12** for a description on how to identify antigen-presenting cells from mouse skin and tissues.

4 Notes

1. A series of antibodies have now been shown to work effectively for whole-mount immunofluorescence microscopy allowing in-depth description of dermal lymphatics, capillaries, APC, and T cell subset localization in both disease and steady states. This has recently been well illustrated by Wang et al. [5].

2. Separating the epidermis from the dermis increases the efficiency of skin digestion and hence increases the yield of some APC subsets. However, it is possible that the peeling process may harm the immune cells located just below the epidermal layer, in particular we have noted a reduction in CD141+ DC yield (own observations).

3. For skin punch biopsies (8 mm) and optimal cell yield, cut biopsy in half prior to addition of Dispase for removal of the epidermal layer. Efficient Dispase digestion of skin punch biopsies can take 1–2 h.

4. For skin punch biopsies, cut the remaining dermal layer into small pieces and leave in Collagenase (0.8 mg/mL, Worthington's Type IV) for approximately 6–8 h.

5. Digestion time and cell viability vary slightly, depending on the anatomical source of the skin. For example, the digestion of skin from mammoplasties is more efficient, with greater cell viability at isolation end point than skin obtained from abdominoplasties (own observations).

6. When performing flow cytometry analysis of collagenase-sensitive markers, cells should be placed back into RF10 at 37 °C for 6–8 h (depending on the antigen of interest) to allow cell surface re-expression.

7. Due to their migratory properties, APCs can simply be obtained by leaving dermal explants in culture for 48–72 h. During this period, the cDCs (CD1c+ DCs and CD141+ DCs), CD14+ cells, and LCs will migrate out of the dermal explants. The advantage of this technique over digestion is that tissue-

a

Lung, commonly used gating strategy

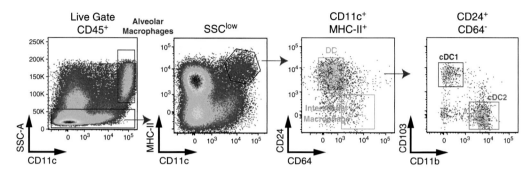

b

Small intestine, commonly used gating strategy

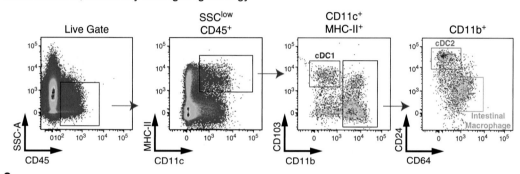

c

Skin, Dermis, commonly used gating strategy

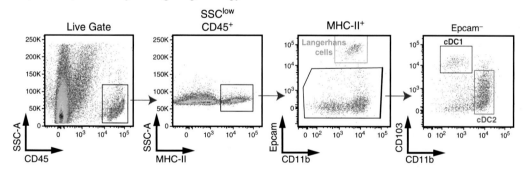

Fig. 2 Gating strategy commonly used to identify mouse tissue antigen-presenting cells [6]. (**a**) By successive gating on DAPI-Live cells and CD45+ cells, CD11c+SSChigh alveolar macrophages (*red*) can be identified. Subsequently gating on SSClowCD11c+MHC-II+ cells, APCs can be identified, which are further split in CD24+ DCs and CD64+ interstitial macrophages (*orange*). DCs can be further subdivided into cDC1 (*purple*) and cDC2 (*blue*) using the markers CD103 and CD11b. (**b**) By successive gating on DAPI-Live cells and CD45+CD11c+MHC-II+SSClow cells, intestinal APCs can be identified. cDC1 (*purple*) can be identified gating on CD103+CD11b- cells and CD11b+ cells can be further separated using CD24 and CD64 in CD24+CD64-CD103+/- cDC2 (*blue*) and CD24-CD64+ intestinal macrophages (*orange*). (**c**) By successive gating on DAPI-Live cells and CD45+MHC-II+ cells, dermal APCs can be identified. Within the APC fraction, Langerhans cells can be separated using Epcam. Subsequently, gating on the Epcam- fraction cDC1 (*purple*) and cDC2 (*blue*) can be separated using the markers CD103 and CD11b

Table 1
Human antigen-presenting cells surface markers

	CD1c⁺ DC	CD141⁺ DC	CD14⁺ cells	Macrophages	Langerhans cells
CD45	+	+	+	+	+
HLA-DR	+	+	+	+	+
Autofluorescence	lo	lo	lo	hi	lo
Side-scatter area	lo	lo	lo	hi	lo
CD14	−	−	+	+	−/lo
CD1a	+/−	lo/−	lo/−	−	+++
CD1c	+	−/+	lo/−	−	+
CD141	+/−	++	+/lo	+/lo	+/lo
CD11c	+	lo/−	+	+	lo/−
CD26	−	+	−	lo/+	lo/−
SIRPα	+	−	+	+	+
CD163	−	−	+	+	−
CD11b	+	−	+	+	lo/−
CD64	+	−	+	+	lo/−

resident macrophages, which are nonmigratory, will not migrate from ex vivo tissue explants, thereby ensuring the isolated CD14⁺ cells are not contaminated with the tissue-resident macrophages. Additionally, obtaining cells by migration bypasses the need to use enzymes such as collagenase, which cleaves some surface markers.

8. GM-CSF is frequently added to migrating ex vivo skin explants [7], to aid in cell recovery.

9. Digestion time depends on how thoroughly the tissue is first dispersed.

10. APC subsets are firstly selected by gating on the leukocyte common antigen CD45. APCs can next be selected by gating on HLA-DR⁺ cells (Fig. 1 and Table 1). Tissue-resident dermal macrophages can be readily identified by their high side-scatter area (SSC-A) and auto-fluorescence (AF). Dermal macrophages also express CD14 and CD163, allowing further positive selection. Meanwhile, by selecting AF$^{-/low}$ cells with a lower SSC-A, dermal CD14⁺ cells, and DC subsets can be identified. By plotting CD14 versus CD1a expression profile, CD14⁺ cells can be readily selected, as their CD1a expression

is low to negative. Here also, Langerhans cells can be identified by their high CD1a surface expression [7–10]. Further positive selection of Langerhans cells includes selecting langerin^{high} and Epcam$^+$ cells. The cDC population (CD141$^+$ DCs and CD1c$^+$ DCs) is located within the CD14$^-$ and CD1a$^{-/int}$ population. The CD141$^+$ DCs are selected by their high CD141 expression and low to negative CD1c expression [9]. CD141$^+$ DCs also highly express CD26 and are SIRPα$^-$ within skin (own observations). Meanwhile, the CD1c$^+$ DCs can be identified by their high CD1c expression. Dermal CD1c$^+$ DCs also highly express SIRPα, CD11c, and CD11b. *See* Fig. 1a for commonly used gating strategy to identify human skin antigen-presenting cell subsets.

11. For flow cytometric analysis (*see* Fig. 2a, b and Table 2) of murine DC and macrophage subsets in the lung, liver, colon, and small intestine, live cells are positively selected for CD45 and CD11c. Alveolar macrophages of the lung can be discriminated from the rest of the CD11c$^+$ cells by selecting high SSC-A cells. Low SSC-A cells are gated MHC-II$^+$ cells. Within this CD45$^+$CD11c$^+$MHC-II$^+$SSC^{low} fraction, CD24$^+$CD64$^-$ (DCs), and CD24$^-$CD64$^+$ (macrophages) cells are found. MHC-II$^+$CD24$^+$CD64$^-$ cells can be separated into DC subsets in the lung, colon, and liver by the mutually exclusive expression of CD103 and CD11b, whereas in the small intestine, a CD103$^+$CD11b$^+$ DC subset exists (*see* Fig. 2b).

Table 2
Mouse antigen-presenting cells surface markers

	cDC2	cDC1	Macrophages	Langerhans cells
CD45	+	+	+	+
MHC class II	+	+	+	+
Autofluorescence	lo	lo	hi	lo
Side-scatter area	lo	lo	hi	lo
CD24	+	+	–	+
CD11C	+	+	+	+
CD64	lo/–	lo/–	+	–
CD103	–	++	+/–	–
CD11b	+	–	+	+
EPCAM	–	–	–	+
SIRPα	+	–	+	–

12. For flow cytometry analysis of antigen-presenting cells from the skin (*see* Fig. 2c), leukocytes are firstly selected by gating on the leukocyte common antigen, CD45. In the epidermal fraction, Langerhans cells can be identified by gating on MHC-II⁺Epcam⁺ cells. In the dermal fraction, APCs are selected by gating on MHC-II⁺ cells. Migratory Langerhans cells are identified by their uniquely high expression of Epcam within the dermal compartment. By plotting CD24 versus CD11b, cDC1 can be readily selected as they are highly positive for CD24 and negative for CD11b. The CD11b⁺ cell fraction, is composed of cells with different levels of CD24 expression, which can be delineated in cDC2 (CD24⁺) and monocyte-derived macrophages (CD64⁺) by plotting CD24 versus CD64.

Acknowledgment

This work was supported by core grants of the Singapore Immunology Network.

References

1. Merad M, Sathe P, Helft J, Miller J, Mortha A (2013) The dendritic cell lineage: ontogeny and function of dendritic cells and their subsets in the steady state and the inflamed setting. Annu Rev Immunol 31:563–604

2. Guilliams M, Ginhoux F, Jakubzick C, Naik SH, Onai N, Schraml BU et al (2014) Dendritic cells, monocytes and macrophages: a unified nomenclature based on ontogeny. Nat Rev Immunol 14:571–578

3. Collin M, McGovern N, Haniffa M (2013) Human dendritic cell subsets. Immunology 140:22–30

4. Hashimoto D, Merad M (2011) Harnessing dendritic cells to improve allogeneic hematopoietic cell transplantation outcome. Semin Immunol 23:50–57

5. Wang X-N, McGovern N, Gunawan M, Richardson C, Windebank M, Siah T-W et al (2013) A three-dimensional atlas of human dermal leukocytes, lymphatics, and blood vessels. J Invest Dermatol 134:965–974

6. Schlitzer A, McGovern N, Teo P, Zelante T, Atarashi K, Low D et al (2013) IRF4 transcription factor-dependent CD11b⁺ dendritic cells in human and mouse control mucosal IL-17 cytokine responses. Immunity 38:970–983

7. Haniffa M, Ginhoux F, Wang XN, Bigley V, Abel M, Dimmick I et al (2009) Differential rates of replacement of human dermal dendritic cells and macrophages during hematopoietic stem cell transplantation. J Exp Med 206:371–385

8. McGovern N, Schlitzer A, Gunawan M, Jardine L, Shin A, Poyner E et al (2014) Human dermal CD14⁺ cells are a transient population of monocyte-derived macrophages. Immunity 41:465–477

9. Haniffa M, Shin A, Bigley V, McGovern N, Teo P, See P et al (2012) Human tissues contain CD141ʰⁱ cross-presenting dendritic cells with functional homology to mouse CD103⁺ nonlymphoid dendritic cells. Immunity 37:60–73

10. Haniffa M, Collin M, Ginhoux F (2013) Ontogeny and functional specialization of dendritic cells in human and mouse. Adv Immunol 120:1–49

Part IV

Ex Vivo Functional Analysis of Dendritic Cells

Chapter 13

Measurement of Export to the Cytosol in Dendritic Cells Using a Cytofluorimetry-Based Assay

Omar I. Vivar, Joao G. Magalhaes, and Sebastian Amigorena

Abstract

The presentation of exogenous antigens on MHC class I molecules, known as cross-presentation, is a key function of dendritic cells (DCs). Cross-presentation via the cytosolic pathway involves antigen export from endocytic compartments to the cytosol. We have recently developed a cytofluorimetry-based assay to examine the kinetics and the efficiency of antigen export to the cytosol in DC populations. In this assay, DCs are loaded with a FRET-sensitive cytosolic substrate of β-lactamase, CCF4. Following uptake of β-lactamase by the DCs, the enzyme undergoes export to the cytosol leading to cleavage of the FRET dye. This cleavage and switch of fluorescence are analyzed by flow cytometry, allowing a quantitative measurement of this event.

Key words β-lactamase, Cross-presentation, Dendritic cell, Cytosolic export, CCF4

1 Introduction

Dendritic cells (DCs) are the principal antigen-presenting cells capable of initiating adaptive immune responses in the body. DCs initiate CD8+ T cell responses via presentation of antigens on MHC class I molecules. The presentation of exogenous antigens on MHC class I molecules, known as cross-presentation, is critical for the induction of immune responses against infectious agents which do not directly infect DCs. Different intracellular pathways of cross-presentation have been reported, namely, the cytosolic and vacuolar pathways [1, 2]. The cytosolic pathway is proteasome dependent, suggesting that internalized proteins undergo export to the cytosol where they are degraded by the proteasome [3]. The resulting peptides are then loaded onto MHC class I molecules by different mechanisms [1].

We have recently developed a cytofluorimetry-based assay to quantify cytosolic export in dendritic cells [4, 5]. In this assay, DCs are loaded with a substrate of β-lactamase, CCF4, which accumulates in the cytosol. CCF4 is a FRET dye which fluoresces at

Elodie Segura and Nobuyuki Onai (eds.), *Dendritic Cell Protocols*, Methods in Molecular Biology, vol. 1423,
DOI 10.1007/978-1-4939-3606-9_13, © Springer Science+Business Media New York 2016

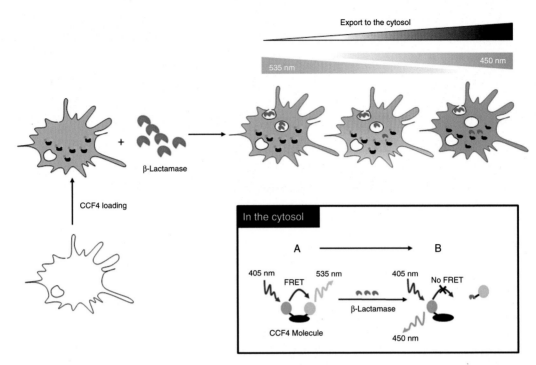

Fig. 1 Principal of the assay. Cells are loaded with CCF4 dye. Cells are then incubated with β-lactamase during different time points. If the β-lactamase undergoes export to the cytosol, the CCF4 dye is cleaved resulting in loss of FRET and the cells switch from 535 nm emission fluorescence (A) to 450 nm fluorescence (B)

535 nm when excited at 405 nm. The CCF4-loaded DCs are then incubated with β-lactamase which upon cytosolic export cleaves CCF4. The cleaved CCF4 no longer undergoes FRET, resulting in a decrease in 535 nm fluorescence and a concomitant increase in 450 nm fluorescence. Using flow cytometry, we can analyze the increase in 450 nm fluorescence allowing us to measure export to the cytosol in real time (*see* Fig. 1).

2   Materials

1. LiveBLAzer™-FRET B/G Loading Kit with CCF4-AM (Life Technologies) (*see* **Note 1**).

2. Loading buffer: 120 mM NaCl, 7 mM KCl, 1.8 mM CaCl₂, 0.8 mM MgCl₂, 5 mM glucose, and 25 mM HEPES, pH 7.3. Store at 4 °C.

3. Probenecid (Life Technologies). Prepare a 250 mM stock solution in Hank's balanced salt solution (HBSS). Store at –20 °C.

4. Penicillinase (β-lactamase) from *Bacillus cereus* (Sigma) (*see* **Note 2**). Prepare a 10 mg/mL stock solution in phosphate-buffered saline (PBS). Aliquot and store at –20 °C.

5. Alexa Fluor -647 labeled β-lactamase (*see* **Note 3**).

6. FACS buffer: PBS, 0.5 % BSA, 2 mM EDTA.

7. 4 % paraformaldehyde.

8. Mouse BD Fc Block™ purified anti-mouse CD16/CD32 mAb 2.4G2 (BD Biosciences).

9. Fixable Viability Dye eFluor 780 (eBioscience).

10. A violet laser-equipped flow cytometer capable of detecting the 500 and 450 nm channels.

3   Methods

1. Prepare CCF4 loading solution: mix 2 μL of the CCF4-AM, 10 μL of probenecid, and 50 μL of solution B (from the LiveBLAzer™-FRET B/G Loading Kit) (*see* **Note 4**) in 938 μL of loading buffer. Protect from light.

2. Collect the cells and retain 1×10^6 unlabeled cells at 4 °C for use in calibrating the flow cytometer.

3. Wash the cells in PBS and pellet them by centrifugation at $300 \times g$ for 5 min at 4 °C.

4. Discard the supernatant and resuspend the cells at a concentration of $10–20 \times 10^6$ cells/mL in CCF4 loading solution. Incubate the cells for 30 min at room temperature protected from light (*see* **Note 5**).

5. Wash the cells once with PBS and pellet them by centrifugation at $300 \times g$ for 5 min at 4 °C. Retain 1×10^6 CCF4 loaded cells at 4 °C in the dark for use as negative control. Retain 1×10^6 CCF4 loaded cells for incubation at 37 °C for the maximum time point used in **step 7** for use as a control of CCF4 retention in the cells.

6. Prepare β-lactamase loading solution: mix 200 μL of β-lactamase stock solution, 20 μL of Alexa Fluor- 647 labeled β-lactamase stock solution (at 10 mg/mL), and 10 μL of probenecid in 780 μL of cell culture medium.

7. Incubate the cells at a concentration of $10–20 \times 10^6$ cells/mL with β-lactamase loading solution at 37 °C for different times (*see* **Note 6**). To stop the reaction, transfer the cells to tubes containing cold PBS. Store the cells on ice until the longest time point is complete (see Fig. 2).

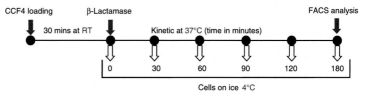

Fig. 2 Timeline of the assay

8. Wash the cells in PBS and pellet them by centrifugation at $300 \times g$ for 5 min at 4 °C, and stain with Fixable Viability Dye eFluor 780 according to the manufacturer's protocol.

9. Wash the cells in FACS buffer and incubate with Mouse BD Fc Block™ to block nonspecific staining. Incubate the cells with antibodies as required (*see* **Note 7**).

10. After staining, wash the cells twice with FACS buffer by pelleting them by centrifugation at $300 \times g$ for 5 min at 4 °C. Fix the cells at a concentration of $10–20 \times 10^6$ cells/mL with 4 % paraformaldehyde for 10 min at 4 °C in the dark. Wash cells twice with FACS buffer by pelleting them by centrifugation at $300 \times g$ for 5 min at 4 °C. Resuspend the cells at 1×10^6 cells in 200 µL of FACS buffer and acquire on the flow cytometer. Typically, a minimum of 1×10^5 cells are acquired for analysis.

11. Analyze the data using a flow cytometry software (*see* Fig. 3). In the example illustrated here, murine bone marrow-derived DCs were stained with anti-CD11c-PE-Cy7 antibody. We first gate on the whole cell population (*see* Fig. 3a) and we then select the single cell population (*see* Fig. 3b). Live cells are then gated in the APC-Cy7-negative population (*see* Fig. 3c), and the DC population is gated on the CD11c$^+$ population (*see* Fig. 3d). Finally, cells in which export to the cytosol has occurred will stain in the 450 nm channel (*see* Fig. 3e) (*see* **Note 8**). The uptake capacity can be followed in the APC channel (*see* Fig. 3f). Uptake in the CD11c$^+$ population is shown.

4 Notes

1. Prepare 500 µM CCF4-AM stock by dissolving the 200 µg vial of CCF4-AM in 365 µL DMSO provided in the kit. Aliquot and store at –20 °C protected from light as indicated by the manufacturer's protocol.

2. This β-lactamase (Sigma, catalog number P0389-25KU) is usually contaminated with endotoxin. If endotoxin contamination will affect the assay, the endotoxin can be removed. In our case, we use the EndoTrap® red Endotoxin Removal Kit according to the manufacturer's recommendations (Hyglos). The levels of endotoxin contamination can be tested using the EndoLISA® ELISA-based Endotoxin Detection Assay according to the manufacturer's recommendations (Hyglos).

3. Label β-lactamase using the Alexa Fluor® 647 Microscale Protein Labeling Kit (Life Technologies) according to the manufacturer's protocol. The labeled β-lactamase will be used to control enzyme uptake by the cells during the assay.

Gating strategy:

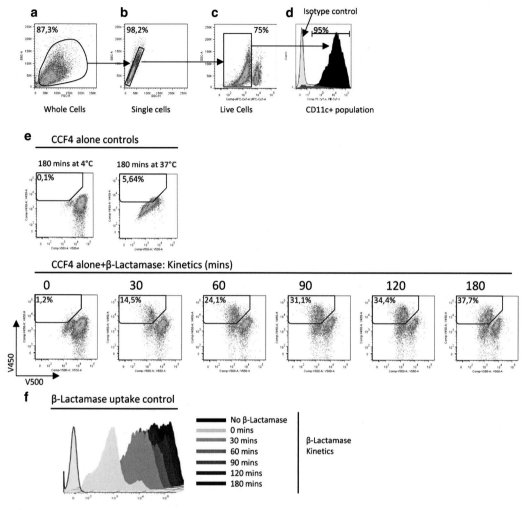

Fig. 3 Method of analysis for the assay. (**a**) An example of an FSC/SSC plot obtained when acquiring cells. The population is gated and can be clearly distinguished from the cell debris. (**b**) An example of an SSC-A/SSC-H plot obtained when acquiring cells. The single cells are gated here. (**c**) An example of an SSC-A/APC-CY7 live/dead staining plot obtained when acquiring cells. The live cells are gated here. (**d**) An example of a histogram after gating on the live cells. Here, we show CD11c-PECy7 staining in the bone marrow-derived DC along with the isotype control. (**e**) Dot plot example of the CCF4 controls and the kinetics of antigen export to the cytosol. The gate represents the percentage of cells in which export to the cytosol of the antigen occurs. (**f**) Typical histogram of β-lactamase-APC-labeled acquisition in the bone marrow-derived DC over time in the CD11c⁺ population. This data can be used when comparing different conditions to ensure that uptake of the antigen is similar

4. The optimal concentration of probenecid will vary depending on the cell type. The user should examine cell viability in order to determine the appropriate concentration resulting in both maximal cell survival and retention of the CCF4 dye inside the cells.

5. The time of incubation in CCF4 loading solution will depend on the cell type. The user should determine the time required to maximally label the cells.

6. Typical time points are 0, 0.5, 1, 2, and 3 h. Kinetics of export to the cytosol are cell type dependent and should be determined by the user.

7. The CCF4-AM substrate and CCF4 cleavage product emission will occupy the 500 nm, 450 nm, and FITC channels. There will also be some spillover into the PE channel and special care should be taken if using this channel.

8. The CCF4 loaded cells kept at 4 °C are used to determine the level of staining of cells in which export to the cytosol has not occurred. The CCF4-loaded cells incubated at 37 °C will be used to control for loss of the CCF4 dye during the incubation time of the assay.

Acknowledgments

This work was financed by Institut Curie, Institut National de la Santé et de la Recherche Médicale (INSERM), the European Research Council (2008 Advanced Grant 233062 PhagoDC), Agence National de Recherche (ANR-11-LABX-0043). O.I.V. received funding from Association pour la Recherche sur le Cancer.

References

1. Joffre OP, Segura E, Savina A, Amigorena S (2012) Cross-presentation by dendritic cells. Nat Rev Immunol 12(8):557–569

2. Mantegazza AR, Magalhaes JG, Amigorena S, Marks MS (2013) Presentation of phagocytosed antigens by MHC class I and II. Traffic 14(2):135–152

3. Kovacsovics-Bankowski M, Rock KL (1995) A phagosome-to-cytosol pathway for exogenous antigens presented on MHC class I molecules. Science 267(5195):243–246

4. Cebrian I, Visentin G, Blanchard N, Jouve M, Bobard A, Moita C, Enninga J, Moita LF, Amigorena S, Savina A (2011) Sec22b regulates phagosomal maturation and antigen crosspresentation by dendritic cells. Cell 147(6):1355–1368

5. Segura E, Durand M, Amigorena S (2013) Similar antigen cross-presentation capacity and phagocytic functions in all freshly isolated human lymphoid organ-resident dendritic cells. J Exp Med 210(5):1035–1047

Chapter 14

Cross-Presentation Assay for Human Dendritic Cells

Elodie Segura

Abstract

The presentation of exogenous antigens on major histocompatibility complex (MHC) class I molecules, termed cross-presentation, is essential for the initiation of cytotoxic immune responses. Numerous studies in mice and human have shown that dendritic cells are the best cross-presenting cells. The protocol described here allows the assessment of the cross-presentation by human dendritic cells of a model antigen (either soluble or cell associated) to antigen-specific CD8 T cells.

Key words Dendritic cells, Cross-presentation, Antigen processing, Human, CD8 T cells

1 Introduction

Antigen-presenting cells activate T lymphocytes via the recognition of antigenic peptides presented on major histocompatibility complex (MHC) molecules at their surface. MHC class I molecules usually present to CD8 T cell peptides derived from endogenous proteins that are degraded in the cytosol by the proteasome. Degradation products are then transported into the endoplasmic reticulum to be loaded onto newly formed MHC class I molecules, which are finally exported to the cell surface. Dendritic cells (DCs) have developed an alternative pathway for the MHC class I-restricted presentation of exogenous antigens to CD8 T cells: the cross-presentation pathway. This process enables the immune system to detect abnormal cells (transformed or infected by a virus or intracellular pathogens) and is essential for the initiation of cytotoxic immune responses.

DCs can internalize antigens through various mechanisms including macropinocytosis, receptor-mediated endocytosis, and phagocytosis. Two main intracellular pathways for the cross-presentation of internalized antigens have been reported: they are usually termed "cytosolic" and "vacuolar" pathways [1]. Cross-presentation through the cytosolic pathway involves the transfer of exogenous proteins from endocytic compartments to the cytosol,

Elodie Segura and Nobuyuki Onai (eds.), *Dendritic Cell Protocols*, Methods in Molecular Biology, vol. 1423,
DOI 10.1007/978-1-4939-3606-9_14, © Springer Science+Business Media New York 2016

where they are degraded by the proteasome. In the vacuolar pathway, internalized antigens are degraded directly in endolysosomes or phagosomes by lysosomal proteases and loaded on preformed MHC class I molecules in the endosomal or phagosomal lumen.

It has been known for a long time that human in vitro-generated monocyte-derived DCs, but not monocyte-derived macrophages, can cross-present antigens [2]. In the past few years, the cross-presentation ability of DC subsets directly isolated from human tissues has attracted considerable attention. Several studies have shown that human pDCs and all classical DC subsets, with the exception of dermal CD14+ DCs, can cross-present several forms of antigen: soluble antigen, cell-associated antigen, or receptor-targeted antigen [3]. By contrast, macrophages isolated from lymphoid organs cannot cross-present [4]. DC targeting for the in vivo delivery of vaccines has proved a powerful tool in the mouse [5–8]. Special emphasis is put on the targeting of cross-presenting DCs in order to elicit strong cytotoxic T cell responses [9, 10]. In order to translate this strategy to humans and to develop novel immunotherapies, it is essential to better understand the cross-presentation capacity of human DCs.

The protocol described here relies on the use of a model antigen and the corresponding antigen-specific CD8 T cell clone. Several such systems have been reported in various cross-presentation studies (*see* Table 1). In this assay, antigen-specific T cell activation is used as a readout for cross-presentation. Therefore, several controls are needed. A control condition without any antigen is essential in order to measure the background activation of the T cell clone (*see* Fig. 1). DCs must also be pulsed with the minimal epitope sequence (preprocessed peptide), in order to

Table 1
Antigen-specific CD8 T clones used for cross-presentation assays in the literature

Antigen	Epitope sequence	HLA restriction	References
gp100	YLEPGPVTA (aa 280-288)	HLA-A2	[12]
gp100	IMDQVPFSV (aa 209-217)	HLA-A2	[13]
Influenza matrix protein	GILGFVFTL (aa58-66)	HLA-A2	[14]
HCMV pp65	NLVPMVATV (aa 495-503)	HLA-A2	[15, 16]
MelanA	EAAGIGILTV (aa 26-35)	HLA-A2	[13]
Hepatitis C virus NS3 (c33c subtype 1a)	CINGVCWTV (aa 1073-1081)	HLA-A2	[17]
Hepatitis B virus surface antigen	FLLTRILTI (aa 183-191)	HLA-A2	[18]

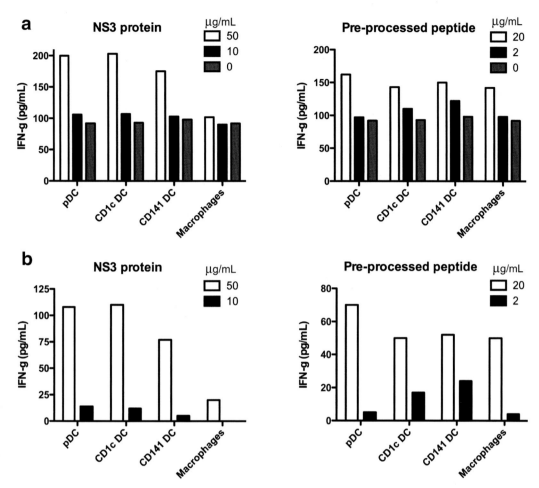

Fig. 1 Cross-presentation of hepatitis C virus NS3 protein by tonsil dendritic cells. Purified tonsil dendritic cells (pDC, CD1c DC, or CD141 DC) or purified tonsil macrophages were incubated with different concentrations of NS3 protein or preprocessed peptide for 3 h before washing and addition of antigen-specific CD8 T cells. After 18 h of culture, supernatants were collected and IFNγ secretion measured by ELISA. (**a**) Raw data. (**b**) Background levels for each cell type were subtracted

verify the reactivity of the T cell clone. In addition, several titrations of the antigen should be tested, especially if several treatments or DC subsets are being compared. Different forms of antigen can be used: soluble, coated onto particles, conjugated to a targeting antibody, cell associated, etc. Because soluble antigen can be degraded in the culture supernatant over time, it is important to wash away the antigen after a short pulse with the DCs. Two variations of the protocol are presented here: using soluble antigen and using necrotic cells that endogenously express the model antigen.

2   Materials

2.1   Cross-Presentation Assay

1. Yssel medium (*see* **Note 1**).
2. 96-well V-bottom plate with lid (Corning).
3. Preprocessed peptide (minimal epitope sequence).
4. Antigen source.
5. Antigen-specific CD8 T cell clone culture.
6. 96-well U-bottom plate.

2.2   Induction of Necrosis in Antigen-Expressing Cells

1. Oxaliplatin (marketed as Eloxatin, Sanofi).
2. Cells expressing the model antigen.

2.3   Assessment of Antigen-Specific T Cell Clone Activation

2.3.1   Analysis of IFNγ Secretion by ELISA

1. ELISA kit for IFNγ.

2.3.2   Analysis of IFNγ Secretion by Intracellular Flow Cytometry

1. Brefeldin A
2. Cytofix/Cytoperm kit (BD Biosciences).
3. Perm/Wash buffer (BD Biosciences).
4. Fluorochrome-conjugated antibodies: anti-IFNγ and anti-CD8.

3   Methods

Because of the high inter-donor variability that is inherent to all experiments on human cells, it is recommended to perform the assay with a minimum of five different donors for dendritic cells. The acquisition of human samples must be approved by an institutional ethics committee.

3.1   Cross-Presentation of Soluble Antigen

All steps must be performed in sterile conditions.

If you wish to test the effect of a compound on cross-presentation (activating drug, proteasome inhibitor, etc.), you should add this compound to the corresponding wells in between **steps 4** and **5**.

1. Collect the dendritic cells you wish to test: you will need 10^4 cells per well, with duplicate wells for each condition (*see* **Note 2**).
2. Resuspend the dendritic cells in Yssel medium (without serum), at a concentration of 2×10^5 cells/mL.
3. Prepare serial dilutions of the antigen and the preprocessed peptide in Yssel medium (without serum), at concentrations twice that of the desired final concentration (*see* **Note 3**).

4. Put 50 µL of antigen (or peptide) solution in each corresponding well in a V-bottom 96-well plate.

5. In each well, add 50 µL of the dendritic cell suspension. Be careful not to carry over any solution from one well to the other (*see* **Note 4**).

6. Cover the plate with the lid and incubate the plate in the incubator (5 % CO_2, 37 °C) during 3 h.

7. Approximately 30 min before the end of the incubation time, warm up Yssel medium (without serum) at 37 °C in a water bath.

8. At the end of the incubation time, add 100 µL of prewarmed Yssel medium in each well of the plate.

9. Centrifuge the plate ($570 \times g$, 7 min, room temperature).

10. With a multichannel pipette, carefully remove the medium from the plate. Be careful not to detach the cells from the bottom of the wells (*see* **Note 5**).

11. With a multichannel pipette, resuspend the cells in 200 µL of prewarmed Yssel medium.

12. Centrifuge the plate ($570 \times g$, 7 min, room temperature).

13. With a multichannel pipette, carefully remove the medium from the plate. Be careful not to detach the cells from the bottom of the wells.

14. With a multichannel pipette, resuspend the cells in 50 µL of prewarmed Yssel medium.

15. Collect T cells from the CD8 T cell clone culture: you will need 2×10^4 cells per well.

16. Wash the T cells, and resuspend them at a concentration of 4×10^5 cells/mL in Yssel medium containing 20 % of fetal calf serum (*see* **Note 6**).

17. Add 50 µL of the T cell suspension in each well.

18. Cover the plate with the lid and incubate the plate in the incubator (5 % CO_2, 37 °C) during 16–24 h.

3.2 Cross-Presentation of Necrotic Cell-Associated Antigen

Cells endogenously expressing the model antigen can be obtained as naturally occurring cells (cells infected with a virus, tumor cells, etc.) or after transfection of a cell line to express the model antigen. Caution should be taken that these cells do not present the antigen on their own MHC class I molecules: if employing a HLA-A2$^+$-restricted T cell clone, HLA-A2$^-$ cells should be used as a source of antigen.

All steps must be performed in sterile conditions.

If you wish to test the effect of a compound on cross-presentation (activating drug, proteasome inhibitor, etc.), you should add this compound to the corresponding wells in between **steps** 7 and **8**.

1. Induce cell death in your cell culture. Add oxaliplatin to the culture medium at a concentration of 10 μg/mL and further culture the cells during 48 h (*see* **Note 7**).

2. Wash necrotic cells a minimum of three times in culture medium, in order to remove oxaliplatin.

3. Resuspend the necrotic cells at a concentration of 1×10^5 cells/mL.

4. Collect the dendritic cells you wish to test: you will need 10^4 cells per well, with duplicate wells for each condition (*see* **Note 8**).

5. Resuspend the dendritic cells in Yssel medium (without serum), at a concentration of 2×10^5 cells/mL.

6. Prepare serial dilutions of the necrotic cells and the preprocessed peptide in Yssel medium (without serum), at concentrations twice that of the final concentration (*see* **Note 9**).

7. Put 50 μL of necrotic cell suspension or peptide solution in each corresponding well in a V-bottom 96-well plate.

8. In each well, add 50 μL of the dendritic cell suspension. Be careful not to carry over any solution from one well to the other (*see* **Note 4**).

9. Cover the plate with the lid and incubate the plate in the incubator (5 % CO_2, 37 °C) during 3 h.

10. Approximately 30 min before the end of the incubation time, warm up Yssel medium (without serum) at 37 °C in a water bath.

11. At the end of the incubation time, add 100 μL of prewarmed Yssel medium in each well of the plate.

12. Centrifuge the plate ($570 \times g$, 7 min, room temperature).

13. With a multichannel pipette, carefully remove the medium from the plate. Be careful not to detach the cells from the bottom of the wells (*see* **Note 5**).

14. With a multichannel pipette, resuspend the cells in 50 μL of prewarmed Yssel medium.

15. Collect T cells from the CD8 T cell clone culture: you will need 2×10^4 cells per well.

16. Wash the T cells, and resuspend them at a concentration of 4×10^5 cells/mL in Yssel medium containing 20 % of fetal calf serum (*see* **Note 6**).

17. Add 50 μL of the T cell suspension in each well.

18. Cover the plate with the lid and incubate the plate in the incubator (5 % CO_2, 37 °C) during 16–24 h.

3.3 Assessment of T Cell Clone Activation   The production of IFNγ is used as a readout for CD8 T cell clone activation. There are two main methods for measuring this IFNγ production: ELISA or intracellular staining for IFNγ. These methods should give overall similar results (*see* Fig. 2).

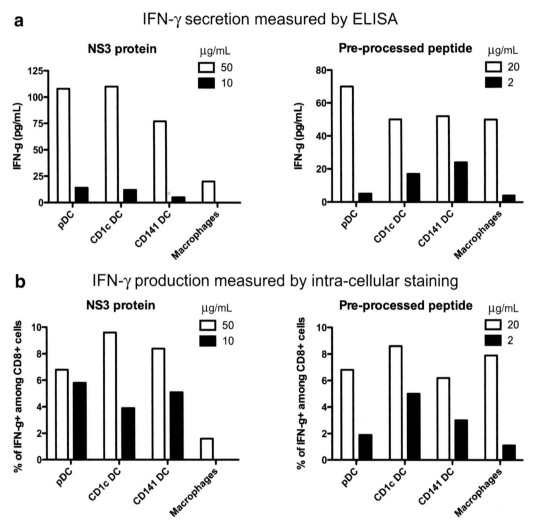

Fig. 2 Cross-presentation of hepatitis C virus NS3 protein by tonsil dendritic cells: comparison of methods for IFNγ detection. Purified tonsil dendritic cells (pDC, CD1c DC, or CD141 DC) or purified tonsil macrophages were incubated with different concentrations of NS3 protein or preprocessed peptide for 3 h before washing and addition of antigen-specific CD8 T cells. After 18 h of culture, supernatants were collected for ELISA. Cells were further incubated with brefeldin A for 5 h, before fixation and permeabilization. Cells were stained with anti-CD8-APC and anti-IFNγPE and analyzed on a MACSQuant flow cytometer. (**a**) IFNγ secretion was measured by ELISA. Background levels for each cell type were subtracted. (**b**) IFNγ production was measured by intracellular staining. Background levels for each cell type were subtracted

1. Centrifuge the plate ($570 \times g$, 7 min, room temperature).

2. With a multichannel pipette, collect the supernatants and place them in a new 96-well U-bottom plate. If not analyzed on the same day, the supernatants can then be kept at –20 °C for further analysis.

3. Measure IFNγ secretion using a standard ELISA, following the manufacturer's protocol.

Steps 1–4 must be performed in sterile conditions.

1. Centrifuge the plate ($570 \times g$, 7 min, room temperature).

2. With a multichannel pipette, remove the supernatants.

3. With a multichannel pipette, resuspend the cells in 100 μL of Yssel medium containing 10 % fetal calf serum and 1 μg/mL of brefeldin A.

4. Cover the plate with the lid and incubate the plate in the incubator (5 % CO_2, 37 °C) during 5 h.

5. Centrifuge the plate ($570 \times g$, 7 min, room temperature).

6. Fix the cells using Cytofix/Cytoperm solution: add 100 μL per well and incubate for 20 min on ice (*see* **Note 10**).

7. Centrifuge the plate ($570 \times g$, 5 min, 4 °C) and wash the cells twice using 200 μL of Perm/Wash buffer per well (*see* **Note 11**).

8. Resuspend the cells in 50 μL per well of Perm/Wash buffer containing anti-CD8 and anti- IFNγ antibodies (*see* **Note 12**).

9. Incubate the plate on ice in the dark during 30 min.

10. Centrifuge the plate ($570 \times g$, 5 min, 4 °C).

11. Wash the cells twice using 200 μL of Perm/Wash buffer per well.

12. Resuspend the cells in 100 μL of Perm/Wash buffer per well and analyze the flow cytometer.

4 Notes

1. Yssel medium is a commercially available IMDM-based serum-free medium [11]. This medium is particularly well suited for the culture of human DCs. Other media can also be used. Unless a serum-free medium is used, small amounts of serum may be added during the DC pulse to preserve cell viability.

2. Conditions for a typical experiment are "no antigen" control, several concentrations of the soluble antigen, and several concentrations of the preprocessed peptide.

3. Dilutions need to be determined by each experimenter: threefold, fivefold, or tenfold dilutions are recommended for the soluble antigen. Tenfold dilutions are recommended for the preprocessed peptide.

4. To avoid the carry-over of medium from one well to the other, the use of a multichannel pipette or a dispenser pipette is recommended.

5. The use of a V-bottom plate makes this step easier. Incline the plate slightly and place the tips along the wall of the wells, without touching the bottom of the well. You should be able to aspirate all of the medium.

6. The use of serum at this stage is important for the optimal activation of the T cell clone. If the T cell clone has been cultured in serum other than fetal calf serum (human serum, for instance), use the same serum at a concentration twice of the final desired concentration.

7. Other treatments can be used to induce necrosis. The necrosis of the cells should be checked by staining for annexin V and DAPI (other viability dyes can also be used). Necrotic cells are positive for both annexin V and DAPI, while apoptotic cells are positive for annexin V only.

8. Conditions for a typical experiment are "no antigen" control, several concentrations of necrotic cells, and several concentrations of the preprocessed peptide.

9. Dilutions need to be determined by each experimenter: twofold or threefold dilutions are recommended for the necrotic cells. Tenfold dilutions are recommended for the preprocessed peptide.

10. Other fixation protocols can be used.

11. Other permeabilization agent-containing buffers can be used.

12. The concentration of each antibody must be determined by each experimenter. Recommended fluorochrome combinations are FITC and APC or PE and APC. This should minimize compensation issues.

Acknowledgments

This work is supported by INSERM, the European Research Council, Labex DCBIOL, and Ligue contre le Cancer.

References

1. Joffre OP, Segura E, Savina A, Amigorena S (2012) Cross-presentation by dendritic cells. Nat Rev Immunol 12(8):557–569

2. Albert ML, Sauter B, Bhardwaj N (1998) Dendritic cells acquire antigen from apoptotic cells and induce class I-restricted CTLs. Nature 392(6671):86–89

3. Segura E, Amigorena S (2014) Cross-presentation by human dendritic cell subsets. Immunol Lett 158(1–2):73–78

4. Segura E, Durand M, Amigorena S (2013) Similar antigen cross-presentation capacity and phagocytic functions in all freshly isolated human lymphoid organ-resident dendritic cells. J Exp Med 210(5):1035–1047

5. Bonifaz L, Bonnyay D, Mahnke K, Rivera M, Nussenzweig MC, Steinman RM (2002) Efficient targeting of protein antigen to the dendritic cell receptor DEC-205 in the steady state leads to antigen presentation on major histocompatibility complex class I products and peripheral CD8+ T cell tolerance. J Exp Med 196(12):1627–1638

6. Bonifaz LC, Bonnyay DP, Charalambous A, Darguste DI, Fujii S, Soares H, Brimnes MK, Moltedo B, Moran TM, Steinman RM (2004) In vivo targeting of antigens to maturing dendritic cells via the DEC-205 receptor improves T cell vaccination. J Exp Med 199(6):815–824

7. Tacken PJ, de Vries IJ, Torensma R, Figdor CG (2007) Dendritic-cell immunotherapy: from ex vivo loading to in vivo targeting. Nat Rev Immunol 7(10):790–802

8. Caminschi I, Lahoud MH, Shortman K (2009) Enhancing immune responses by targeting antigen to DC. Eur J Immunol 39(4):931–938

9. Sancho D, Mourao-Sa D, Joffre OP, Schulz O, Rogers NC, Pennington DJ, Carlyle JR, Reis e Sousa C (2008) Tumor therapy in mice via antigen targeting to a novel, DC-restricted C-type lectin. J Clin Invest 118(6):2098–2110

10. Idoyaga J, Lubkin A, Fiorese C, Lahoud MH, Caminschi I, Huang Y, Rodriguez A, Clausen BE, Park CG, Trumpfheller C, Steinman RM (2011) Comparable T helper 1 (Th1) and CD8 T-cell immunity by targeting HIV gag p24 to CD8 dendritic cells within antibodies to Langerin, DEC205, and Clec9A. Proc Natl Acad Sci U S A 108(6):2384–2389

11. Yssel H, De Vries JE, Koken M, Van Blitterswijk W, Spits H (1984) Serum-free medium for generation and propagation of functional human cytotoxic and helper T cell clones. J Immunol Methods 72(1):219–227

12. Tel J, Schreibelt G, Sittig SP, Mathan TS, Buschow SI, Cruz LJ, Lambeck AJ, Figdor CG, de Vries IJ (2013) Human plasmacytoid dendritic cells efficiently cross-present exoge-nous Ags to CD8+ T cells despite lower Ag uptake than myeloid dendritic cell subsets. Blood 121(3):459–467

13. Faure F, Mantegazza A, Sadaka C, Sedlik C, Jotereau F, Amigorena S (2009) Long-lasting cross-presentation of tumor antigen in human DC. Eur J Immunol 39(2):380–390

14. Mittag D, Proietto AI, Loudovaris T, Mannering SI, Vremec D, Shortman K, Wu L, Harrison LC (2011) Human dendritic cell subsets from spleen and blood are similar in phenotype and function but modified by donor health status. J Immunol 186(11):6207–6217

15. Bachem A, Guttler S, Hartung E, Ebstein F, Schaefer M, Tannert A, Salama A, Movassaghi K, Opitz C, Mages HW, Henn V, Kloetzel PM, Gurka S, Kroczek RA (2010) Superior antigen cross-presentation and XCR1 expression define human CD11c+CD141+ cells as homologues of mouse CD8+ dendritic cells. J Exp Med 207(6):1273–1281

16. Jongbloed SL, Kassianos AJ, McDonald KJ, Clark GJ, Ju X, Angel CE, Chen CJ, Dunbar PR, Wadley RB, Jeet V, Vulink AJ, Hart DN, Radford KJ (2010) Human CD141+ (BDCA-3)+ dendritic cells (DCs) represent a unique myeloid DC subset that cross-presents necrotic cell antigens. J Exp Med 207(6):1247–1260

17. Accapezzato D, Visco V, Francavilla V, Molette C, Donato T, Paroli M, Mondelli MU, Doria M, Torrisi MR, Barnaba V (2005) Chloroquine enhances human CD8+ T cell responses against soluble antigens in vivo. J Exp Med 202(6):817–828

18. Haniffa M, Shin A, Bigley V, McGovern N, Teo P, See P, Wasan PS, Wang XN, Malinarich F, Malleret B, Larbi A, Tan P, Zhao H, Poidinger M, Pagan S, Cookson S, Dickinson R, Dimmick I, Jarrett RF, Renia L, Tam J, Song C, Connolly J, Chan JK, Gehring A, Bertoletti A, Collin M, Ginhoux F (2012) Human tissues contain CD141(hi) cross-presenting dendritic cells with functional homology to mouse CD103(+) nonlymphoid dendritic cells. Immunity 37(1):60–73

Chapter 15

Analysis of Intracellular Trafficking of Dendritic Cell Receptors for Antigen Targeting

Haiyin Liu, Claire Dumont, Angus P.R. Johnston, and Justine D. Mintern

Abstract

Antibody-targeted vaccination aims to efficiently deliver antigen to dendritic cells by targeting specific receptors at the cell surface. The choice of receptor depends on different factors, including their capacity to induce internalization of the delivered antigen/adjuvant cargo. Assays currently used to monitor internalization in dendritic cells have several limitations. We have developed a novel DNA-based probe that allows for simple and robust high-throughput analysis of internalization. Designed for flow cytometry, the probe can also be used for fluorescence microscopy to clearly distinguish internalized from surface-bound material. Here, we describe the steps for modifying material (antibodies, proteins) with the probe, undertaking the assay, and analyzing the data obtained from flow cytometry.

Key words Dendritic cell, Endocytosis, Internalization, Fluorescence, Sensor, Flow cytometry, Antibodies, Binding, Receptor

1 Introduction

Antibody-targeted vaccine delivery of antigens directly to dendritic cells (DCs) is a promising approach to improve the effectiveness of vaccines [1–3]. To this end, compatible vaccine carriers (e.g., nanoparticles, liposomes) are loaded with antigen and adjuvants and functionalized with targeting antibodies that bind DC-specific receptors. Expected benefits include reducing the required antigen dose, avoiding antigen loss due to uptake by nontargeted cells, and the possibility of simultaneously co-delivering adjuvants to the same cell. In addition, receptor targeting may enable continued antigen delivery to activated DCs, which have been shown to downregulate macropinocytic uptake while maintaining receptor-mediated endocytosis [4]. This is of particular interest for vaccination in circumstances where individuals may be chronically exposed to low-level infection.

To date, a number of different DC receptors have been targeted with antigen, including DEC-205, CD11c, and CLEC9A [1, 5–7].

Elodie Segura and Nobuyuki Onai (eds.), *Dendritic Cell Protocols*, Methods in Molecular Biology, vol. 1423,
DOI 10.1007/978-1-4939-3606-9_15, © Springer Science+Business Media New York 2016

While any DC receptor has the potential to be targeted with vaccine antigen, so far there has been mixed success in improving immunogenicity in vivo by receptor targeting, with no clear consensus as to which receptors elicit superior outcomes. This is, at least in part, due to the limited knowledge of their biological roles and trafficking characteristics [8–10]. Recent findings have indicated that different DC receptors have the potential to vary significantly in their capacity to internalize antigen and that this may impact their ability to elicit the desired T cell response [11, 12]. Given that cargo internalization is essential for efficient antigen processing, a more comprehensive understanding of DC receptor trafficking will be instrumental in defining the criteria for selecting superior receptors for vaccine targeting.

Current methods to study molecule internalization at the single-cell level include imaging-based techniques (e.g., fluorescence microscopy, high-content imaging, imaging flow cytometry) as well as flow cytometry analysis of biochemical internalization assays [13, 14]. While imaging-based methods can provide valuable additional information on subcellular localization, it can be difficult to clearly define "surface-bound" or "internalized" material merely based on signal location, especially at low resolution. Furthermore, high-content imaging and imaging flow cytometry often require specialized instrumentation that is not widely accessible. On the other hand, flow cytometry-based internalization assays are widely used owing to the relatively simple protocols, high sample throughput, and low equipment requirements. Presently, a large number of flow cytometry-based assays are available that enable detection of molecule internalization via changes in fluorescent signal [15, 16]. Many commonly used assays, however, have shortcomings that limit their usefulness for application with primary DCs. One method widely used is acid wash stripping of surface proteins. This has obvious limitations for use with primary cells as it can impact cell viability and can also nonspecifically remove the phenotype markers required to distinguish cell subsets in a mixed population [17]. The biotin pulse-chase assay equates a reduction in surface antibody to internalization, but neglects to account for loss of molecules through other means such as shedding (whether it be specific or not) from the cell surface [18]. Finally, pH-based probes such as pHrodo™ rely on a drop in pH after endocytosis, rendering them less sensitive in DCs due to their weakly acidifying endosomes [19].

To overcome the limitations of current internalization assays for DC analysis, we have recently established a specific hybridization internalization (SHIP) assay. This assay is compatible with DCs, does not affect phenotyping, and can detect protein or particle internalization in flow cytometry assays or fluorescence microscopy [17, 20]. For the SHIP assay, an antibody of interest is modified with a short fluorophore-labeled ssDNA oligonucleotide

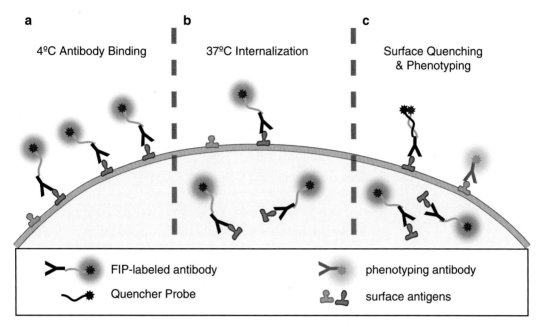

Fig. 1 Schematic for SHIP Assay. (**a**) Antibodies labeled with the fluorescent internalization probe (FIP) are bound to cell surface receptors at 4 °C. (**b**) Incubation at 37 °C triggers internalization of receptors, including bound FIP-antibody conjugates. (**c**) A quencher probe (QP) is added that hybridizes to the ssDNA of the FIP and specifically quenches FIP fluorescence. Fluorescence of phenotyping stains will be unaffected

(fluorescent internalization probe, FIP) and allowed to bind and be taken up by cells. A second ssDNA strand with the complementary sequence conjugated to a quencher dye (quencher probe, QP) is added. This hybridizes to the FIP, quenching only the signal of surface-bound FIP while leaving phenotyping markers unaffected (*see* Fig. 1). Since the QP is unable to penetrate the cell membrane, internalized material can be quantified as the remaining FIP fluorescence. Internalization can be normalized to the total fluorescence of unquenched sample (total bound material) to account for signal loss due to shedding from the cell surface and to compare internalization efficiency between differentially expressed receptors. Furthermore, by using multiple fluorophore/ssDNA sequence and fluorochrome combinations in the same assay, different receptors can be tracked simultaneously in the same sample. Longer wavelength dyes (such as Cy5, AF647, etc.) are generally favored for the assay, as they have a high signal to noise ratio, due to the lower autofluorescence of cells at the higher wavelengths.

In this chapter, we will detail the FIP labeling procedure for antibodies and describe the procedure and data analysis of a SHIP assay by flow cytometry.

2 Materials

2.1 FIP Labeling

1. Antibody (or protein ligand) for surface protein/receptor to be studied.

2. 10 mM sodium bicarbonate, pH 8.0.

3. 10 mM phosphate-buffered saline (PBS), pH 7.4.

4. NHS ester-DIBO alkyne (Life Technologies), dissolved at 1 mg/ml in anhydrous DMSO (*see* **Note 1**).

5. Zeba Spin Desalting Columns (7K MWCO 0.5 ml) (Thermo Fisher).

6. Cy5-FIP-Azide (5′ Cy5-TCAGTTCAGGACCCTCGGCT-N_3 3′, Integrated DNA Technologies), dissolved at 150 μM in nuclease-free water and stored at –20 °C (*see* **Note 2**).

7. Amicon Ultra Centrifugal Filters (30K MWCO 0.5 ml) (Merck Millipore).

8. 1.5 ml microcentrifuge tubes (both normal and low binding if available).

9. UV-Vis spectrophotometer.

2.2 SHIP Assay

1. Prepared dendritic cells in suspension.

2. Wash buffer: 1× balanced salt solution with 5 mM EDTA, 2 % FCS.

3. FIP-labeled antibody in PBS.

4. Quencher probe (5′ AGC CGA GGG TCC TGA ACT GA-BHQ2 3′, Integrated DNA Technologies) dissolved at 600 μM in nuclease-free water and stored at –20 °C (*see* **Note 3**).

5. DC medium (RPMI, 10 % FCS, 100 μM 2-mercaptoethanol, 50 U/l penicillin, 50 μg/l streptomycin, 1 % GlutaMAX).

6. Dendritic cell phenotyping antibodies.

7. Multichannel pipette (200 μl).

8. 10 ml tubes.

9. 96-well round bottom cell culture plates.

10. 0.5 mg/ml propidium iodide in wash buffer.

11. Quenching buffer: 1 mM quencher probe, 0.5 mg/ml propidium iodide in wash buffer.

12. Centrifuge for pelleting cells ($670 \times g$), with rotor bucket for plates.

13. Cell incubator at 37 °C, 10 % CO_2.

14. Flow cytometer with at least 488 and 640 nm lasers, ideally with 96-well plate sampler.

15. Flow cytometry analysis software.

3   Methods

This section details the procedure for labeling 200 μg of antibody with Cy5-FIP to achieve a degree of labeling (DOL) of approximately 1–1.5 probe molecules per antibody. Proteins other than immunoglobulin may also be used, and the reaction can be scaled up or down for the desired amount of protein to be labeled (*see* **Note 4**).

3.1   FIP Labeling

1. Prepare 200 μg antibody (IgG) solution at 2 mg/ml in 10 mM sodium bicarbonate (*see* **Note 5**).
2. Add 7 μl NHS-DIBO alkyne (10× molar excess to IgG) (*see* **Note 6**).
3. Incubate for 2 h at 4 °C (*see* **Note 7**).
4. Place 7K Zeba column in a 1.5 ml tube and centrifuge ($1500 \times g$ 1 min) and discard flow-through.
5. Add 300 μl PBS, centrifuge ($1500 \times g$, 1 min), and discard flow-through. Repeat twice.
6. Place column in a 1.5 ml low-binding tube and add the protein/DIBO solution.
7. Centrifuge ($1500 \times g$, 2 min) and collect eluate containing DIBO-labeled IgG.
8. To the eluate, add 20 μl Cy5-FIP-Azide (2× molar excess to IgG) and incubate for at least 2 h, ideally overnight, at 4 °C.
9. Equilibrate Amicon filter by adding 500 μl PBS and centrifuging briefly ($13,000 \times g$, 1 min) and discard flow-through and retentate.
10. Dilute protein/DIBO solution into 500 μl total volume with PBS and centrifuge in the Amicon filter unit ($13,000 \times g$, 5 min).
11. Wash the sample repeatedly ($13,000 \times g$, 5 min) with PBS, until flow-through is clear (approximately three washes).
12. Retrieve sample by placing the filter upside down in a clean supplied 2 ml tube and centrifuge at $1000 \times g$ for 2 min.
13. Measure protein concentration and DOL with a spectrophotometer (*see* **Note 8**).
14. Store in dark at 4 °C for short term, or aliquot and store at −20 °C until use.

3.2   SHIP Assay (*See* **Note 9**)

The internalization assay can be performed either as a two-point experiment with an "internalized" sample (incubated at 37 °C for a desired amount of time to allow internalization) and a "non-internalized" control (incubated on ice or with 150 mM sodium azide) or as a time course experiment to study internalization kinetics. For the latter, the required number of cells is first stained in bulk on ice with the FIP-labeled antibody and then placed at 37 °C, with samples removed and placed on ice to arrest internalization at different time points.

1. Dilute isolated DCs in cold washing buffer at 5×10^6 cells/ml.

2. Pipette cells (calculate 200,000 cells per time point excluding controls and pipetting allowance) into a 10 ml tube, add FIP-labeled antibody (*see* **Note 10**), and incubate on ice for 30 min.

3. Pellet cells ($670 \times g$, 7 min) and wash away unbound antibody with 10 ml washing buffer, followed by 5 ml DC medium.

4. Resuspend cells in DC medium at a concentration of 10^6 cells/ml. Remove 200 µl sample for 0 min time point and place into a chilled 96-well round bottom plate on ice.

5. Place tube with remaining cells into an incubator (37 °C, 10 % CO_2), with the cap loosened.

6. At each time point, remove 200 µl and place cells into the chilled 96-well round bottom plate.

7. At the end of the time course, pellet the cells by centrifuging the 96-well plate ($670 \times g$, 2 min). Resuspend cells in 200 µl washing buffer per well (using a multichannel pipette) and pellet cells again ($670 \times g$, 2 min).

8. Cells can be stained with desired phenotyping antibodies, diluted in 50 µl washing buffer per sample, for 30 min on ice.

9. Pellet cells ($670 \times g$, 2 min), wash 1× with 200 µl washing buffer, and pellet cells again ($670 \times g$, 2 min).

10. Resuspend cell pellets in 200 µl. Divide samples into 2×100 µl, before pelleting cells for the last time ($670 \times g$, 2 min).

11. Resuspend one half of the samples ("unquenched") in 100 µl wash buffer containing 0.5 mg/ml propidium iodide. The other half ("quenched") of samples are resuspended in 100 µl quenching buffer. Plates are to be kept chilled and dark until analysis (*see* **Note 11**).

12. Acquire samples (at least 10,000 cells per sample) on a flow cytometer with 488 nm excitation and a 610/20 band-pass filter for propidium iodide, 640 nm excitation and 670/14 filter for Cy5-FIP, as well as appropriate laser/filter sets for the phenotyping antibodies.

3.3   Data Analysis

Analysis of SHIP assay data is demonstrated here with an antibody specific for the H-2K^{bkqrs} alleles of the class I major histocompatibility complex (MHC I). Immature CD11c$^+$ spleen DCs were isolated from C57BL/6 mice, and phenotyping was performed with anti-CD11c-PE and anti-CD8α-AF488.

1. Using flow cytometry analysis software, gate the appropriate FSC/SSC population for lymphocytes, excluding nonviable propidium iodide[high] cells (*see* Fig. 2a, b).

2. Gate-desired cell subsets—in this example, we have gated CD11c+conventional DCs, in addition to CD11c+ CD8+ and CD11c+ CD8− DC subsets (*see* Fig. 2c). Following 90 min of internalization, MHC I-FIP signal can be detected in cells exposed to QP (*see* Fig. 2d).

3. Generate the geometric mean of the Cy5 signal for each sample for the different cell subsets and import the table into a spreadsheet program. To compare the fluorescent signal of different proteins, the geometric mean signal should be divided by the DOL (*see* **Note 8**).

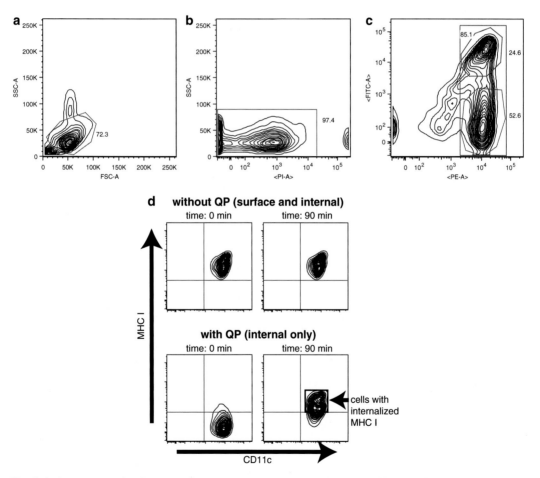

Fig. 2 Gating strategy for SHIP assay. (**a**) Lymphocytes were gated in a FSC and SSC plot, (**b**) nonviable propidium iodide[high] cells were excluded, and (**c**) CD11c+ CD8+ and CD11c+ CD8− DC subsets were identified with CD11c-PE and CD8α-AF488 staining. (**d**) DCs were stained with anti-MHC I-FIP at 4 °C only (0 min) or stained at 4 °C and incubated at 37 °C for 90 min. Cells were then treated with an antisense DNA probe conjugated to a quencher (QP, *top*) or not (*bottom*)

4. Subtract the Cy5 value of background sample (cells stained only with phenotyping antibodies and propidium iodide) from all values and plot against time. The total amount of internalized antibody over the time course is equivalent to the fluorescent signal of "quenched" samples (*see* Fig. 3a). This can be used to compare the absolute amounts of cargo (e.g., antigen or vaccine) that is delivered to a cell by different receptors.

5. To assess the efficiency of delivery (i.e., internalization relative to the total amount of bound antibody), the % internalization can be calculated as the ratio of "quenched" signal to "unquenched" sample signal elicited at time 0 (*see* Fig. 3b):

$$\% \text{ internalisation} = \frac{\text{Quenched}_{t=n} - \text{Quenched}_{t=0}}{\text{Unquenched}_{t=0} - \text{Quenched}_{t=0}}$$

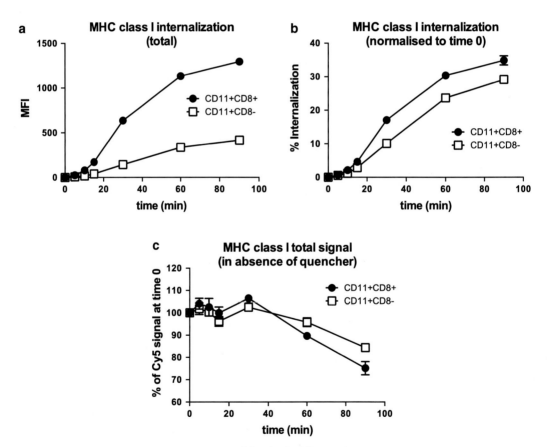

Fig. 3 Analysis of MHC class I internalization in DC subsets. SHIP assay was performed with FIP-labeled anti MHC I antibody using immature spleen DCs isolated from C57BL/6 mice. Total internalization (**a**) was calculated as the geometric mean of absolute FIP fluorescence signal after addition of QP. Normalized internalization (**b**) was calculated as ratio of FIP fluorescence of QP-exposed samples (at specified times) and unquenched sample at time 0, assuming no internalization at time 0. MHC class I total signal (**c**) was calculated from absolute FIP fluorescence without addition of QP. Graphs represent triplicate data from 10,000 cells, geometric mean ± SD

where Quenched$t$=n is the fluorescence signal (after the addition of quencher) from cells at time n, Quenched$t_{=0}$ is the fluorescence signal (after the addition of quencher) from cells kept at 4 °C to prevent internalization, and Unquenched$t_{=0}$ is the fluorescence signal from cells kept at 4 °C. In this example, total MHC I internalization is higher in CD8$^+$ DCs, while the proportion of molecules internalized is very similar for both CD8$^+$ and CD8$^-$ DC subsets.

6. A reduction in signal from "unquenched" samples over time indicates the loss of antibody from the cell (*see* **Note 12**). In this example, there is a gradual loss of total anti-MHC I signal indicating loss of antibody and/or MHC I from the cells, potentially by surface molecule shedding (*see* Fig. 3c).

4   Notes

1. NHS-DIBO alkyne is highly susceptible to hydrolysis and should be protected from moisture (i.e., from condensation after thawing) to prevent loss of reactivity.

2. All fluorophore-containing reagents (e.g., Cy5-FIP-Azide, FIP-antibody conjugate) should be protected from light exposure, whenever possible.

3. The quencher dye (e.g., Black Hole Quencher 1 or 2, QXL, etc.) should be matched to the fluorophore conjugated to FIP.

4. Protein concentrations should be between 1 and 10 mg/ml for optimal labeling efficiency. When using lower concentrations, raising the relative amount of NHS-DIBO alkyne (up to 20× molar excess) may yield better labeling ratios.

5. The antibody solution should have a pH between 8.0 and 8.3 to balance NHS-amine reactivity and hydrolysis [21]. The reaction buffer should contain no amines which may react with NHS-DIBO alkyne, e.g., Tris is a poor buffer as the NHS-DIBO will react with Tris, leading to poor yields. If this cannot be avoided, we recommend a buffer exchange into 10 mM sodium bicarbonate, pH 8.0.

6. For very lowly expressed receptors, it is possible to improve fluorescent signal by increasing the amount of NHS-DIBO/Cy5-FIP-Azide in the labeling reaction. However, higher degrees of labeling may affect antibody function and binding, and the effect of labeling should always be tested.

7. Avoid incubating NHS-DIBO alkyne with the protein for more than 2 h, to minimize nonspecific reactions that may affect protein solubility or activity.

8. To determine the correct protein concentration via spectrophotometry, the ssDNA absorbance at 280 nm must be calculated and subtracted from the measured A_{280} value. This correction factor must be obtained empirically by measuring the absorbance of the free FIP at the λ_{max} of the fluorophore and at 280 nm. The correction factor is determined by the equation:

$$CF = \frac{A_{280}}{A_{dye}}.$$

where A_{dye} is the absorbance of unconjugated FIP at the absorbance maximum of the fluorophore and A_{280} is the absorbance of unconjugated FIP at 280 nm.

For Cy5-FIP, the $\lambda_{max} = 649$ nm and $CF_{Cy5\text{-}FIP} \approx 0.55$.

The DOL of protein-FIP conjugate is calculated by dividing the molar concentrations of fluorophore and protein:

$$DOL = \frac{[\text{fluorophore}]}{[\text{protein}]}$$

$$= \frac{\left(\dfrac{A_{dye}}{\varepsilon_{dye}}\right)}{\left(\dfrac{A_{280} - CF \times A_{dye}}{\varepsilon_{protein}}\right)}$$

$$= \frac{A_{dye} \times \varepsilon_{protein}}{\varepsilon_{dye}\left(A_{280} - CF \times A_{dye}\right)}$$

where A_{dye} is the absorbance of the dye at its absorbance maximum, ε_{dye} is the extinction coefficient of the dye at its absorbance maximum, A_{280} is the absorbance at 280 nm, $\varepsilon_{protein}$ is the extinction coefficient of the protein at 280 nm, and CF is the empirically obtained correction factor.

9. Keep buffers/tubes/equipment chilled at all times to prevent unintended internalization.

10. The optimal concentration of FIP-labeled antibody should be titrated for each receptor; typical amounts range from 0.2 to 1.0 µg/100,000 cells.

11. Cells may be fixed with 4% PFA, but note that the quencher probe must be allowed to hybridize for at least 2 min before fixation.

12. If significant loss of unquenched signal over time is detected, signal loss due to FIP degradation should be excluded. This can be done by comparing the fluorescent signal with time to that elicited by an antibody labeled with a conventional fluorophore (e.g., Alexa Fluor 647).

References

1. Bonifaz LC, Bonnyay DP, Charalambous A et al (2004) In vivo targeting of antigens to maturing dendritic cells via the DEC-205 receptor improves T cell vaccination. J Exp Med 199:815–824

2. Caminschi I, Shortman K (2012) Boosting antibody responses by targeting antigens to dendritic cells. Trends Immunol 33:71–77

3. Ueno H, Klechevsky E, Schmitt N et al (2011) Targeting human dendritic cell subsets for improved vaccines. Semin Immunol 23:21–27

4. Platt CD, Ma JK, Chalouni C et al (2010) Mature dendritic cells use endocytic receptors to capture and present antigens. Proc Natl Acad Sci U S A 107:4287–4292

5. Cheong C, Choi J-H, Vitale L et al (2010) Improved cellular and humoral immune responses in vivo following targeting of HIV Gag to dendritic cells within human anti-human DEC205 monoclonal antibody. Blood 116:3828–3838

6. Wei H, Wang S, Zhang D et al (2009) Targeted delivery of tumor antigens to activated dendritic cells via CD11c molecules induces potent antitumor immunity in mice. Clin Cancer Res 15:4612–4621

7. Caminschi I, Proietto AI, Ahmet F et al (2008) The dendritic cell subtype-restricted C-type lectin Clec9A is a target for vaccine enhancement. Blood 112:3264–3273

8. Idoyaga J, Lubkin A, Fiorese C et al (2011) Comparable T helper 1 (Th1) and CD8 T-cell immunity by targeting HIV gag p24 to CD8 dendritic cells within antibodies to Langerin, DEC205, and Clec9A. Proc Natl Acad Sci U S A 108:2384–2389

9. Kreutz M, Tacken PJ, Figdor CG (2013) Targeting dendritic cells--why bother? Blood 121:2836–2844

10. Kastenmüller W, Kastenmüller K, Kurts C et al (2014) Dendritic cell-targeted vaccines - hope or hype? Nat Rev Immunol 14(10):705–11

11. Cohn L, Chatterjee B, Esselborn F et al (2013) Antigen delivery to early endosomes eliminates the superiority of human blood BDCA3+ dendritic cells at cross presentation. J Exp Med 210:1049–1063

12. Cruz LJ, Rosalia RA, Kleinovink JW et al (2014) Targeting nanoparticles to CD40, DEC-205 or CD11c molecules on dendritic cells for efficient CD8(+) T cell response: a comparative study. J Control Release 192:209–218

13. Göstring L, Chew MT, Orlova A et al (2010) Quantification of internalization of EGFR-binding Affibody molecules: methodological aspects. Int J Oncol 36:757–763

14. Naslavsky N, Weigert R, Donaldson JG (2003) Convergence of non-clathrin- and clathrin-derived endosomes involves Arf6 inactivation and changes in phosphoinositides. Mol Biol Cell 14:417–431

15. Van Amersfoort ES, Van Strijp JA (1994) Evaluation of a flow cytometric fluorescence quenching assay of phagocytosis of sensitized sheep erythrocytes by polymorphonuclear leukocytes. Cytometry 17:294–301

16. Miksa M, Komura H, Wu R et al (2009) A novel method to determine the engulfment of apoptotic cells by macrophages using pHrodo succinimidyl ester. J Immunol Methods 342:71–77

17. Liu H, Johnston APR (2013) A programmable sensor to probe the internalization of proteins and nanoparticles in live cells. Angew Chem Int Ed Engl 52:5744–5748

18. Shi M, Dennis K, Peschon JJ et al (2001) Antibody-induced shedding of CD44 from adherent cells is linked to the assembly of the cytoskeleton. J Immunol 167:123–131

19. Savina A, Peres A, Cebrian I et al (2009) The small GTPase Rac2 controls phagosomal alkalinization and antigen crosspresentation selectively in CD8(+) dendritic cells. Immunity 30:544–555

20. Ana-Sosa-Batiz F, Johnston APR, Liu H et al (2014) HIV-specific antibody-dependent phagocytosis matures during HIV infection. Immunol Cell Biol 92:679–687

21. Kalkhof S, Sinz A (2008) Chances and pitfalls of chemical cross-linking with amine-reactive N-hydroxysuccinimide esters. Anal Bioanal Chem 392:305–312

Chapter 16

Characterization of Dendritic Cell Subsets Through Gene Expression Analysis

Thien-Phong Vu Manh and Marc Dalod

Abstract

Dendritic cells (DCs) are immune sentinels of the body and play a key role in the orchestration of the communication between the innate and the adaptive immune systems. DCs can polarize innate and adaptive immunity toward a variety of functions, sometimes with opposite roles in the overall control of immune responses (e.g., tolerance or immunosuppression versus immunity) or in the balance between various defense mechanisms promoting the control of different types of pathogens (e.g., antiviral versus antibacterial versus anti-worm immunity). These multiple DC functions result both from the plasticity of individual DC to exert different activities and from the existence of various DC subsets specialized in distinct functions. Functional genomics represents a powerful, unbiased, approach to better characterize these two levels of DC plasticity and to decipher its molecular regulation. Indeed, more and more experimental immunologists are generating high-throughput data in order to better characterize different states of DC based, for example, on their belonging to a specific subpopulation and/or on their exposure to specific stimuli and/or on their ability to exert a specific function. However, the interpretation of this wealth of data is severely hampered by the bottleneck of their bioinformatics analysis. Indeed, most experimental immunologists lack advanced computational or bioinformatics expertise and do not know how to translate raw gene expression data into potential biological meaning. Moreover, subcontracting such analyses is generally disappointing or financially not sustainable, since companies generally propose canonical analysis pipelines that are often unadapted for the structure of the data to analyze or for the precise type of questions asked. Hence, there is an important need of democratization of the bioinformatics analyses of gene expression profiling studies, in order to accelerate interpretation of the results by the researchers at the origin of the research project, of the data and who know best the underlying biology. This chapter will focus on the analysis of DC subset transcriptomes as measured by microarrays. We will show that simple bioinformatics procedures, applied one after the other in the framework of a pipeline, can lead to the characterization of DC subsets. We will develop two tutorials based on the reanalysis of public gene expression data. The first tutorial aims at illustrating a strategy for establishing the identity of DC subsets studied in a novel context, here their in vitro generation in cultures of human CD34+ hematopoietic progenitors. The second tutorial aims at illustrating how to perform a posteriori bioinformatics analyses in order to evaluate the risk of contamination or of improper identification of DC subsets during preparation of biological samples, such that this information is taken into account in the final interpretation of the data and can eventually help to redesign the sampling strategy.

Key words Dendritic cell subsets, Microarray analysis for beginners, Workflow analysis, Cell identity characterization, Transcriptomic signatures, Gene set enrichment approach

Elodie Segura and Nobuyuki Onai (eds.), *Dendritic Cell Protocols*, Methods in Molecular Biology, vol. 1423,
DOI 10.1007/978-1-4939-3606-9_16, © Springer Science+Business Media New York 2016

1 Introduction

Over the past few decades, major advances in the field of both molecular biology and genomic technologies and in the ability of personal computers to handle a high quantity of information have led to the generation of huge amounts of omics data that should help to better understand all biological processes, including the functions of DC subsets. However, data analysis and interpretation represents a major bottleneck to achieve this aim, mainly because it requires a dual expertise in bioinformatics and in state-of-the-art knowledge of DC biology and more generally immunology. Hence, it is necessary on one hand to democratize bioinformatics analyses in order to render them accessible to experimental biologists lacking advanced computational and statistical skills and, on the other hand, to adapt methodologies and softwares for customization of analyses in order to adequately fit on a case-per-case basis to the structure of the dataset analyzed and to the scientific questions addressed. The ambition of this chapter is to help in advancing toward this general goal. In order to favor the independency of experimental biologists for the analysis of their data, the emphasis has been set on free computational tools that we believe are accurate and easy to use with hardly any background knowledge in bioinformatics analyses. We focused on the analysis of microarray data, because raw results for this technology are easier to preprocess than RNA-seq data for which consensus preprocessing protocols have not yet emerged. However, after data preprocessing, the workflow described to ensure of the identity of the cell subset under study or to answer other biological questions can, at least in part, be transposed to the analysis of preprocessed RNA-seq data. This chapter is not meant to be exhaustive, neither in the approaches nor in the parameter settings of the programs and algorithms used. To readers interested in gaining more knowledge in these types of bioinformatics analysis, we recommend consulting the online user guides for each of the programs used in this chapter and the published papers we referenced. In addition, alternative protocols of microarray data analysis or more generally systems biology approaches have been described by others [1–6].

A microarray experiment starts with its precise designing (*see* Fig. 1), which must be done already bearing in mind the different questions that it is meant to address, and thus the types of bioinformatics analyses that will be needed. These key issues will determine the different conditions to include, the numbers of independent replicates to generate for each condition, the type of sampling protocol to be used to obtain the desired biological material, and the eventual necessity to perform specific quality controls for validation of this sampling procedure in order to ensure the proper identity of the samples before using them for microarray data generation. Data generation then encompasses many different steps, including RNA

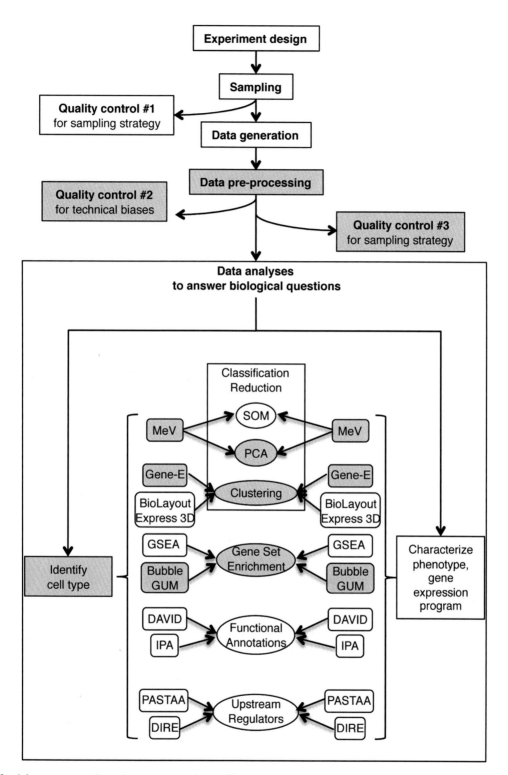

Fig. 1 A gene expression microarray experiment. The procedure encompasses various steps, starting from the experimental design and ending with the bioinformatics analysis and interpretation of the results in order to answer biological questions. The emphasis is set to focus on the last steps starting from the data preprocessing. The steps that are discussed in this chapter are in *grey*. Methods are *circled*; softwares are surrounded by *rounded-angle rectangles*. *SOM* self-organizing map, *PCA* principal component analysis, *GSEA* gene set enrichment analysis, *IPA* ingenuity pathway analysis, *DIRE* distant regulatory elements of co-regulated genes, *PASTAA* predicting associated transcription factors from annotated affinities

extraction, synthesis and labeling of complementary nucleic acid sequences, their hybridization on the arrays, and the scanning of the arrays. Each of these steps needs thorough quality controls and has been discussed in detail elsewhere [7, 8]. Moreover, laboratories most often subcontract data generation as it is a robust and well-controlled process. Thus, we will not discuss these steps in detail. Rather, we will focus on bioinformatics analysis starting from the preprocessing of the raw data to the interpretation of the results, because it generally represents a major bottleneck for experimental biologists. It necessitates custom analyses which are not easily subcontracted at reasonable prices. This chapter describes a complete workflow for gene expression analysis protocols which can allow ensuring of the identity and purity of DC subsets. This workflow encompasses analyses through hierarchical clustering, principal component analysis (PCA), generation of cell subset transcriptomic signatures, and gene set enrichment (GSEA). This workflow can be adapted to use microarray analyses for establishing the identity of potentially any candidate cell type, provided that it can be compared to a variety of other referent cell types, including i) an index one hypothesized to be the most likely homologous to the candidate cell type, ii) other cell types that are the closest known relatives to the index cell type, as well as iii) other cell types that could be potential contaminants. This workflow is illustrated through two step-by-step tutorials reanalyzing data from published studies. The protocol and bioinformatics tool described here for identifying transcriptomic signatures can also be used as an alternative to more classical methods of gene clustering, in order to identify sets of genes associated with specific subsets or activation states of DCs, then allowing to look for potential candidate functions and molecular regulatory mechanisms associated with the expression of these sets of genes. For these types of analyses that we are not covering here, the reader can refer to other articles [9–11].

The first tutorial corresponds to our own work where the biological question was to rigorously establish the identity of DC subsets derived in vitro [12]. Specifically, we will analyze the gene expression profiles of different human cell subsets generated in vitro in cultures of CD34+ hematopoietic progenitor cells initially expanded with a cytokine/growth factor cocktail called FS3T as it was composed of FLT3-L, SCF, IL-3, and TPO. This culture protocol is detailed in another chapter of this book (Chapter 2). The FS3T CD34+ cultures included CD11c^{low}CD141^{high} cells coined FS3T_XCR1+ DCs as their selective expression of XCR1 and a few other cell surface markers suggested that they could be homologs to blood XCR1+ DCs. The FS3T CD34+ cultures also contained CD11c^{high}CD141+ cells, coined FS3T_XCR1− DCs, which phenotype suggested homologies with DCs derived in vitro from monocytes (MoDCs). In order to better characterize FS3T_XCR1+ DCs and FS3T_XCR1− DCs, we performed microarray

Table 1
Affymetrix HuGene 1.0 ST gene chip samples from Balan et al. considered in the first tutorial

GEO array accession	Cell description	Cell type
GSM1386452	UN FS3T XCR1+ CD34− #1	FS3T_XCR1+
GSM1386455	UN FS3T XCR1+ CD34− #2	FS3T_XCR1+
GSM1386457	UN FS3T XCR1− CD34− #1	FS3T_XCR1−
GSM1386460	UN FS3T XCR1− CD34− #2	FS3T_XCR1−
GSM1386463	UN FS3T XCR1+ CD34− #3	FS3T_XCR1+
GSM1386466	UN FS3T XCR1+ CD34− #4	FS3T_XCR1+
GSM1386469	UN FS3T XCR1− CD34− #3	FS3T_XCR1−
GSM1386472	UN FS3T XCR1− CD34− #4	FS3T_XCR1−
GSM1386475	UN FS3T XCR1+ CD34− #5	FS3T_XCR1+
GSM1386478	UN FS3T XCR1+ CD34− #6	FS3T_XCR1+
GSM1386481	UN FS3T XCR1− CD34− #5	FS3T_XCR1−
GSM1386485	UN FS3T XCR1− CD34− #6	FS3T_XCR1−
GSM1386487	UN CD1c_bDC #1	CD1c+ bDC
GSM1386491	UN XCR1_bDC #1	XCR1+ bDC
GSM1386495	UN CD1c_bDC #2	CD1c+ bDC
GSM1386499	UN XCR1_bDC #2	XCR1+ bDC
GSM1386503	UN bpDC	pDC
GSM1386504	UN CD34-MoDC #1	MoDC
GSM1386509	UN CD34-MoDC #3	MoDC
GSM1386513	UN CD34-MoDC #4	MoDC

experiments to compare their global gene expression programs to that of DC subsets directly isolated ex vivo from the peripheral blood of healthy donors (XCR1+ bDCs, CD1c+ bDCs, and pDCs) and to MoDCs (*see* Table 1) [12].

The second tutorial corresponds to the work of others aiming at identifying candidate genes potentially associated to the functional specialization of a mouse DC subpopulation [13]. The authors of this study have identified in C56BL/6J mice a subset of steady-state splenic CD8α+XCR1+ DCs that are more potent for phagocytosis of blood-borne apoptotic cells and cross presentation of the associated antigens. In the absence of danger signals allowing their immunogenic maturation, these DCs are most efficient for the induction of immunological tolerance to cell-associated antigens, via the inactivation or clonal deletion of the corresponding CD8[+] T cells. These DCs can be distinguished from the other CD8α[+]XCR1[+] DCs based either on their selective expression of the CD103 molecule (CD11c[+]CD8α[+]CD103[+] cells) or on their selective staining consecutive to their engulfment of fluorescent apoptotic cells (CD11c[+]CD8α[+]APO[+] cells). In this study, mice were intravenously injected or not with labeled apoptotic cells. Splenic DCs from these mice were enriched with anti-CD11c microbeads and stained with anti-CD11c and anti-CD8α antibodies for cell sorting. CD11c[+]CD8α[+]APO[+] and CD11c[+]CD8α[+]APO[−] DCs were sorted from injected mice. In parallel, CD11c[+]CD8α[+]CD103[+] and CD11c[+]CD8α[+]CD103[−] DCs were sorted from non-injected mice. The gene expression data of these four DC subsets was generated. The analysis of these arrays aimed at identifying genes that play a role in the high efficiency of CD8α[+]APO[+] and CD8α[+]CD103[+] for cross presentation. However, a different population of cells, a subset of pDCs, shares the CD11c[+]CD8α[+]APO[−] and CD11c[+]CD8α[+]CD103[−] phenotypes with the subpopulation of CD8α[+]XCR1[+] cDCs that the authors aimed to study here. Hence, in this experiment, proper identification of candidate genes regulating cross presentation in CD8α[+]XCR1[+] cDCs, as genes more strongly expressed in CD103[+] or APO[+] cells as compared to their CD103[−] or APO[−] counterparts, might be confounded by a potential contamination of the latter by pDCs. To examine to which extent this might be the case, which has not been addressed in the original study, we will thus reanalyze this data together with external reference populations encompassing not only XCR1[+] cDCs but also pDCs (*see* Table 2). This second dataset will thus be

Table 2
Affymetrix Mouse 430 2.0 gene chip samples considered in the second tutorial, with the description (Sample Title) as it appears on the RefDIC server

RefDIC sample ID	Sample title
RSM01586	CD8a+
RMXXDC002003	CD24posi-B6
RMXXDC001003	PDC-B6
RSM01658	C57BL/6 PDCA-1+DC
RMSPDC049001	CD8a+APO−
RMSPDC050001	CD8a+Apo+
RMSPDC051001	CD8a+CD103−
RMSPDC052001	CD8a+CD103+

used as a concrete example to show how a posteriori bioinformatics analyses can be used to check purity of sorted DC subsets as an additional quality control of the data, to be aware of contaminants by other cell populations, and how this should be taken into account in the interpretation of the data.

2   Materials

In this section, we introduce the public expression datasets and the softwares which we use to develop the workflow. We also provide details about where they can be downloaded (*see* **Note 1**).

2.1   Accessing the Expression Dataset for the First Tutorial

1. Download Balan et al. raw expression CEL files (*see* Table 1) from the Gene Expression Omnibus (GEO) website [14]. The GEO accession number is GSE57671: http://www.ncbi.nlm.nih.gov/geo/query/acc.cgi?acc=GSE57671

2. Download the raw expression CEL files corresponding to human immunocyte samples (*see* **Note 2**), from two different sources:
   – ArrayExpress [15]: http://www.ebi.ac.uk/arrayexpress/ (accession number E-TABM-34 for the DC subsets and CD16$^+$ monocyte data).
   – GEO website: http://www.ncbi.nlm.nih.gov/geo/query/acc.cgi?acc=GSE72642 for the remaining cell populations.

2.2   Accessing the Expression Dataset for the Second Tutorial

1. Download Qiu et al. raw expression CEL files (*see* Table 2) from the RefDIC database (Reference Database of Immune Cells) [16]: http://refdic.rcai.riken.jp/welcome.cgi

2. Download from the GEO website the raw expression CEL files corresponding to murine immunocyte samples listed in the Supplemental Material 1 from Vu Manh et al. [17] (*see* **Note 3**).

2.3   Preprocessing of the Data and Various Statistical Analyses

Download R from http://www.r-project.org, install, and open it.

2.4   Classification by Hierarchical Clustering

Download Gene-E from the Broad Institute web server: http://www.broadinstitute.org/cancer/software/GENE-E

2.5   Handling of the Expression Datasets

Download LibreOffice from http://www.libreoffice.org/ (*see* **Note 4**).

2.6   Data Reduction by Principal Component Analysis

Download Mev from the TIGR web server [18]: http://www.tm4.org

2.7 Gene Set-Based
Approaches

Download BubbleGUM [19] from the CIML web server:
    http://www.ciml.univ-mrs.fr/applications/BubbleGUM/
index.html

---

3 Methods

3.1 Normalization
and Background
Correction of the Data
for the First Tutorial

Once the raw data has been generated/received, it is necessary to normalize it (*see* **Note 5**). Background can also be subtracted from signal intensities, although this is not systematically performed (*see* **Note 6**).

The following lines describe how to preprocess the expression data from Balan et al. using the R statistical environment (*see* **Note 7**).

1. Put all CEL files into a single separate folder.

2. Open R.

3. At the R prompt (only the lines beginning with ">" should be inserted into R.), define where the CEL files are located (*see* **Note 8**):
   ```
   >setwd("path_to_the_directory_containing_
   the_CEL_files")
   ```

4. Install and load the "oligo" package from Bioconductor [20]:
   ```
   >source("http://bioconductor.org/
   biocLite.R")
   >biocLite("oligo")
   >library(oligo)
   ```

5. Read the CEL files into an object named "raw_eset"
   ```
   >celFiles <- list.celfiles()
   >raw_eset<- read.celfiles(celFiles)
   ```

6. Apply RMA [21] to normalize the data (*see* **Note 9**). The normalized data is in an object named "rma_eset."
   ```
   >rma_eset<- rma(raw_eset)
   ```

3.2 Quality Control
Analyses of the Data
for the First Tutorial

1. Check the chip hybridization images. For that, we use the R function fitProbeLevelModel which output can be accessed by other functions, such as image, RLE, and NUSE. Change the value of "which" to examine all the chip hybridizations and look for any irregularities (*see* Fig. 2a) (*see* **Note 10**).
   ```
   >pset<-fitProbeLevelModel(raw_eset)
   >image(pset, which=1)
   >image(pset, which=2)
   ...
   >image(pset, which=20)
   ```

2. Generate the RLE (Relative Log Expression) (*see* Fig. 2b) and NUSE (Normalized Unscaled Standard Error) (*see* Fig. 2c) plots that provide sensitive measures to evaluate array quality (*see* **Note 11**).

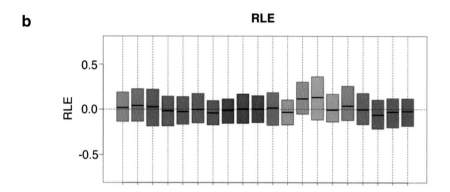

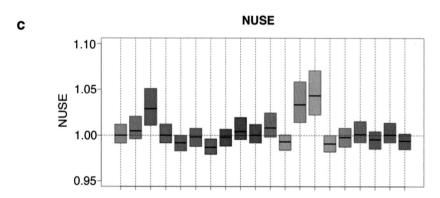

Fig. 2 Quality control #2 for technical biases. (**a**) Examination of the chips to look for irregularities. (**b**) Relative Log Expression (RLE) and (**c**) Normalized Unscaled Standard Error (NUSE), to look for low-quality arrays

```
>RLE(pset, main="RLE")
>NUSE(pset, main="NUSE")
```

3. Make the box plot of the raw data, then of the RMA normalized data (*see* Fig. 3a), to check that quantiles have correctly been adjusted between arrays (*see* **Note 12**).

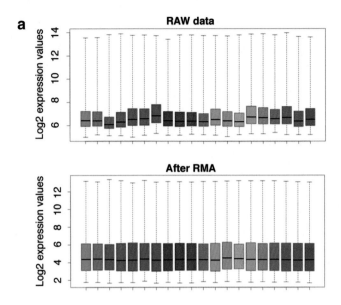

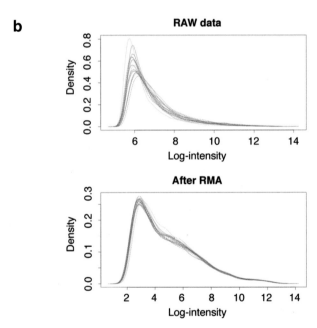

Fig. 3 Quality control #2 for technical biases. (**a**) Boxplots of the expression data, before (*top*) and after (*bottom*) RMA normalization. (**b**) Density plots of the expression data, before (*top*) and after (*bottom*) RMA normalization

```
>boxplot(raw_eset, main="RAW data")
>boxplot(rma_eset, main="After RMA")
```

4. Make the density plot of the raw data, then of the RMA normalized data (*see* Fig. 3b), to check that distributions of the signal intensities are comparable across arrays:

```
>hist(raw_eset, main="RAW data")
>hist(rma_eset, main="After RMA")
```

5. If no major problem has been found in the dataset, as it is the case with the current dataset, create the final log2-scaled normalized expression data file (*see* **Note 13**). It will be created in the current folder:

```
>write.exprs(rma_eset,   file="RMA_normal-
ized_data_file.txt", sep="\t", quote=FALSE)
```

3.3 Classification of the Data of the First Tutorial by Hierarchical Clustering

Unsupervised classification, such as hierarchical clustering or self-organizing map, is a widely used method in microarray data analysis, which aims at classifying the genes and/or the samples into classes/clusters without prior knowledge (*see* **Note 14**).

3.3.1 Classification by Hierarchical Clustering on All Probe Sets

To perform a hierarchical clustering analysis on the entire dataset from Balan et al.:

1. Open Gene-E.

2. Open the RMA_normalized data_file.txt into Gene-E.

3. Select the first cell (upper left corner) containing an expression value and press OK. A heatmap displaying relative expression appears.

4. Optional: Associate a class (.cls) file to this heatmap in order to label the samples with distinct colors according to the population type (*see* **Note 15**).

   An example of a class file, named class.cls, matching with the expression data file, is represented in Fig. 4. To associate it to the heatmap in Gene-E, click File -> Annotate Columns, and choose the class.cls file.

5. Click on the "hierarchical clustering" button.

6. Choose to cluster the columns (samples) (rows/genes can also be clustered, either independently or together with the samples).

7. Choose the One minus Pearson correlation metric.

```
20   6   1
#    FS3T_XCR1pos        FS3T_XCR1neg   CD1c_DC        XCR1_DC
        pDC      MoDC
FS3T_XCR1pos   FS3T_XCR1pos    FS3T_XCR1neg
        FS3T_XCR1neg    FS3T_XCR1pos    FS3T_XCR1pos
        FS3T_XCR1neg    FS3T_XCR1neg    FS3T_XCR1pos
        FS3T_XCR1pos    FS3T_XCR1neg    FS3T_XCR1neg
        CD1c_DC         XCR1_DC         CD1c_DC        XCR1_DC
        pDC      MoDC    MoDC     MoDC
```

Fig. 4 Class file associating a color code to each cell population in the expression dataset

8. Choose the "average linkage" method (*see* **Note 16**).

9. Press Ok to get the hierarchical clustering result (*see* Fig. 5a).

10. In the resulting tree/dendrogram, verify that the biological replicates cluster together (*see* **Note 17**).

11. Observe the main branches of the dendrogram to check whether there is any batch effect (*see* **Note 18**).

12. If no problem was observed for **steps 10** and **11**, proceed to the interpretation of the results.

Rather than performing the hierarchical clustering on all probe sets, one may want to filter out the noise from the expression dataset (*see* **Note 19**). The coming parts describe how to perform such a filter.

3.3.2 Remove the Probes Always Below the Background Level

In order to remove the probes that are not expressed in any sample, the principle is to evaluate the background level and discard the probes that are never expressed above that level:

1. Estimate the background level using the density plot (*see* Fig. 3b, bottom) (*see* **Note 20**). Assuming that in a cell, most of the genes are not expressed, one can determine that the background encompasses the first peak of the plot; thus, its upper limit is 4 in a log2 scale.

2. Open the RMA normalized expression dataset in LibreOffice Calc (or Microsoft® Excel).

3. For each probe, calculate in a new cell its maximal (MAX) value across all samples. To calculate the MAX value of all probes across all samples, calculate it for the first probe: Select the first empty cell at the end of the data related to the first probe. For the file entitled RMA_normalized_data_file.txt, this cell should be V2. In V2, type the exact formula: =MAX(B2:U2) (*see* **Note 21**).

   Then, copy V2; select the cells below V2, i.e., from V3 down to the last line of the table; and paste.

4. Select the entire table, click on "Data," and then "sort" according to the calculated MAX value.

5. Erase the lines with probes having a MAX value smaller than 4.

6. Save the file as a tab-separated "RMA_normalized_over_bkg_data_file.txt."

Fig. 5 (continued) Pvclust computes for each cluster the AU (Approximately Unbiased) *p*-value, by multiscale bootstrap resampling. The higher the AU value is (displayed in percentage), the more robust is the cluster. *Numbers* displayed for each node correspond to the AU *p*-values. Clusters having no value have an AU equal to 100. For all three results, MoDC replicates are mixed with the FS3T_XCR1⁻ DC in one of the two main branches, suggesting that these two cell types might be very similar. The second main branch encompasses XCR1⁺ DC, the CD1c⁺ DC, the pDC, and the FS3T_XCR1⁺ cells, which cluster differently depending on the method and dataset used

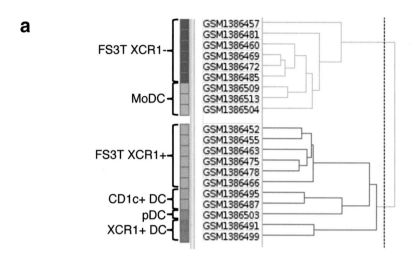

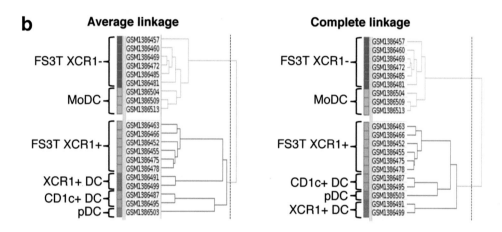

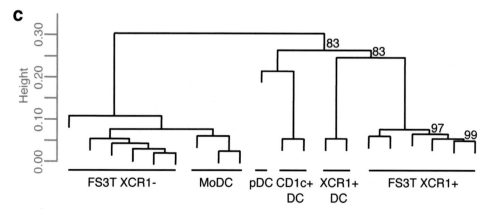

Fig. 5 Hierarchical clustering (HC) of the expression data. (**a**) HC on all the probe sets. (**b**) HC of the filtered expression data, using the One minus Pearson correlation metric and the Average (*left*) vs. Complete (*right*) Linkage methods. (**c**) Bootstrap HC of the filtered expression data using the Average linkage method.

3.3.3 Remove the Probes with No or Little Expression Changes Across Cell Populations

1. In "RMA_normalized_over_bkg_data_file.txt," with LibreOffice Calc (or Microsoft® Excel), select the cell W2 and type the following formula:

    =IF(MAX(MIN(B2:C2,F2:G2,J2:K2),MIN(D2:E2,H2:I2,L2:M2),MIN(N2,P2),MIN(O2,Q2),MIN(R2),MIN(S2:U2))-MIN(MAX(B2:C2,F2:G2,J2:K2),MAX(D2:E2,H2:I2,L2:M2),MAX(N2,P2),MAX(O2,Q2),MAX(R2),MAX(S2:U2))>1,1,0).

    This formula aims to attribute 1 to the probes that have a minimal fold change of 2 (1 in a log2 scale) between at least two cell populations and 0 to the probes that do not have such variation (*see* **Note 22**).

2. Copy W2; select the cells below W2, i.e., from W3 down to the last line of the table; and paste.

3. Select the entire table, click on "Data," and then "sort" according to this calculated score.

4. Erase the lines with probes having this score equal to 0.

5. Save the file as a tab-separated "RMA_normalized_over_bkg_minFC2x_data_file.txt" and close it.

This newly generated file now contains 5549 probes.

6. Perform a hierarchical clustering of the newly generated file using Gene-E as described in Subheading 3.3.1, with the same One minus Pearson distance and Average linkage method (*see* Fig. 5b, left panel).

7. Perform with Gene-E another hierarchical clustering of the same dataset, this time using as criteria the One minus Pearson distance and the Complete linkage method (*see* Fig. 5b, right panel).

8. Look at the differences between the two dendrograms obtained (*see* **Note 23**).

3.3.4 Apply a Multiscale Bootstrap Resampling to Hierarchical Clustering

Resampling methods, such as the bootstrap [22], allow assigning measures of robustness to dendrograms (*see* **Note 24**). Here, we will apply a multiscale bootstrap resampling, meaning that the bootstrap samples will have different sizes (from 50 to 140 % of the original sample data).

1. In R, download and install pvclust, the package performing multiscale bootstrap:

```
>source("http://bioconductor.org/
biocLite.R")
>biocLite("pvclust")
>library(pvclust)
```

2. Read the expression data file into an object named Data:

```
>Data<-read.table("RMA_normalized_over_
bkg_minFC2x_data_file.txt",          sep="\t",
header=T, row=1)
```

3. Select the columns of Data containing the expression values.

```
>Exp.Data<-Data(,1:20)
```

4. Perform the multiscale bootstrap resampling (by default, generates 1000 randomly peaked gene samples for each dataset sample ranged from 50 to 140 % of the genes in the actual dataset, every 10 %, i.e., 1000 bootstrap samples of 50 %, 1000 bootstrap samples of 60 %, etc., 1000 bootstrap samples of 140 % of the original data).

```
>result    <-   pvclust(Exp.Data,   method.
dist="cor",          method.hclust="average",
nboot=1000)
```

5. Show the final tree with % of times each node was found reproduced in the bootstrap samples (*see* Fig. 5c):

```
>plot(result)
```

3.4 Data Reduction by Principal Component Analysis on the Data for the First Tutorial

Principal component analysis (PCA) [23] aims at classifying the samples or the genes, by reducing the intrinsic complexity of an expression dataset representing tens of samples and thousands of probe sets (*see* **Note 25**).

The procedure to perform a PCA is as follows:

1. Launch MeV [18] and load the RMA_normalized_over_bkg_minFC2x_data_file.txt file.

2. Select the first cell (upper left corner) containing an expression value and press "Load." A heatmap displaying relative expression appears.

3. Click on "Data Reduction" -> "Principal Component Analysis."

4. Choose "Cluster Samples" as the Sample Selection parameter (since we want to observe the similarity/dissimilarity between samples).

5. Choose "Mean" as the Centering mode parameter and press "OK."

   Once the analysis has been performed:

6. Examine the relative contributions of the PC, in "Analysis Results" -> "PCA" -> "Eigenvalues" -> "Values" (*see* Fig. 6).

7. Look at the distribution of the samples in a 2D graph where PC1, PC2, and PC3 are compared with each other in a pairwise manner: Click on "Analysis Results" -> "PCA" -> "Projections on PC axes" -> "Components 1, 2, 3" -> "2D Views".

   – Click on "1, 2" to see the samples in a PC1 vs. PC2 space (*see* Fig. 7a).

   – Click on "1, 3" to see the samples in a PC1 vs. PC3 space (*see* Fig. 7b).

   – Click on "2, 3" to see the samples in a PC2 vs. PC3 space (*see* Fig. 7c).

Right clicking on these buttons provides some options, including the one of displaying the sample names.

Principal Component 1	04.205	47.824 %
Principal Component 2	01.471	16.730 %
Principal Component 3	00.600	06.827 %
Principal Component 4	00.415	04.723 %
Principal Component 5	00.388	04.412 %
Principal Component 6	00.298	03.385 %
Principal Component 7	00.238	02.712 %
Principal Component 8	00.169	01.923 %
Principal Component 9	00.165	01.871 %
Principal Component 10	00.137	01.554 %
Principal Component 11	00.134	01.519 %
Principal Component 12	00.106	01.211 %
Principal Component 13	00.106	01.200 %
Principal Component 14	00.097	01.108 %
Principal Component 15	00.088	01.006 %
Principal Component 16	00.068	00.779 %
Principal Component 17	00.061	00.693 %
Principal Component 18	00.046	00.525 %
Principal Component 19	00.000	00.000 %

First 2 components: 64.553 %
First 3 components: 71.380%

Fig. 6 Eigenvalues (*second column*) computed by MeV, from which each principal component contribution (*third column*) is calculated. The first PC encounters for 47.8 % of the total variations in the dataset, the second and the third PC for 16.7 % and 6.8 %, respectively. Thus, if looking at the first three PC, one can have an approximation of the data which reflects 71.3 % of the total variations in the dataset

3.5 Gene Set-Based Approach Applied to the Data for the First Tutorial

In order to gain more insight about the identity of the in vitro-derived cell populations from Balan et al., we will use a gene set-based approach (*see* **Note 26**).

Our strategy will consist in assessing, using the gene set enrichment analysis (GSEA) algorithm [24], the enrichment of sets of genes of interest on the cell populations studied in Balan et al. To better characterize the cell populations derived in vitro, we need to generate our own gene sets (*see* **Note 27**): We will extract the transcriptomic signatures of cell populations that may be equivalent or not to the in vitro-derived cells we want to characterize and test their enrichment on the transcriptomic data from Balan et al. (*see* **Note 28**). The cell populations from which we will extract the signatures include a wide range of human immunocytes from an independent expression dataset (*see* **Note 2**). To do so, we will use BubbleGUM [19] (*see* **Note 29**).

3.5.1 Generate Cell-Specific Transcriptomic Signatures

The human immunocyte gene expression data were normalized with RMA, and quality control was performed as described above:

1. Open BubbleGUM.

2. Choose the amount of memory the program will use. This depends on the resources of your computer.

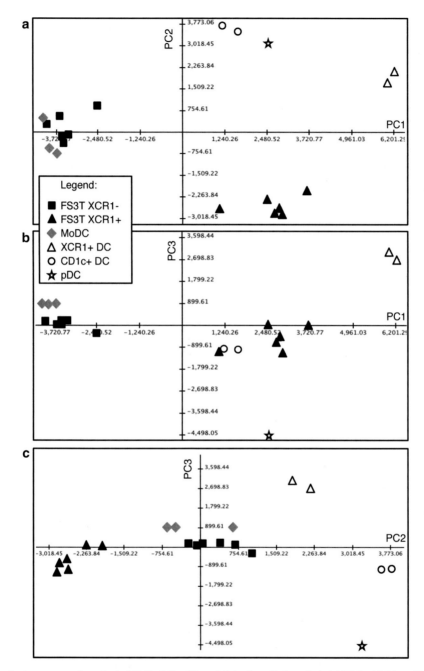

Fig. 7 Principal component analysis (PCA) of the filtered data from Balan et al. (**a**) PCA of PC1 vs. PC2. (**b**) PCA of PC1 vs. PC3. (**c**) PCA of PC2 vs. PC3. PC1 separates the samples into two main groups: the XCR1+, CD1c+, and FS3T_XCR1+ DC on the one hand and the MoDC and the FS3T_XCR1− DC on the other hand. This separation is similar to the one observed by hierarchical clustering (represented by the two main branches)

3. Choose the GeneSign tab (default tab).

4. Upload the GCT and CLS files (*see* Fig. 8) (*see* **Note 30**): The populations/classes appear in the "Available classes" window.

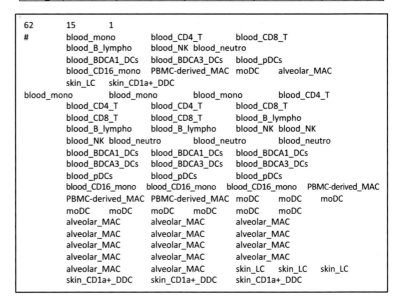

#1.2					
54675	62				
NAME	Description	FRSharp_blood_ monocytes 1	FRSharp_blood_ monocytes 2	FRSharp_blood_ monocytes 3	FRSharp_blood_ CD4 T 1
1007_s_at	DDR1	6.294	6.188	6.244	6.395
1053_at	RFC2	7.416	7.362	7.468	7.184
117_at	HSPA6	9.178	9.163	9.310	6.793
121_at	PAX8	7.698	7.826	7.731	7.607
1255_g_at	GUCA1A	4.174	3.739	4.069	4.115
1294_at	UBE1L	9.656	9.661	9.848	9.724
1316_at	THRA	6.343	6.092	6.184	6.530

```
62          15          1
#           blood_mono          blood_CD4_T          blood_CD8_T
            blood_B_lympho      blood_NK  blood_neutro
            blood_BDCA1_DCs     blood_BDCA3_DCs      blood_pDCs
            blood_CD16_mono     PBMC-derived_MAC  moDC       alveolar_MAC
            skin_LC    skin_CD1a+_DDC
blood_mono          blood_mono          blood_mono          blood_CD4_T
            blood_CD4_T         blood_CD4_T          blood_CD8_T
            blood_CD8_T         blood_CD8_T          blood_B_lympho
            blood_B_lympho      blood_B_lympho       blood_NK  blood_NK
            blood_NK  blood_neutro          blood_neutro          blood_neutro
            blood_BDCA1_DCs     blood_BDCA1_DCs      blood_BDCA1_DCs
            blood_BDCA3_DCs     blood_BDCA3_DCs      blood_BDCA3_DCs
            blood_pDCs          blood_pDCs           blood_pDCs
            blood_CD16_mono     blood_CD16_mono  blood_CD16_mono   PBMC-derived_MAC
            PBMC-derived_MAC  PBMC-derived_MAC  moDC       moDC       moDC
            moDC       moDC       moDC       moDC       moDC       moDC
            alveolar_MAC        alveolar_MAC         alveolar_MAC
            alveolar_MAC        alveolar_MAC         alveolar_MAC
            alveolar_MAC        alveolar_MAC         alveolar_MAC
            alveolar_MAC        alveolar_MAC         alveolar_MAC
            alveolar_MAC        alveolar_MAC  skin_LC    skin_LC    skin_LC
            skin_CD1a+_DDC      skin_CD1a+_DDC  skin_CD1a+_DDC
```

Fig. 8 GCT (*top*) and CLS (*bottom*) files uploaded into GeneSign. The GCT file contains the expression values of each probe in each sample. The CLS file describes the samples included in the GCT file

5. Check the "Data is in log2 scale" box (RMA produces log2 expression values).

6. Select all the immunocyte populations as Test classes, and keep the default option for the Reference classes: "All remaining populations" (*see* **Note 31**).

7. Select the Min(test)/Max(ref) as the analysis method (*see* **Note 32**).

8. Set the minimal linear fold change to 1 and press Start.
   Once the signatures are generated (one per tab in the result window):

9. Select the menu "Cart -> Add all signatures."

10. Open the Cart and "Export" the signatures into a GMT file, using as identifiers the "Description" field containing the gene symbols.

This GMT-formatted signature file will be used by BubbleMap, the second module of BubbleGUM, to assess the enrichment of the newly generated signatures on the expression data from Balan et al.

3.5.2 Assess the Enrichment of the Transcriptomic Signatures

1. In BubbleGUM, click on the BubbleMap tab.

2. Upload the four files required to perform the analysis (*see* **Note 33**).

3. Choose the parameters shown in Fig. 9 and press Start.
   BubbleMap performs GSEA on all pairwise comparisons between the populations present in the GCT-formatted expression data file, using the signatures extracted from GeneSign.

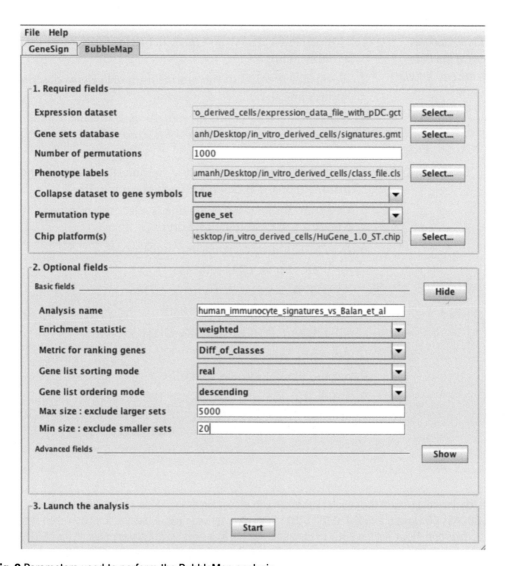

Fig. 9 Parameters used to perform the BubbleMap analysis

4. Once the BubbleMap is obtained (*see* **Note 34**), set the FDR threshold to 0.05 (*see* **Note 35**).

   One can select directly in the BubbleMap the comparisons and the gene sets of interest, in order to restrict the interpretation to the most relevant enrichments:

5. Select the signatures of MoDC, BDCA3$^+$ DC, pDC, and neutrophils (as a negative control).

6. Select the pairwise comparisons involving the FS3T_XCR1$^+$ DC, FS3T_XCR1$^-$ DC, pDC, and MoDC populations.

7. Click on "Apply selection."

A new tab appears, with the restricted BubbleMap (*see* Fig. 10), where specific bubble/enrichment patterns are observable (aligned blue bubbles) (*see* **Note 36**).

3.6 Computational Analysis of the Data for the Second Tutorial

1. Normalize the data from Qiu et al. [13] with RMA as described in Subheading 3.1.

2. Normalize the data of the murine immune cells as described in Subheading 3.1.

3. Generate the transcriptomic signatures of murine immune cells as described in Subheading 3.5.1 (use 1.5 as minimal linear fold change).

4. Assess the enrichment of these murine signatures on the data from Qiu et al., as described in Subheading 3.5.2.

5. Once the BubbleMap is obtained, select the murine pDC and XCR1$^+$ transcriptomic signatures.

6. Select the following pairwise comparisons:
   - PDC-B6 vs. CD24posi-B6
   - C57BL/6 PDCA-1$^+$DC vs. CD8α^+
   - CD8α^+APO$^-$ vs. CD8α^+Apo$^+$
   - CD8α^+CD103$^-$ vs. CD8α^+CD103$^+$

7. Apply the Selection.

8. Set the FDR threshold to 0.05.

9. Verify that the pDC or XCR1$^+$ DC transcriptomic signatures are enriched as expected in the pairwise comparisons of pDC vs. CD8α^+ DC and pDC vs. CD24$^+$ DC.

Fig. 10 (continued) in any of the populations taken into account in this analysis, (3) the systematic enrichment of the BDCA3$^+$ DC signature in the FS3T_XCR1$^+$ in in vitro-derived cells, whatever the population it is compared to (*blue box*). However, the BDCA1$^+$ signature was not found systematically enriched in the FS3T_XCR1$^+$ population (*orange box*), (4) the systematic enrichment of the MoDC signature in the FS3T_XCR1$^-$ in in vitro-derived cells, whatever the population it is compared to, except the MoDC themselves (*dark green box*)

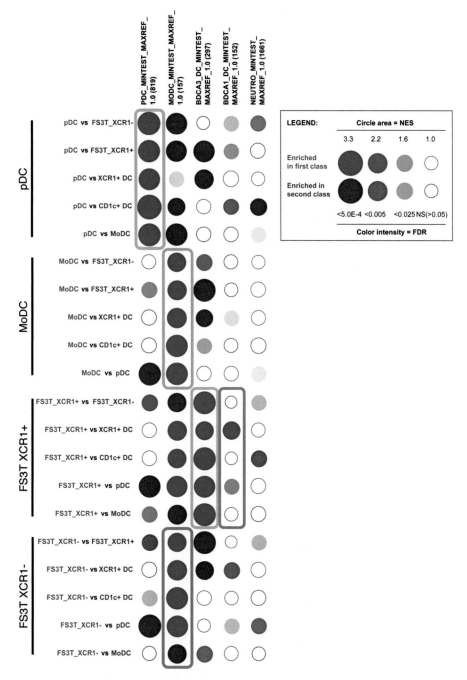

Fig. 10 Enrichment patterns observed in BubbleMap. Each line corresponds to a pairwise comparison; each column corresponds to a transcriptomic signature. Results are represented as bubbles, the size and intensity of color increasing as the enrichment was stronger and more significant, in a color matching that of the cell population in which the signature was enriched. Specifically, the bubble size is proportional to the absolute value of the normalized enrichment score, and the color intensity of bubbles is indicative of the false discovery rate statistical *q* value. These results display (1) positive controls: The systematic enrichment of the pDC and MoDC signatures in the pDC and MoDC populations, respectively, whatever the population they are compared to (*light green box* and *purple box*), (2) negative controls: The neutrophil signature is not systematically enriched

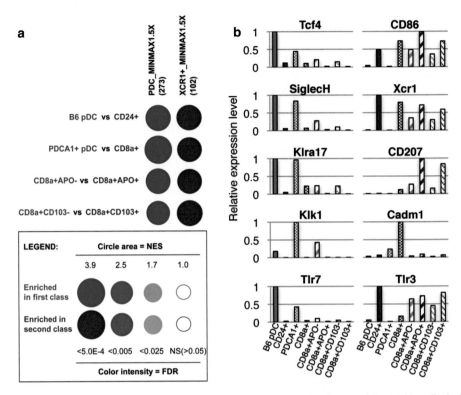

Fig. 11 Assessing the contamination of CD8α⁺APO⁻ and CD8α⁺CD103⁻. (a) Output of BubbleMap displaying the enrichment of the murine pDC and XCR1⁺ DC transcriptomic signatures. The murine pDC signature is found significantly enriched in CD8α⁺APO⁻ and CD8α⁺CD103⁻ as compared to CD8α⁺APO⁺ and CD8α⁺CD103⁺, respectively, to the same level as the enrichment that is observed when comparing pDC and CD8α⁺ DCs or pDC and CD24⁺ DC, thus meaning that the CD8α⁺APO⁻ and the CD8α⁺CD103⁻ are contaminated by pDCs. (b) Histograms displaying the expression profile of pDC (*left column*) or XCR1⁺ DC (*right column*)-specific genes across all the samples normalized together. The relative gene expression values were scaled to the maximal value across the samples

10. Examine whether the pDC or XCR1⁺ DC transcriptomic signatures are enriched in the pairwise comparisons of CD8α⁺APO⁻ DCs vs. CD8α⁺APO⁺ DCs and CD8α⁺CD103⁻ DCs vs. CD8α⁺CD103⁺ DCs (*see* Fig. 11a) (*see* **Note 37**).

In this second tutorial, we have highlighted through the reanalysis of public gene expression data that bioinformatics analyses can be crucial for the a posteriori quality control of the purity of the cell populations of interest that were sorted in an experiment. This analysis can thus help to improve and refine the sorting strategy that was used.

4 Notes

1. Special attention has been put on the fact that all these softwares run on any operating systems, provided that Java is installed. If it is not the case, please visit www.java.com

2. This dataset, screened on Affymetrix Human Genome U133 Plus 2.0 gene chips, will be used to generate cell-specific transcriptomic signatures that will then be assessed for enrichment on the data from Balan et al. It encompasses expression data for blood CD14+ and CD16+ monocytes, different types of macrophages (alveolar, PBMC derived, monocyte derived), blood CD4+ and CD8+ T cells, B cells, NK cells, neutrophils, MoDC, BDCA1+ and BDCA3+ DC, and pDC. Since the generation of a cell-specific transcriptomic signature is based on the higher expression of genes in the cell type of interest relative to other cell types, the more cell types are taken into account in the compendium, the more accurate the transcriptomic signatures are.

3. This dataset, screened on Affymetrix MoGene 1.0 ST gene chips, will be used to generate cell-specific transcriptomic signatures that will then be assessed for enrichment on the data from Qiu et al. It encompasses expression profiles of splenic CD8α+ and CD11b+ DC, B cells, CD4+ and CD8+ T cells, NK cells, CD8+ and CD8− pDC, blood monocytes, neutrophils, CD8α+, and CD11b+ DC from the cutaneous lymph nodes, macrophages from the lung and the peritoneum, Langerhans cells, and in vitro-derived MoDC.

4. LibreOffice is a free alternative suite to Microsoft® Office. If you already have Microsoft® Office installed on your computer, as it is the case for many researchers, skip this step.

5. The normalization process aims at removing the technical variations that systematically occur in microarray experiments, while preserving the biological variations which are of interest for the investigator. The technical variations can result from subtle differences in many experimental steps, including in the preparation of the microarrays, in the procedures for the isolation of the cells, in the extraction of the RNA from the cells, in the synthesis and labeling of cRNA probes, in their hybridization on the microarrays, and in the scanning of the microarrays [25]. Normalization consists in adjusting the signal intensity of each probe on each array in such a way that the distributions of all signal intensities are uniform across all arrays of the same experiment (or the mean ratio is equal to 1 in two-color array experiments). This process is necessary to be able to compare the expression of a probe set across arrays. A commonly used methodology to achieve this is quantile normalization.

6. The background level can be computed as the mean intensity signal where no cDNA is attached to the array. However, many arrays include negative control probes that do not match any known mRNA sequence. Defining background signal as the mean intensity signal of these negative control probes appears to be more appropriate since it accounts for nonspecific hybridization signals. It is important to remove the background

signal when possible, since it will provide a better calculation of fold changes.

7. If you work with Windows or Mac, there is a simpler point and click alternative although less flexible option to using R for RMA normalization: RMAExpress. A complete documentation is available on the website (http://rmaexpress.bmbolstad.com/).

8. With Linux the path to your file should resemble to:

   "/home/vumanh/Documents/CEL_files/."

   With Mac: "/Users/vumanh/Documents/CEL_files/."

   With Microsoft® Windows: "C:\Users\vumanh\Documents\ CEL_files\"

9. The gene chips that were used for this study are Affymetrix HuGene 1.0 ST. The described command lines can be used to normalize most Affymetrix expression gene chips. For Affymetrix chips, we recommend to use the Robust Multichip Average (RMA) normalization [21] although other normalization methods exist, such as MAS 5.0 and dChip [26] (*see* ref. 27 for the comparison of these methods). RMA includes (1) background adjustment, (2) log2 transformation, and (3) quantile normalization. Then, a linear model is fit to the normalized data to obtain an expression measure for each probe set on each gene chip.

10. Irregularities such as dots, marks, or lines could be responsible for incorrect signal intensity attribution. If you observe any, keep in mind that subsequent unexpected results related to the corresponding arrays may potentially be due to these marks and thus might lead to the decision of discarding these arrays.

11. In RLE box plot, the boxes should be similar in range and be centered to 0. If a box is not centered to 0 or has a greater spread than the others, this reflects a lower quality of the array. In NUSE box plot (*see* Fig. 2c), if a box is elevated as compared to the others and/or has a greater spread than the others, this also reflects a lower quality of the array. Typically, a box centered above 1.1 corresponds to an array with quality problems. Arrays with low quality should be removed from the analysis. However, if the sample is very important and either the spread of the box in RLE or the shift of the box in NUSE is not critical, keep that sample in the analysis and proceed to further quality control analyses. Always keep in mind that any unexpected result (i.e., density plot different from the others, the sample does not cluster with its replicates) may be due to the quality of that array.

12. The box plots (*see* Fig. 3a), representing for each array the extreme expression values (minimum and maximum) as well as the first, second (median), and third quartiles, should be

aligned after RMA normalization. If this is not the case, there was a problem during the normalization procedure, and repeat the procedure.

13. In log2 scale, distribution of the signal intensities is more uniform than in linear scale where the intensities are mainly in low values; thus, log transformation facilitates visual inspection of the data. Moreover, using log transformed values reduces the dependency on the magnitude of the values and improves variance estimation [28].

14. This classification is based on the similarity of the gene/sample expression profiles. It is particularly useful when one does not know what relationships are to be expected from the genes/samples. The notion of similarity of expression profiles requires that a distance or dissimilarity measure between two expression profiles be defined. Two commonly used metrics for measuring correlation are the Euclidean and the Pearson correlations. The type of correlation metric used depends largely on what is measured. For gene expression, Pearson correlation seems to be more appropriate.

15. Description of the CLS file format is available: www.broadinstitute.org/cancer/software/gsea/wiki/index.php/Data_formats

16. When performing a hierarchical clustering analysis, the linkage method decides which classes to connect by computing a distance between the classes. Once two classes have been connected, the distances between the new class and the other classes are recomputed. Four main methods exist [28]:

    – Single linkage: It uses the shortest distance (i.e., the smallest dissimilarity measure) between all pairs of members (genes or samples) in different classes and tends to produce long chains of genes/samples.

    – Complete linkage: It uses the maximum distance and tends to produce compact, spherical clusters.

    – Average linkage: It uses the average of all distances between genes/samples in the different classes.

    – Centroid linkage: It uses the distances between class centroids (e.g., class means).

    Average and centroid linkages tend to produce compromise clusters between the single and complete linkage. An alternative and interesting method is Wards' method, which joins clusters so as to minimize the within-cluster variance, but this method is not implemented in Gene-E.

17. Classification methods, such as hierarchical clustering or principal component analysis, constitute an extra quality control analysis, since the biological replicates of the samples are expected to cluster together. If it is not the case, it could mean

that there was a problem during the previous steps of the microarray experiment or that there is an intrinsic variability among biological replicates that is higher than the variability between the tested phenotypic conditions/cell types.

18. When dealing with microarray samples coming from different laboratories, related to different tissues, or processed at different times, one can observe batch effects especially through classification methods such as hierarchical clustering and principal component analysis. It is important to detect if there is any since it will have a strong impact on the interpretation of the results. Through hierarchical clustering, a strong batch effect is typically observed by the separation of the microarray samples according to their batch of origin into as many main branches as the number of batches taken into consideration in the study. Through principal component analysis, a batch effect is typically observed in the projection of one of the principal components. If this batch effect is strong as compared to the variations linked to the biological differences of the samples, the associated axis will be among the first principal components. Algorithms of batch correction exist, as, for example, ComBat [29], which is particularly robust when combining batches of microarray samples of small sizes.

19. It is highly recommended to filter the expression dataset in order to remove (1) the probes that do not or barely change their expression across the cell populations and (2) the probes that are not expressed in any of the populations considered, because these probes introduce noise to the actual expression changes, thus to the calculation of the distances between samples.

20. Estimation of the background in a microarray experiment is still a matter of debate. The goal here is not to attribute the final expression level of each probe considering the background level, which is something already taken into account by the RMA normalization, but to roughly estimate what the threshold is for considering a probe as expressed or not, relative to the background level; one simple solution consists in examining the density plot after normalization (*see* Fig. 3b, bottom).

21. This should give as a result the maximal value of the first probe (with Affymetrix ID:7892501). If this returns an error, check that the format of the cells (the ones selected in the formula and also V2) is set to "General" so that calculations can be correctly performed.

22. In other words, this formula means: If one subtracts the minimum of the maximal values (among the replicates) of each population to the maximum of the minimal values (among the replicates) of each population and the result is higher than 1 in

a log2 scale (=2 in a linear scale), then return 1, otherwise return 0. This method uses an arbitrary cutoff (that should be adapted depending on the dynamic range of the dataset) of two-folds as a minimal fold change to consider that a probe set is differentially expressed between at least two populations. Alternative methods exist, such as the one also based on an arbitrary cutoff and used by the Immgen Consortium, which consists in focusing on the 15 % most variable genes across all populations [30].

23. The discrepancy in the hierarchical clustering results depending on the chosen parameters shows how subjective it can be to infer a biological interpretation only based on a dendrogram. Other complementary analyses are necessary to support the interpretation.

24. Because a hierarchical clustering will always build a hierarchy of clusters, regardless of the sample size, the data quality, or the experimental design, it is important to provide a measure of the reproducibility of the clustering. The principle of bootstrap is to use the data of the current sample study as a basis to generate many (generally few thousands) bootstrap samples by resampling with replacement the current sample data, in order to approximate its sampling distribution.

25. Principal component analysis [23] is a statistical technique that aims at finding patterns in high-dimensional data, such as an expression dataset with thousands of lines and tens of columns, where the dispersion matrix (also called variance-covariance matrix) is too large to interpret properly the data, due to the large number of variables. Once it has found these patterns of similarity or dissimilarity, it is able to compress the data and decrease the complexity of the dataset by reducing the number of dimensions into few interpretable linear combinations, without much loss of information. Each linear combination is a principal component (PC). There are as many PCs as necessary to account for 100 % of the variations present in the data. The PCs are ranked from the first to the last according to the descending order of variability gathered in each PC.

26. The gene set approach differs from the single gene approach because it is based on the observation that many phenotypes are due to the moderate dysregulation of sets of genes that are closely related rather than to the strong regulation of few individual genes. Hence, the genes of a gene set could be found statistically dysregulated as a group, although each individual gene of the gene set is not detected as significantly dysregulated. One popular method to assess the global dysregulation of a gene set of interest is the gene set enrichment analysis (GSEA) [24]. GSEA tests statistically whether an a priori

defined gene set is differentially expressed between two phenotypic conditions.

27. Public databases of gene sets exist and are based on experiments held by various laboratories in the world or on annotation databases. Assessing the enrichment of these public gene sets on microarray data can be informative and is complementary with the use of annotation enrichment tools such as DAVID [11] or ingenuity pathway analysis (IPA) which aim at characterizing functionally the most significantly dysregulated genes in a study, at the biological process or pathway level. However, the knowledge gathered in the public gene set databases as well as in the public annotation databases could be irrelevant for the specific question that is addressed in one's project. Typically, the characterization of DC subsets needs precise sets of genes that define the transcriptional fingerprint of each DC subset. This information is so far lacking in the public databases.

28. A cell-specific transcriptomic signature is defined as the list of genes that are more highly expressed in the cell population of interest as compared to many others.

29. BubbleGUM [19] is composed of two modules, GeneSign which automatically extracts transcriptomic signatures from an expression dataset and BubbleMap which performs serial GSEA using the signatures generated by GeneSign (or third-party gene sets).

30. In order to generate transcriptomic signatures for each of the immunocytes mentioned above, using GeneSign, two files are necessary:

    – A normalized expression data file, in GCT format

    – A class file, in CLS format, describing the populations present in the GCT file

    Examples are provided in Fig. 8. The GCT file corresponds to the RMA normalized text file generated using R, slightly modified. It was completed with two extra lines on top of the text file, and an extra column, entitled "Description" and containing the gene symbols, is necessary. When dealing with your own data, this information is usually provided by the facility which generated the data. Otherwise, the chip annotation of most chips commonly used can be downloaded from the Broad Institute MSigDB [24] web server: http://www.broadinstitute.org/gsea/msigdb/index.jsp (an account is necessary); Go to Download, click on the Array Annotations ftp server, and get the annotations corresponding to the gene chip used, in this particular example, Affymetrix Human Genome U133 Plus 2.0. Beware that the two first columns of the GCT file should be exactly entitled "NAME" and "Description."

Description of the GCT and CLS file formats is available: www.broadinstitute.org/cancer/software/gsea/wiki/index. php/Data_formats

The class file (CLS) should not contain special symbols such as + or –. One can use "pos" or "neg" as exemplified in Fig. 4.

31. GeneSign offers the option to generate absolute and relative transcriptomic signatures. Absolute signatures correspond to the selection of genes that are more highly expressed in a cell population of interest as compared to all other cell populations considered in the compendium, whereas relative signatures correspond to the selection of genes that are more highly expressed in a cell population of interest as compared to a restricted list of the remaining cell populations.

32. Contrary to the other methods proposed by GeneSign, the Min(test)/Max(ref) does not calculate p-values and computes the immunocyte transcriptomic signatures in few seconds. The other methods can also be used to extract the signatures, but the Min(test)/Max(ref) methods turns out to be fairly stringent [19].

33. To perform a BubbleMap analysis, four files are necessary (the same as the ones required to run a GSEA):

    – An expression data file (the preprocessed file generated earlier from Balan et al. [12]: RMA_normalized_data_file. txt), in a GCT format (*see* **Note 30**).

    – A class file, in CLS format, describing the populations present in the GCT file (*see* **Note 30**).

    – A gene set/signature file, in GMT format, generated earlier by GeneSign.

    – A chip annotation file, in CHIP format, containing the annotations related to the HuGene 1.0 ST gene chip (corresponding to the data from Balan et al.). This file should be downloaded from the MsigDB website (*see* **Note 30**).

    Description of these file formats is available here: www. broadinstitute.org/cancer/software/gsea/wiki/index.php/ Data_formats

34. The results are displayed as a BubbleMap. Each bubble is a GSEA result and reflects the GSEA enrichment plot: The color of the bubble corresponds to the population from the pairwise comparison in which the signature tested is enriched. The bubble's size is proportional to the normalized enrichment score (NES) calculated by GSEA. The color intensity corresponds to the significance of the GSEA-modified Kolmogorov-Smirnov statistical test (FDR). A nonsignificant enrichment is displayed as an empty bubble.

35. When interpreting the BubbleMap results, beware of the false discovery rate (FDR) cutoff used. The FDR is the estimated probability that a gene set with a given normalized enrichment score (NES) represents a false-positive finding. The developers of GSEA recommend to set this threshold to a maximum of 25 % in the case where the number of samples is high. However, if the number of samples for each condition is low as it is the case in the present study, the permutations have to be performed on the gene sets rather than on the phenotypes and a more stringent FDR cutoff (5 %) could be more appropriate.

36. The combination of approaches used in this tutorial, encompassing hierarchical clustering, principal component analysis, generation of homemade signatures, and the assessment of their enrichment, leads to the conclusion that the FS3T_XCR1$^-$ DCs are equivalent to MoDC and the FS3T_XCR1$^+$ DCs are equivalent to BDCA3$^+$ DC.

37. The murine pDC signature is found significantly enriched in CD8α^+APO$^-$ and CD8α^+CD103$^-$ as compared to CD8α^+APO$^+$ and CD8α^+CD103$^+$, respectively, to the same level as the enrichment that is observed when comparing pDC and CD8α^+ DCs or pDC and CD24$^+$ DC (*see* Fig. 11a). This result shows that the CD8α^+APO$^-$ and the CD8α^+CD103$^-$ are contaminated by pDCs, as confirmed by looking at the expression profiles of individual genes known to be expressed specifically in pDC and not in XCR1$^+$ DC, such as *Tcf4*, *Siglech*, *Klra17*, *Klk1*, and *Tlr7* (*see* Fig. 11b, left column). Hence, even if it is probably modest, the specific contamination of APO$^-$ cells and CD103$^-$ cells by pDCs and potentially other cell types does contribute to their lower expression of genes that the initial report identified based on their higher expression in APO$^+$ cells and CD103$^+$ cells and proposed to be candidates for molecular regulation of cross presentation, including *Langerin* (*Cd207*), *Cd86*, and *Itgae* (*Cd103*). Indeed, it is striking to observe that the whole CD8α^+XCR1$^+$ DC signature is less expressed in APO$^-$ cells and CD103$^-$ cells, as illustrated by the expression profiles of individual genes known to be expressed in all *bona fide* CD8α^+ cDCs such as *Xcr1* and *Tlr3* in Fig. 11b (right column). It is unlikely that each of these individual genes is more highly expressed in the subset of the CD8α^+XCR1$^+$ DCs, the most efficient for cross presentation as compared to the other CD8α^+XCR1$^+$ DCs. Thus, rigorous identification through this approach of candidate genes potentially involved in the regulation of cross presentation would require doing this experiment again, with a more rigorous protocol for cell subset sampling, including, for example, XCR1 expression as an additional phenotypic marker to ensure rigorously comparing different subsets of pure XCR1$^+$ DCs, as opposed to the current dataset

comparing pure XCR1$^+$ DCs with a mixed population of XCR1$^+$ DCs and pDCs. More generally, this dataset illustrates the necessity to define consensus guidelines for rigorous and universal phenotypic identification of DC subsets across tissues, across laboratories, and even across experiments performed in the same laboratories [31], since the analysis of the expression profiles of genes known to be expressed specifically in pDCs or in *bona fide* CD8α$^+$ cDC shows striking differences between cells defined as pDC-B6 vs. PDCA1$^+$DC on one hand and CD8α$^+$ vs. CD24$^+$ DCs on the other hand (*see* Fig. 11).

Acknowledgements

The laboratory of Marc Dalod receives funding from the European Research Council under the European Community's Seventh Framework Programme (FP7/2007–2013 Grant Agreement no. 281225, including salary support for T.-P.V.M.); from the I$_2$HD collaborative project between CIML, AVIESAN, and SANOFI; from Institut National du Cancer (INCa grant #2011-155); from Agence Nationale de la Recherche (project PhyloGenDC, ANR-09-BLAN-0073-02); and from FRM (Equipe labellisée to M.D.), as well as institutional support from Inserm and CNRS. We also acknowledge support from the DCBIOL Labex (ANR-11-LABEX-0043, grant ANR-10-IDEX-0001-02PSL*) and the A*MIDEX project (ANR-11-IDEX-0001-02) funded by the French Government's "Investissements d'Avenir" program managed by the ANR.

References

1. Pascual V, Chaussabel D, Banchereau J (2010) A genomic approach to human autoimmune diseases. Annu Rev Immunol 28:535–571

2. Chaussabel D, Baldwin N (2014) Democratizing systems immunology with modular transcriptional repertoire analyses. Nat Rev Immunol 14(4):271–280

3. Pandey G, Cohain A, Miller J, Merad M (2013) Decoding dendritic cell function through module and network analysis. J Immunol Methods 387(1–2):71–80

4. Mabbott NA, Kenneth Baillie J, Hume DA, Freeman TC (2010) Meta-analysis of lineage-specific gene expression signatures in mouse leukocyte populations. Immunobiology 215(9–10):724–736

5. Zak DE, Tam VC, Aderem A (2014) Systems-level analysis of innate immunity. Annu Rev Immunol 32:547–577

6. Pulendran B, Li S, Nakaya HI (2010) Systems vaccinology. Immunity 33(4):516–529

7. Weeraratna AT, Taub DD (2007) Microarray data analysis: an overview of design, methodology, and analysis. Methods Mol Biol 377:1–16

8. Imbeaud S, Auffray C (2005) 'The 39 steps' in gene expression profiling: critical issues and proposed best practices for microarray experiments. Drug Discov Today 10(17): 1175–1182

9. Theocharidis A, van Dongen S, Enright AJ, Freeman TC (2009) Network visualization and analysis of gene expression data using BioLayout Express (3D). Nat Protoc 4(10): 1535–1550

10. Reich M, Liefeld T, Gould J, Lerner J, Tamayo P, Mesirov JP (2006) GenePattern 2.0. Nat Genet 38(5):500–501

11. da Huang W, Sherman BT, Lempicki RA (2009) Systematic and integrative analysis of large gene lists using DAVID bioinformatics resources. Nat Protoc 4(1):44–57

12. Balan S, Ollion V, Colletti N, Chelbi R, Montanana-Sanchis F, Liu H, Vu Manh TP, Sanchez C, Savoret J, Perrot I, Doffin AC, Fossum E, Bechlian D, Chabannon C, Bogen B, Asselin-Paturel C, Shaw M, Soos T, Caux C, Valladeau-Guilemond J, Dalod M (2014) Human XCR1+ dendritic cells derived in vitro from CD34+ progenitors closely resemble blood dendritic cells, including their adjuvant responsiveness, contrary to monocyte-derived dendritic cells. J Immunol 193(4):1622–1635

13. Qiu CH, Miyake Y, Kaise H, Kitamura H, Ohara O, Tanaka M (2009) Novel subset of CD8a+dendritic cells localized in the marginal zone is responsible for tolerance to cell-associated antigens. J Immunol 182(7):4127–4136

14. Edgar R, Domrachev M, Lash AE (2002) Gene Expression Omnibus: NCBI gene expression and hybridization array data repository. Nucleic Acids Res 30(1):207–210

15. Brazma A, Parkinson H, Sarkans U, Shojatalab M, Vilo J, Abeygunawardena N, Holloway E, Kapushesky M, Kemmeren P, Lara GG, Oezcimen A, Rocca-Serra P, Sansone SA (2003) ArrayExpress—a public repository for microarray gene expression data at the EBI. Nucleic Acids Res 31(1):68–71

16. Hijikata A, Kitamura H, Kimura Y, Yokoyama R, Aiba Y, Bao Y, Fujita S, Hase K, Hori S, Ishii Y, Kanagawa O, Kawamoto H, Kawano K, Koseki H, Kubo M, Kurita-Miki A, Kurosaki T, Masuda K, Nakata M, Oboki K, Ohno H, Okamoto M, Okayama Y, O-Wang J, Saito H, Saito T, Sakuma M, Sato K, Seino K, Setoguchi R, Tamura Y, Tanaka M, Taniguchi M, Taniuchi I, Teng A, Watanabe T, Watarai H, Yamasaki S, Ohara O (2007) Construction of an open-access database that integrates cross-reference information from the transcriptome and proteome of immune cells. Bioinformatics 23(21):2934–2941

17. Vu Manh TP, Marty H, Sibille P, Le Vern Y, Kaspers B, Dalod M, Schwartz-Cornil I, Quere P (2014) Existence of conventional dendritic cells in Gallus gallus revealed by comparative gene expression profiling. J Immunol 192(10):4510–4517

18. Saeed AI, Sharov V, White J, Li J, Liang W, Bhagabati N, Braisted J, Klapa M, Currier T, Thiagarajan M, Sturn A, Snuffin M, Rezantsev A, Popov D, Ryltsov A, Kostukovich E, Borisovsky I, Liu Z, Vinsavich A, Trush V, Quackenbush J (2003) TM4: a free, open-source system for microarray data man-agement and analysis. Biotechniques 34(2):374–378

19. Spinelli L, Carpentier S, Montañana Sanchis F, Dalod M, Vu Manh TP (2015) BubbleGUM: automatic extraction of phenotype molecular signatures and comprehensive visualization of multiple Gene Set Enrichment Analyses. BMC Genomics 16(1):814

20. Gentleman RC, Carey VJ, Bates DM, Bolstad B, Dettling M, Dudoit S, Ellis B, Gautier L, Ge Y, Gentry J, Hornik K, Hothorn T, Huber W, Iacus S, Irizarry R, Leisch F, Li C, Maechler M, Rossini AJ, Sawitzki G, Smith C, Smyth G, Tierney L, Yang JY, Zhang J (2004) Bioconductor: open software development for computational biology and bioinformatics. Genome Biol 5(10):R80

21. Bolstad BM, Irizarry RA, Astrand M, Speed TP (2003) A comparison of normalization meth-ods for high density oligonucleotide array data based on variance and bias. Bioinformatics 19(2):185–193

22. Efron B, Tibshirani R (1991) Statistical data analysis in the computer age. Science 253(5018):390–395

23. Pearson K (1901) On lines and planes of clos-est fit to systems of points in space. Philos Mag 2:559–572

24. Subramanian A, Tamayo P, Mootha VK, Mukherjee S, Ebert BL, Gillette MA, Paulovich A, Pomeroy SL, Golub TR, Lander ES, Mesirov JP (2005) Gene set enrichment analy-sis: a knowledge-based approach for interpret-ing genome-wide expression profiles. Proc Natl Acad Sci U S A 102(43):15545–15550

25. Churchill GA (2002) Fundamentals of experi-mental design for cDNA microarrays. Nat Genet 32(Suppl):490–495

26. Li C, Wong WH (2001) Model-based analysis of oligonucleotide arrays: expression index computation and outlier detection. Proc Natl Acad Sci U S A 98(1):31–36

27. Irizarry RA, Bolstad BM, Collin F, Cope LM, Hobbs B, Speed TP (2003) Summaries of Affymetrix GeneChip probe level data. Nucleic Acids Res 31(4):e15

28. Amaratunga D, Cabrera J, Shkedy Z (2014) Exploration of analysis of DNA microarray and other high-dimensional data. In: Amaratunga D, Cabrera J, Shkedy Z (eds) Exploration of analysis of DNA microarray and other high-dimensional data, Wiley series in probability and statistics. Wiley, Hoboken, NJ

29. Johnson WE, Li C, Rabinovic A (2007) Adjusting batch effects in microarray expres-sion data using empirical Bayes methods. Biostatistics 8(1):118–127

30. Bezman NA, Kim CC, Sun JC, Min-Oo G, Hendricks DW, Kamimura Y, Best JA, Goldrath AW, Lanier LL (2012) Molecular definition of the identity and activation of natural killer cells. Nat Immunol 13(10): 1000–1009

31. Guilliams M, Henri S, Tamoutounour S, Ardouin L, Schwartz-Cornil I, Dalod M, Malissen B (2010) From skin dendritic cells to a simplified classification of human and mouse dendritic cell subsets. Eur J Immunol 40(8):2089–2094

Part V

In Vivo Analysis of Dendritic Cells

Chapter 17

In Vivo Ablation of a Dendritic Cell Subset Expressing the Chemokine Receptor XCR1

Hiroaki Hemmi, Katsuaki Hoshino, and Tsuneyasu Kaisho

Abstract

Dendritic cells (DCs) are one of the key populations controlling immune responses. To establish a cell depletion system in vivo, human diphtheria toxin (DT) receptor (DTR) is transduced to the mice in which DTR is expressed under the control of a specific promoter. In these mice, DTR-expressing cells are inducibly depleted after DT injection. Using this system, analysis of mouse models in which DTR was expressed under the CD11c promoter has contributed to our knowledge of DC biology by depleting CD11c$^+$ cells. Other mouse models to inducibly eliminate specific DC subsets upon DT treatment have been also generated. Here, we describe a new mouse model in which the XCR1$^+$ DC subset is inducibly and transiently depleted in vivo.

Keyword Dendritic cells, Dendritic cell subset, Inducible ablation of dendritic cell subsets, Diphtheria toxin, Diphtheria toxin receptor

1 Introduction

Dendritic cells (DCs) are professional antigen-presenting cells linking innate and adaptive immunity [1]. DCs are heterogeneous and consist of several subsets [2]. For example, in the murine spleen, DCs can be identified as CD11c$^+$ cells, which can be divided into plasmacytoid DCs (pDCs) and conventional DCs (cDCs). cDCs can be further divided into CD8α^+CD11b$^-$ and CD8α^-CD11b$^+$ cDCs. CD103, instead of CD8α^+, is often used to analyze these DC subsets in the lymph nodes or peripheral tissues. These DC subsets exhibit subset-specific functions. pDCs are specialized to produce type I interferon by sensing virus- or host-derived nucleic acids and contribute to protective immune responses against viral infection or to the pathogenesis of certain autoimmune diseases. CD8α^+CD11b$^-$ cDCs and CD103$^+$CD11b$^-$ cDCs have a high ability to uptake dying or apoptotic cells and to cross-present antigens to support the generation of cytotoxic CD8$^+$

Elodie Segura and Nobuyuki Onai (eds.), *Dendritic Cell Protocols*, Methods in Molecular Biology, vol. 1423,
DOI 10.1007/978-1-4939-3606-9_17, © Springer Science+Business Media New York 2016

T cells. This cross presenting activity is important for immune responses against viral infection or tumors.

In order to elucidate the in vivo roles of certain kinds of cells, cell ablation is very useful. In particular, the expression of human diphtheria toxin receptor (DTR) in the mice has often been used [3]. Diphtheria toxin (DT), when taken up by cells, can inhibit protein synthesis and lead to cell death. Human cells are sensitive to DT, whereas murine cells are resistant to DT, because human DTR, also called as heparin-binding EGF-like growth factor (HB-EGF), binds DT with more than 10^3-fold higher affinity than murine HB-EGF. Several gene-manipulated mice have been established, in which expression of human DTR is driven by gene promoters specifically expressed in all DCs or in DC subsets (Table 1) [4–15, 18]. If DT is injected in these mice, DTR-expressing cells are temporally eliminated, while the other murine cells remain. Analysis of these mice has uncovered in vivo functions of DCs and DC subsets.

We have generated the XCR1-DTRvenus mice by knock in of a gene coding for a fusion protein consisting of a human DTR and the fluorescent protein venus (DTRvenus) into the XCR1 gene locus (see Fig. 1, and see ref. 9). The fusion protein gene is expressed

Table 1
Conditional ablation of DCs and DC subsets

	Cells ablated	Tracing of cells by fluorescence*	References
CD11c-DTR	CD11c^{high} DC (CD8α/CD103$^+$CD11b$^-$ and CD8α/CD103$^-$CD11b$^+$), metallophilic and marginal zone macrophages, activated CD8 T cells	Possible	[4–6]
CD11c.DOG	CD11c^{high} DC (CD8α/CD103$^+$CD11b$^-$ and CD8α/CD103$^-$CD11b$^+$), activated T cells?	Impossible	[7]
zDC(Zbtb46)-DTR	cDC (CD8α/CD103$^+$CD11b$^-$ and CD8α/CD103$^-$CD11b$^+$), activated monocytes?	Impossible	[8]
XCR1-DTR	CD8α/CD103$^+$CD11b$^-$ cDC (XCR1$^+$)	Possible	[9]
Langerin-DTR	CD8α/CD103$^+$CD11b$^-$ cDC (Langerin$^+$), Langerhans cells	Possible	[10, 11]
CD205-DTR	CD8α/CD103$^+$CD11b$^-$ cDC (CD205$^+$), a part of CD8α/CD103$^-$CD11b$^+$ cDC, radioresistant cells (cortical thymic epithelial cells and Langerhans cells)	Impossible	[12]
Clec9a-DTR	CD8α/CD103$^+$CD11b$^-$ cDC (CD205$^+$), pDC	Impossible	[13]
BDCA2-DTR	pDC	Impossible	[14]
SiglecH-DTR	pDC	Impossible	[15]

*CD11c-YFP, Zbtb46-GFP, and XCR1-venus mice can be used for tracing of CD11c$^+$, Zbtb46$^+$, and XCR1$^+$ cells, respectively, although not for cell ablation [9, 16, 17].

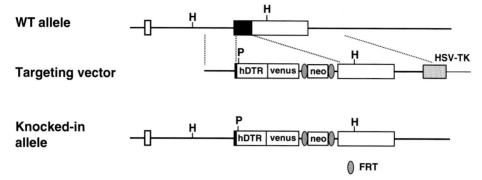

Fig. 1 Generation of XCR1-DTRvenus mice. Schematic diagrams of the mouse XCR1-gene WT allele, a targeting vector, and DTRvenus knocked-in allele. *Filled* and *open boxes* denote coding and noncoding exons of the XCR1 gene, respectively. H, *Hinc*II; P, *Pst*I

in the majority of splenic CD8α⁺CD11b⁻ cDCs and in CD103⁺CD11b⁻ cDCs in various lymph nodes and peripheral tissues, such as skin and intestine (*see* Fig. 2, and *see* ref. 9). This mouse model is based on the selective expression of the chemokine receptor XCR1 in those DC subsets [19, 20]. Here we describe how to deplete XCR1⁺ DCs in the XCR1-DTRvenus mice.

2 Materials

2.1 Mice

1. Heterozygote (XCR1⁺ᐟᴰᵀᴿᵛᵉⁿᵘˢ) mice (indicated as XCR1-DTRvenus) (*see* **Note 1**).

2. Control mice: littermates or C57BL/6J mice.

2.2 Reagents

1. Diphtheria toxin (DT) (Sigma-Aldrich): reconstitute at 1 mg/mL (according to the manufacturer's instruction, reconstitute with distilled water) and aliquot (*see* **Note 2**). Avoid repeated freezing and thawing. Dilute with PBS before injection.

2. PBS.

3. FACS buffer: PBS, 2 % FCS (*see* **Note 3**).

4. Antibodies: Fc block (anti-CD16/32 antibody), fluorochrome-conjugated antibodies against mouse MHC class II (M5/114.15.2), CD8α (53–6.7), CD11b (M1/70), CD11c (N418), and CD103 (M290).

5. Collagenase D (Roche): prepare a 10× stock solution of 4000 Mandl unit/mL in Hanks' balanced salt solution (HBSS) (*see* **Note 4**). To prepare a working solution (400 unit/mL), dilute the stock solution with HBSS or RPMI1640.

6. 0.5 M EDTA (pH 8.0).

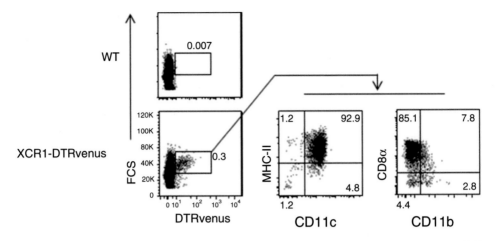

Fig. 2 Expression of venus in XCR1-DTRvenus mice. Single-cell suspensions from the spleens of WT and XCR1-DTRvenus mice were prepared as described in the Method section, stained with the indicated antibodies, and analyzed by a flow cytometer. Majority of venus-expressing cells were MHC class II (MHC-II)$^+$CD11c$^+$ and CD8α$^+$CD11b$^-$ cells

3   Materials

1. Precision weighing scale.
2. Syringe and needle.
3. Forceps and surgery scissors.
4. Petri dish.
5. 15 mL tubes.
6. 70–100 μm mesh (or cell strainers).

4   Methods

4.1   Ablation of XCR1$^+$ Cells in XCR1-DTRvenus Mice

1. Measure the body weight (BW) of each mouse.
2. Inject mice i.p. with DT (25 ng/g BW). For control mice, inject DT to WT mice or inject XCR1-DTRvenus mice with PBS.

4.2   Evaluation of Deletion Efficiency

1. Sacrifice mice and take out spleen and other lymphoid tissues of interest. Place them in a Petri dish.
2. Perfuse the spleen with 5 mL of collagenase D working solution using syringe and needle, mince it, and incubate at 37 °C for 25 min (*see* **Note 5**).
3. Add 0.5 M EDTA (1/10 volume of collagenase D solution), pipet the cells up and down, and incubate them at 37 °C for a further 5 min.

Splenic MHC-II+CD11c+ cells

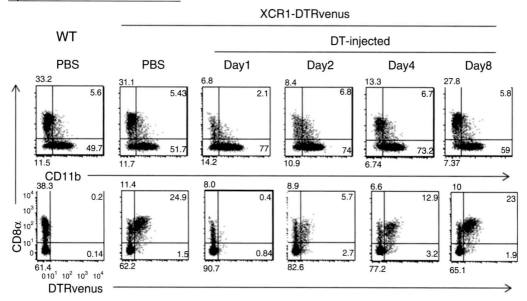

Fig. 3 Ablation of venus-expressing cells in XCR1-DTRvenus mice. XCR1-DTRvenus mice were injected with DT at day 0, and splenocytes were analyzed on CD8α, CD11b, and venus in MHC class II (MHC-II)+CD11c+ cells by a flow cytometer at the indicated days. CD8α+CD11b− cells, i.e., venus-expressing cells, were deleted on day 1 and day 2 after DT injection. These cells begin to recover at day 4 or later

4. Collect the cells into a 15 mL tube and pass them through a 70–100 μm mesh to remove debris.

5. Wash the cells with cold PBS or FACS buffer, and count cell numbers.

6. Add anti-CD16/32 antibody to block the binding of antibodies to Fc receptor, and stain the cells with anti-CD8α, anti-MHC class II, and anti-CD11c antibodies for 20–30 min (*see* **Note 6**).

7. Wash the cells with FACS buffer.

8. Analyze the cells using a flow cytometer. Kinetics of cell ablation is shown in Fig. 3 (*see* **Note 7**).

5 Notes

1. XCR1-DTRvenus mice are available upon request.

2. Diphtheria toxin is harmful to human (minimum lethal dose <100 ng/kg). Handle with care and avoid needlestick injury.

3. Optional: 0.05 % NaN$_3$ might be added as a preservative.

4. We usually prepare collagenase D stock solution (10× concentration) and keep it at −30 °C. Avoid repeated freezing and

thawing. Roche supplies collagenase D based on the Wunsch unit. Approx. 0.15 Wunsch units equal 200 Mandl units.

5. To prepare single-cell suspension to analyze DCs, spleen is digested with collagenase D according to a protocol published elsewhere [21] with small modifications.

6. Venus is detected in the same channel as FITC. So FITC or equivalent fluorochrome-labeled antibodies cannot be used for the staining.

7. If cell ablation for longer time is required, DT can be injected repeatedly every 3 days. We have not formally tested how neutralizing antibodies against DT are generated.

Acknowledgments

This work was supported by the Kishimoto Foundation; Grant-in-Aid for Scientific Research from Japan Society for the Promotion of Science; Grant-in-Aid for Scientific Research on Innovative Areas from the Ministry of Education, Culture, Sports, Science and Technology; and the Uehara Memorial Foundation.

References

1. Steinman RM (2012) Decisions about dendritic cells: past, present, and future. Annu Rev Immunol 30:1–22. doi:10.1146/annurev-immunol-100311-102839

2. Mildner A, Jung S (2014) Development and function of dendritic cell subsets. Immunity 40:642–656. doi:10.1016/j.immuni.2014.04.016

3. Saito M, Iwawaki T, Taya C, Yonekawa H, Noda M, Inui Y, Mekada E, Kimata Y, Tsuru A, Kohno K (2001) Diphtheria toxin receptor-mediated conditional and targeted cell ablation in transgenic mice. Nat Biotechnol 19:746–750

4. Jung S, Unutmaz D, Wong P, Sano G, De los Santos K, Sparwasser T, Wu S, Vuthoori S, Ko K, Zavala F, Pamer EG, Littman DR, Lang RA (2002) In vivo depletion of CD11c + dendritic cells abrogates priming of CD8+ T cells by exogenous cell-associated antigens. Immunity 17:211–220

5. Probst HC, Tschannen K, Odermatt B, Schwendener R, Zinkernagel RM, Van Den Broek M (2005) Histological analysis of CD11c-DTR/GFP mice after in vivo depletion of dendritic cells. Clin Exp Immunol 141:398–404

6. Sathaliyawala T, O'Gorman WE, Greter M, Bogunovic M, Konjufca V, Hou ZE, Nolan GP, Miller MJ, Merad M, Reizis B (2010) Mammalian target of rapamycin controls dendritic cell development downstream of Flt3 ligand signaling. Immunity 33:597–606

7. Hochweller K, Striegler J, Hämmerling GJ, Garbi N (2008) A novel CD11c.DTR transgenic mouse for depletion of dendritic cells reveals their requirement for homeostatic proliferation of natural killer cells. Eur J Immunol 38:2776–2783. doi:10.1002/eji.200838659

8. Meredith MM, Liu K, Darrasse-Jeze G, Kamphorst AO, Schreiber HA, Guermonprez P, Idoyaga J, Cheong C, Yao KH, Niec RE, Nussenzweig MC (2012) Expression of the zinc finger transcription factor zDC (Zbtb46, Btbd4) defines the classical dendritic cell lineage. J Exp Med 209:1153–1165. doi:10.1084/jem.20112675

9. Yamazaki C, Sugiyama M, Ohta T, Hemmi H, Hamada E, Sasaki I, Fukuda Y, Yano T, Nobuoka M, Hirashima T, Iizuka A, Sato K, Tanaka T, Hoshino K, Kaisho T (2013) Critical roles of a dendritic cell subset expressing a chemokine receptor, XCR1. J Immunol 190:6071–6082. doi:10.4049/jimmunol.1202798

10. Kissenpfennig A, Henri S, Dubois B, Laplace-Builhé C, Perrin P, Romani N, Tripp CH, Douillard P, Leserman L, Kaiserlian D, Saeland S, Davoust J, Malissen B (2005) Dynamics and function of Langerhans cells in vivo: dermal dendritic cells colonize lymph node areas distinct from slower migrating Langerhans cells. Immunity 22:643–654

11. Bennett CL, van Rijn E, Jung S, Inaba K, Steinman RM, Kapsenberg ML, Clausen BE (2005) Inducible ablation of mouse Langerhans cells diminishes but fails to abrogate contact hypersensitivity. J Cell Biol 169:569–576

12. Fukaya T, Murakami R, Takagi H, Sato K, Sato Y, Otsuka H, Ohno M, Hijikata A, Ohara O, Hikida M, Malissen B, Sato K (2012) Conditional ablation of CD205+ conventional dendritic cells impacts the regulation of T-cell immunity and homeostasis in vivo. Proc Natl Acad Sci U S A 109:11288–11293. doi:10.1073/pnas.1202208109

13. Piva L, Tetlak P, Claser C, Karjalainen K, Renia L, Ruedl C (2012) Cutting edge: Clec9A+ dendritic cells mediate the development of experimental cerebral malaria. J Immunol 189:1128–1132. doi:10.4049/jimmunol.1201171

14. Swiecki M, Gilfillan S, Vermi W, Wang Y, Colonna M (2010) Plasmacytoid dendritic cell ablation impacts early interferon responses and antiviral NK and CD8(+) T cell accrual. Immunity 33:955–966. doi:10.1016/j.immuni.2010.11.020

15. Takagi H, Fukaya T, Eizumi K, Sato Y, Sato K, Shibazaki A, Otsuka H, Hijikata A, Watanabe T, Ohara O, Kaisho T, Malissen B, Sato K (2011) Plasmacytoid dendritic cells are crucial for the initiation of inflammation and T cell immunity in vivo. Immunity 35:958–971. doi:10.1016/j.immuni.2011.10.014

16. Lindquist RL, Shakhar G, Dudziak D, Wardemann H, Eisenreich T, Dustin ML, Nussenzweig MC (2004) Visualizing dendritic cell networks in vivo. Nat Immunol 5:1243–1250

17. Satpathy AT, KC W, Albring JC, Edelson BT, Kretzer NM, Bhattacharya D, Murphy TL, Murphy KM (2012) Zbtb46 expression distinguishes classical dendritic cells and their committed progenitors from other immune lineages. J Exp Med 209:1135–1152. doi:10.1084/jem.20120030

18. van Blijswijk J, Schraml BU, Reis e Sousa C (2013) Advantages and limitations of mouse models to deplete dendritic cells. Eur J Immunol 43:22–26. doi:10.1002/eji.201243022

19. Yamazaki C, Miyamoto R, Hoshino K, Fukuda Y, Sasaki I, Saito M, Ishiguchi H, Yano T, Sugiyama T, Hemmi H, Tanaka T, Hamada E, Hirashima T, Amakawa R, Fukuhara S, Nomura S, Ito T, Kaisho T (2010) Conservation of a chemokine system, XCR1 and its ligand, XCL1, between human and mice. Biochem Biophys Res Commun 397:756–761. doi:10.1016/j.bbrc.2010.06.029

20. Dorner BG, Dorner MB, Zhou X, Opitz C, Mora A, Güttler S, Hutloff A, Mages HW, Ranke K, Schaefer M, Jack RS, Henn V, Kroczek RA (2009) Selective expression of the chemokine receptor XCR1 on cross-presenting dendritic cells determines cooperation with CD8+ T cells. Immunity 31:823–833. doi:10.1016/j.immuni.2009.08.027

21. Inaba K, Swiggard WJ, Steinman RM, Romani N, Schuler G, Brinster C (2009) Isolation of dendritic cells. Curr Protoc Immunol Chapter 3:Unit 3.7. doi: 10.1002/0471142735.im0307s86

Chapter 18

In Vivo Analysis of Intestinal Mononuclear Phagocytes

Caterina Curato, Biana Bernshtein, Tegest Aychek, and Steffen Jung

Abstract

The study of the intestinal dendritic cell (DC) compartment, its homeostasis, regulation, and response to challenges calls for the investigation within the physiological tissue context comprising the unique anatomic constellation of the epithelial single cell layer and the luminal microbiota, as well as neighboring immune and nonimmune cells. Here we provide protocols we developed that use a combination of conditional cell ablation, conditional compartment mutagenesis, and adoptive precursor transfers to study DC and other intestinal mononuclear phagocytes in in vivo context. We will highlight pitfalls and strengths of these approaches.

Key words Mononuclear phagocytes, Dendritic cells, Intestine, Cell ablation, Conditional mutagenesis

1 Introduction

The study of the intestinal DC compartment has significantly contributed to our understanding of non-lymphoid organ DC. It has largely been focused on the small and large intestine, e.g., the ileum and colon, respectively. Murine colonic and ileal DC subpopulations are currently defined by expression of the integrins CD11c (αX), CD11b (αM), and CD103 (αEβ7), as well as CD24 [1] (*see* Fig. 1, Table 1). CD103$^+$ and CD11b$^-$ DC, like classical splenic XCR1$^+$ CD8α^+ DC, depend for their development on the transcription factors Irf8, Id2, and Batf3 [16]. These cells migrate to the mesenteric lymph nodes to prime and polarize naïve T cells and are, as their splenic counterpart, specialized in cross-presentation [4]. Ileal CD103$^+$ and CD11b$^+$ DC were shown to be required for efficient generation of Th17 cells [6–9]. In the colon, CD103$^+$ and CD11b$^+$ DC are rare but might be functionally replaced by a population of CD11b$^+$ DC that lack CD103 expression [11]. The ontogeny of CD103$^+$ and CD103$^-$ CD11b$^+$ DC remains less well defined, as these cells are likely heterogeneous and comprise aside from preDC-derived DC also a varying fraction

Elodie Segura and Nobuyuki Onai (eds.), *Dendritic Cell Protocols*, Methods in Molecular Biology, vol. 1423,
DOI 10.1007/978-1-4939-3606-9_18, © Springer Science+Business Media New York 2016

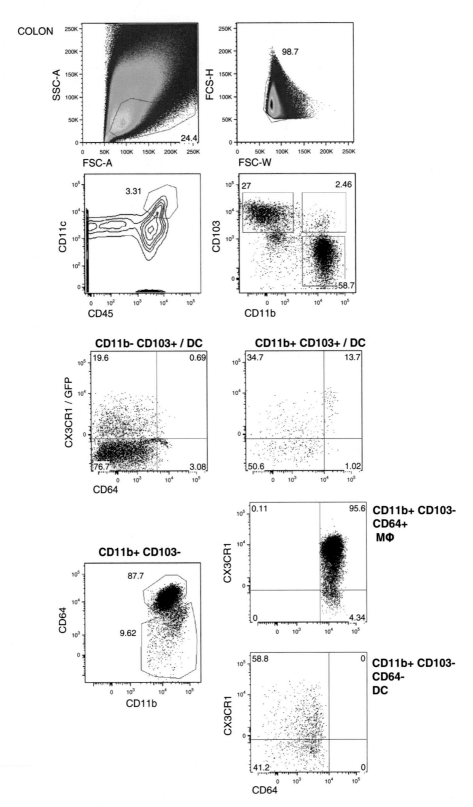

Fig. 1 The intestinal mononuclear phagocyte compartment. Flow cytometric analysis of mononuclear phagocytes in the colon of a CX3CR1^{GFP} mouse. Intestinal DC is currently defined by the expression of the integrins CD11c, CD11b, and CD103 (as well as CD24, not shown). Intestinal macrophages are best defined as CD11b$^+$, CD64$^+$, and CX$_3$CR1^{hi} cells

Table 1
Macrophages and DC subsets described in the mouse intestinal tissue

Intestinal mononuclear phagocyte		Growth/transcription/ environmental factor dependence	Lymphoid organ DC equivalent	Functional specialization	Selected references
DC	CD103+ CD11b−	Flt3L Irf8, Id2, Batf3	XCR1+ CD8α+	Cross-presentation	Edelson et al. [2] Ginhoux et al. [3] Cerovic et al. [4]
	CD103+ CD11b+	Csf-2 (GM-CSF) Retinoic acid (SI) Flt3L (partially) Irf4 Notch2		Priming of TH17 cells	Bogunovic et al. [5] Lewis et al. [6] Welty et al. [7] Schlitzer et al. [8] Persson et al. [9] Klebanoff et al. [10] Cerovic et al. [4]
	CD103− CD11b+	Flt3L		Priming of IL-17 and INFγ- producing T cells	Bogunovic et al. [5] Cerovic et al. [11] Scott et al. [12]
	CD103− CD11b−	Ftl3L		Priming of TH17 (in vitro)	Cerovic et al. [11]
Macrophages	CD64+ CX₃CR1+	Csf-1 (M-CSF)			Varol et al. [13] Bogunovic et al. [5]
Monocytes	CD64+ Ly6C+			Recent monocyte infiltrates	Guilliams et al. [14] Zigmond et al. [15]

of monocyte-derived cells [17]. Different intestinal DC subsets might play distinct functional roles during oral tolerance and in driving immune response after vaccination or intestinal stimulation with Toll-like receptor ligands [1, 16].

As DC in other organs, also intestinal DC are embedded in tight cellular networks, including close relatives, the tissue-resident macrophages [13]. Intestinal macrophages are currently best defined as CD11b+ cells that express the Fc-gamma receptor 1 (FcγRI) CD64 and the chemokine receptor CX_3CR1 [18, 19]. Indeed the latter two markers have been instrumental in highlighting the monocytic origin of these cells, a feature that distinguishes adult intestinal macrophages from most other tissue macrophages that are embryonic derived [20]. Shortly after birth, embryo-derived intestinal macrophages are replaced by monocyte-derived cells that are conditioned in the healthy gut to become noninflammatory [15, 21]. Intestinal macrophages are in steady state nonmigratory and critically involved in the maintenance of gut homeostasis [22]. Under dysbiosis and chronic inflammation, these cells can however be mobilized to the mesenteric lymph nodes [23, 24]. As tissue-resident immune sentinels, mononuclear phagocytes are key players in the establishment and maintenance of immune homeostasis and the orchestration of innate and adaptive immune defense. Specific activities of the intestinal DC subsets and macrophages and, in particular, the likely important cross talk between these mononuclear phagocytes remain poorly understood.

Murine cells are resistant to diphtheria toxin (DT), since they lack a high-affinity receptor for this bacteria exotoxin [25]. These cells can however be rendered DT sensitive by transgenic expression of a primate diphtheria toxin receptor (DTR). By directing DTR expression to distinct cell types using defined promoter elements, this approach can be used for conditional ablation of specific cells [26, 27]. We have pioneered this strategy using CD11c-DTR mice that harbor the DTR transgene under control of the *itgax* promoter thus targeting CD11c-expressing mononuclear phagocytes, including DC and selected macrophage populations [28]. Here we provide the basic protocols for the use of the system, but would also like to refer to additional and more detailed references on the use of the CD11c-DTR mice highlighting their strengths and pitfalls [29]. In addition, we will elaborate here on further applications of the system, beyond the mere cell ablation. Specifically, we will focus on protocols that highlight its value in understanding the in vivo biology of DC subsets in the broader context of intestinal immunity, which involves interactions with surrounding cells, such as tissue-resident macrophages.

Conditional mutagenesis of the intestinal mononuclear phagocyte compartment. Below we will outline how the DT-based conditional cell ablation strategy can be exploited to perform conditional compartment mutagenesis. As an example, we illustrate our recent definition of the cellular source of an essential cytokine (e.g., IL-23) during intestinal inflammation. Hematopoietic IL-23 is a critical factor in innate immune responses [30]. To define the cell type that produces IL-23 in the infected colon, we employed a mixed BM chimera approach. Lethally irradiated wild-type (WT) mice were reconstituted with equal mixtures of IL-23-deficient BM and BM from CD11c-DTR mice. The resulting mixed [*mutant*/CD11c-DTR>WT] chimeras harbor two types of CD11c-expressing cells: mutant, DT-resistant and DT-sensitive, and DTR transgenic cells. Upon DT injection, the mice are left with only mutant CD11c⁺ mononuclear phagocytes, whereas other cellular compartments, such as lymphocytes, are left with mutant and WT cells. Ablation of the CD11c-DTR transgenic cells can be confirmed by flow cytometric analysis using respective CD45 markers for grafts and recipients (*see* Fig. 2). In our experiment, we found that DT-treated [*il23a*$^{-/-}$/CD11c-DTR>WT] chimeras succumbed to infection and thus established, in line with other studies [30, 31], the critical role of mononuclear phagocytes, as IL-23 source [32]. By using animals with more restricted expression of the DTR transgene in DC subsets or macrophages, this approach can be refined. In our study,

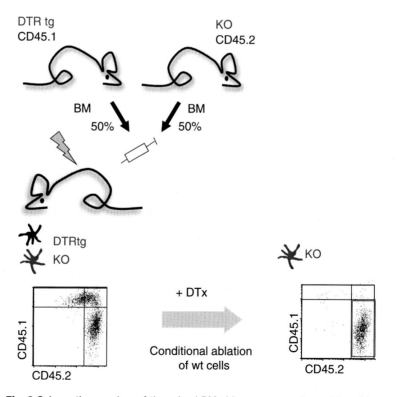

Fig. 2 Schematic overview of the mixed BM chimera approach used to achieve conditional compartment mutagenesis. Note that non-DTR-expressing cell compartments remain composes half of WT cells

for instance, we introduced a system that targets CX₃CR1⁺ cells and could thereby reveal a critical additional role of IL-23 in the cross talk between intestinal macrophages and DC (Aychek et al., *in press*).

In vivo *reconstitution of intestinal mononuclear phagocytes.* Upon ablation of CD11c^{hi} cells in DT-treated [DTR>WT] BM chimeras, these cells can be reconstituted by engraftment of the DT-treated animals with defined non-DTR transgenic precursors. We and others have used this approach in the past to define the origin of intestinal DC and macrophages [5, 13, 33]. Specifically, preDC were shown to reconstitute CD103⁺ DC; CX₃CR1⁺ macrophages on the other hand originated from adoptively transferred Ly6C⁺, but not Ly6C⁻ monocytes, that underwent clonal expansion to occupy the niches left free by the ablation [13]. Ly6C^{hi} monocytes express CX₃CR1 and their migration and differentiation process in the intestine can hence be readily monitored by using CX₃CR1$^{GFP/+}$ monocyte donors [34]. Grafted cell populations can be retrieved and analyzed flow cytometric and histologically in the in vivo context. Moreover, by engraftment with a mix of mutant and WT monocytes, this approach allows to study the impact of specific mutations on the function of monocyte-derived cells along their WT counterparts in the tissue of the same recipients (*see* Fig. 3).

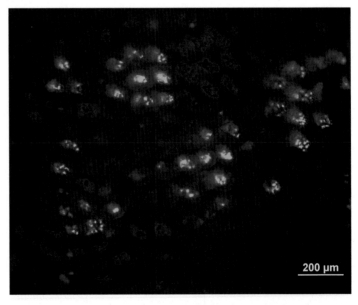

Fig. 3 In vivo reconstitution of intestinal mononuclear phagocytes. Histological analysis of intestine engrafted with GFP-expressing WT monocytes (*green*) and RFP-expressing mutant monocytes (*red*). Upon graft, GFP⁺ WT and RFP⁺ mutant monocytes equally colonize the intestinal tissue and clonally expand (*see* also Varol et al. 2009)

Collectively these examples illustrate the available tools that enable targeting of specific cell types in a given tissue context to probe functions of the mononuclear phagocytes during intestinal homeostasis and inflammation. Below we provide basic protocols that outline the use of these systems highlighting the strengths and limitations of the model.

2 Materials

2.1 Animals

Our approaches involve the use of the following mouse strains (*see* **Note 1**):

1. CD11c-DTR transgenic mice (B6.FVB-Tg (Itgax-DTR/GFP)57Lan/J, Jackson laboratory, stock number 004509) that carry a human diphtheria toxin receptor (DTR) – GFP transgene under control of the murine CD11c promoter [28] (*see* **Note 2**).

2. CX$_3$CR1^{GFP} mutant mice (B6.129P- Cx3cr1tm1Litt/J, Jackson laboratory, stock number 005582) harboring a targeted replacement of the cx3cr1 gene by an enhanced green fluorescent protein (EGFP) reporter gene [34] (*see* **Note 3**). CX$_3$CR1^{GFP} donor mice have been crossed to mice bearing the CD45.1 allotype (B6.SJL- Ptprca Pepcb/BoyJ, Jackson laboratory, stock number 002014).

2.2 Reagents

1. Diphtheria toxin (DT; Sigma-Aldrich, Germany): Dissolve DT in PBS at a concentration of 1 mg/mL and store in 10 µL single-use aliquots at –80 °C. Dilute aliquots in PBS prior to use.

2. Collagenase type VIII (Sigma-Aldrich): 1 mg/mL.

3. PBS.

4. PBS without Ca^{2+} and Mg^{2+} (PBS$^{-/-}$).

5. DTT (Sigma-Aldrich): 1 mM DTT in PBS$^{-/-}$ to a final concentration of 0.1543 mg/mL.

6. FACS buffer: PBS, 1 % fetal calf serum (FCS), 2 mM Na$_2$EDTA, and 0.05 % sodium azide.

7. Fluorochrome-conjugated antibodies: CD11c (clone N418), CD11b (clone M1/70), CD45.1 (clone A20), CD45.2 (clone 104), CD103 (clone 2E7), CD64 (clone X54-5/7.1), and Ly6C (clone HK1.4), store at 4 °C. Dilute in FACS buffer.

8. FACS-blocking reagents: anti-mouse CD16/32-FC (Fc) block (clone 93), store at 4 °C.

9. Magnetic cell sorting (MACS) buffer: PBS$^{-/-}$, 1% FCS, and 2 mM Na$_2$EDTA, filter sterile and degas (*see* **Note 4**).

10. Predigestion solution: Hank's balanced salt solution (HBSS), 2 mM EDTA, 1 mM DTT, and 5 % FCS.

11. Digestion solution: PBS, 1.5 mg/mL collagenase type VIII, and 5 % FCS.

2.3 Materials

1. Weighing scale.

2. Syringes (1 and 3 mL) and needles (27 and 21 G).

3. Dissection tools.

4. Irradiator.

5. Cell sorter.

6. 15 and 50 mL tubes.

7. 100 μm mesh strainers.

8. Tube rotator.

3 Methods

3.1 In Vivo Ablation of Intestinal Mononuclear Phagocytes

Conditional ablation of CD11c^{high} cells is achieved using a DT challenge of DTR transgenic mice [28] (*see* **Note 5**). After entry into the cytoplasm, the DT A subunit inactivates elongation factor 2 by catalyzing its ADP-ribosylation. The resulting blockade of protein synthesis triggers programmed cell death, with dying cells being silently removed by phagocytes. Notably, CD11c-DTR mice as such do not allow for prolonged DC ablation due to the fact that repetitive DT application results in adverse side effects [5, 34]. If the experimental setup allows, this problem can be overcome by use of mixed bone marrow (BM) chimeras generated by reconstitution of lethally irradiated wild-type mice with CD11c-DTR tg BM [35, 36] (*see* Subheading 3.1.2. below). An alternative way to circumvent DT-induced lethality is the development of protocols that aim at local ablation by restricted toxin delivery, as exemplified by the ablation of alveolar macrophages through intratracheal installation of DT [37, 38] and the ablation of DC from skin-draining LNs [39].

3.1.1 Conditional Short-Term Ablation of CD11c^{high} DC

1. Weigh each mouse.

2. For systemic short-term depletion of CD11c^{high} DC, inject mice once intraperitoneally (IP) with 4 ng DT/g body weight.

3. Use mice between 8 h and 2 days after IP injection (*see* **Note 6**).

3.1.2 Conditional Long-Term Ablation of CD11c^{high} DC

In contrast to CD11c-DTR transgenic mice, syngeneic [DTR > wt] BM chimeras, generated by reconstitution of lethally irradiated wt recipient mice with CD11c-DTR BM, can be treated with DT for prolonged periods of time without adverse side effects [35, 36].

1. Irradiate C57BL/6 wt recipient mice with a single lethal total body irradiation of 950 rad.

2. One day after, prepare BM from CD11c-DTR transgenic mice.

3. Isolate femora and tibiae from donor mice.

4. Remove the surrounding soft tissue to expose bone surface.

5. Open bone at both ends.

6. Using a 27 G needle fitted to a 1 mL syringe filled with 1 mL of cold PBS$^{-/-}$, flush out the marrow out of the bone cavity into a 15 mL tube. Repeat this step several times until the bone is empty and becomes white.

7. Gently resuspend BM cells using a 3 mL syringe fitted with a 21 G needle until the suspension is homogenous.

8. Wash the cells with 10 mL of PBS$^{-/-}$ by centrifugation at 350×*g* at 4 °C for 7 min.

9. Following their isolation, resuspend BM cells in PBS$^{-/-}$ and count them.

10. Inject 5×10^6 cells into each recipient mouse intravenously (IV) (*see* **Note 7**).

11. Allow chimeric mice to rest for 8 weeks before use (*see* **Note 8**).

12. Weigh each mouse.

13. For systemic long-term DC depletion, inject [DTR>wt] BM chimeras every other day i.p. with 8 ng DT/g body weight.

3.1.3 Conditional Long-Term Ablation of Mutated DC Compartments

The generation of mixed BM chimeras reconstituted with mutant BM and CD11c-DTR transgenic BM can be a powerful means to investigate molecular contributions of DC [40]. Upon DT injection, these chimeras are left only with mutant mononuclear phagocytes, while other cell compartments are left with wt and mutant cells (*see* Fig. 2). For instance, a mixed BM chimera approach was developed to distinguish the cellular source of essential factors (e.g., IL-23) in the context of intestinal inflammation (Aychek et al., *in press*).

1. Generate BM chimeras as described in Subheading 3.1.2 (**steps 1–10**), using wt recipient mice and reconstituting them with a 50:50 mixture of BM from CD11c-DTR transgenic mice and BM from mutant mice of interest.

2. Allow chimeric mice to rest for 8 weeks before use (*see* **Note 8**).

3. Weigh each mouse.

4. For systemic long-term mononuclear phagocyte depletion, inject BM chimeras every other day i.p. with 8 ng DT/g body weight.

3.2 Adoptive Precursor Transfers into Mononuclear Phagocyte-Depleted Mice

To study the in vivo origins of mononuclear phagocytes, recipient mice are engrafted with $CX_3CR1^{GFP/+}$ mononuclear phagocyte precursors, e.g., MDP or monocytes. CX_3CR1 is expressed by MDP [41], as well as BM and blood monocytes [34, 42], which are therefore green fluorescent in $CX_3CR1^{GFP/+}$ mice. The long half-life of the GFP label allows the detection of the adoptively transferred GFP-expressing precursors and their descendants in recipient mice, even if the progeny ceased to express CX_3CR1/GFP [33, 43] (*see* **Note 9**).

1. Generate [DTR > wt] BM chimeras as described in Subheading 3.1.2 (**steps 1–10**), using wt recipient mice and reconstituting them with BM from CD11c-DTR transgenic mice.

2. Allow chimeric mice to rest for 8 weeks before use (*see* **Note 8**).

3. Weigh each mouse.

4. Inject chimeric mice i.p. with a double dose of DT (16 ng DT/g body weight) on days 2 and 1 prior graft.

5. Isolate total BM cells from CD45.1+ $CX_3CR1^{GFP/+}$ mice, as described in Subheading 3.1.2 (**steps 3–9**).

6. Label BM cells with anti-CD115, anti-CD117, anti-Ly6C, and anti-CD11b antibodies (*see* **Note 10**).

7. Sort CD115+ CD117- Ly6C^{hi} GFP^{int} CD11b+ cells into a new 15 mL tube using a cell sorter (*see* **Note 11**).

8. Transfer 2×10^6 CD115+ CD117- Ly6C^{hi} GFP^{int} CD11b+ monocytes i.v. into each DT-treated [DTR > wt] BM chimeric mouse.

9. Inject mice every other day i.p. with 8 ng DT/g body weight till further analysis for a maximum of 14 days (*see* Fig. 3).

3.3 Isolation of Intestinal Lamina Propria Mononuclear Phagocytes

1. Sacrifice mice and isolate the intestine (*see* **Note 12**).

2. Remove mesenteric tissue (lymph nodes) and fat.

3. Flush the intestine with PBS$^{-/-}$ to remove all its fecal content.

4. Open the intestine longitudinally and cut into 0.5 cm pieces. Place the pieces in a 50 mL tube.

5. Incubate the pieces in 5 mL of predigestion solution for 30 min at 37 °C under rotation (250 rpm) (*see* **Note 13**).

6. At the end of the incubation, vortex the solution for 10s and filter through a 100 μm mesh strainer. The filtered cell suspension contains the intestinal epithelial cellular fraction.

7. Transfer remaining pieces of undigested intestinal tissue into a 50 mL tube and add 5 mL of digestion solution.

8. Incubate for 40 min at 37 °C under rotation (250 rpm) in a thermal incubator.

9. After incubation, vortex the cell solution intensively for 30s and pass through a 100 μm mesh strainer into a 15 mL tube.

10. Wash the tube with PBS$^{-/-}$ in order to collect all released cells.

11. Centrifuge the cells at $350 \times g$ for 7 min at 4 °C.

12. Proceed with cell analysis.

---

4   Notes

1. All mice are maintained under specific pathogen-free (SPF) conditions and handled under institutional protocols according to international guidelines.

2. CD11c-DTR transgenic mice are generally used as heterozygotes. Note that homozygote CD11c-DTR transgenic mice are more sensitive to DT. For the generation of BM chimeras, it is recommended to use CD11c-DTR homozygous mice. Offspring of intercrosses of heterozygote CD11c-DTR mice with wt mice is screened by PCR on tail DNA using the following primers (DTR1, 5′-GCCACCATGAAGCTGCTGCCG-3′; DTR2, 5′- CGGGTGGGAATTAGTCATGCC-3′). The PCR protocol: 94 °C 4 min [35 cycles: 30 s 95 °C, 1 min 58.5 °C, 30 s 72 °C] and 15 s 72 °C, holds 4 °C yielding a product of 625 base pairs. Tail DNA is prepared as follows: Tails are lysed in Eppendorf tube in 500 μL lysis buffer (100 mM Tris–HCl pH 8.5, 5 mM EDTA, 0.2 % SDS, 200 mM NaCl, 100 μg/mL proteinase K) by agitated overnight incubation at 55 °C; samples are spun at 13,000 rpm ($375g$) for 10 min to obtain firm pellet; supernatant is removed and transferred to fresh Eppendorf tube and supplemented with 500 μL isopropanol; samples are mixed until viscosity is gone; DNA is recovered as aggregated precipitate (using pipette tips) and transferred to a tube containing 400 μL of TE; and samples are incubated at 37 or 55 °C, until fully dissolved.

3. CX$_3$CR1 is the receptor of the membrane-tethered chemokine CX$_3$CL1 (fractalkine (Fkn)) [44]. In the transfers we generally use heterozygote mutant mice (CX$_3$CR1$^{GFP/+}$) that show a reduced surface receptor expression but exhibit no overt phenotype [34].

4. Do not add sodium azide.

5. Notably, other cells except DC, which make use of the *itgax* promoter are deleted in CD11c-DTR transgenic mice upon administration of DT. These include alveolar macrophages [37, 38], splenic metallophilic and marginal zone macrophages [45, 46], activated CD8$^+$ T cells (unpublished observation), a subsets of pDC [39], and subsets of NK cells [47]. Cells, which are sensitive to DT, can be detected by flow cytometric or histological analysis for expression of the DTR-GFP fusion protein [39]. On the other hand, the 5.5 kb *itgax* promoter fragment [48] is not active in all CD11c-expressing cells,

including monocyte-derived splenic CD11c^{int} cells [33], CD11c^{int} pDC [39], and most NK cells [47], which transcribe their endogenous *Itgax* alleles but are resistant to DT [39].

6. 8 h after IP injection, toxin-induced depletion of CD11c^{high} DC is detected in the spleen, intestine, BM, and LNs. cDC depletion persists for 2 days, after which cDC numbers are gradually restored [28].

7. We inject cells into recipient mice via the tail vein in a volume of 200 μL (volumes lower than 200 μL can also be injected). Prior to injection, recipient mice are moderately heated with an infrared lamp in order to promote vasodilatation to ease tail vein injection.

8. Irradiated mice receive transiently (for 1 week) an antibiotic (Ciproxin 0.2%) in their drinking water (20 μg/mL) to prevent bacterial infections during the period of immunodeficiency. The water bottle should be either dark or covered with aluminum foil to protect the light-sensitive drug.

9. In addition to the GFP label, it is advisable to include allotypic markers that identify the grafted cells, such as CD45. We therefore regularly use CX$_3$CR1$^{GFP/+}$ mice on a CD45.1 background as donors.

10. Antibodies are used according to the manufacturer's protocols. All antibodies should be tested and titrated in advance to determine the best staining conditions, before analysis of samples from recipient mice. The suggested total staining volume is 20 μL for up to 2×10^6 cells and 50 μL for 5×10^6 cells. A staining time of 10–15 min on ice is usually sufficient. When working with less than 10^7 cells, use the indicated volumes. When working with higher cell numbers, scale up all of the reagents and total volumes, accordingly.

11. Cell sorting was performed with FACS Aria III (BD Biosciences). Flow cytometric analyses were acquired with LSR Fortessa (BD Biosciences, United States) and analyzed with FlowJo software (Treestar).

12. By choosing proximal, central, or distal portions of the intestine, the analysis can be focused on the *duodenum, jejunum, ileum,* and *colon,* respectively.

13. This incubation removes the epithelial cell layer [44].

Acknowledgments

We would like to thank the past and present members of the Jung lab for sharing their protocols. This work was supported by the European Research Council (340345).

References

1. Farache J, Zigmond E, Shakhar G, Jung S (2013) Contributions of dendritic cells and macrophages to intestinal homeostasis and immune defense. Immunol Cell Biol 91(3):232–239

2. Edelson BT, KC W, Juang R, Kohyama M, Benoit LA et al (2010) Peripheral CD103+ dendritic cells form a unified subset developmentally related to CD8α+ conventional dendritic cells. J Exp Med 207(4):823–836

3. Ginhoux F, Liu K, Helft J, Bogunovic M, Greter M et al (2009) The origin and development of nonlymphoid tissue CD103+ DCs. J Exp Med 206(13):3115–3130

4. Cerovic V, Houston SA, Westlund J, Utriainen L, Davison ES et al (2015) Lymph-borne CD8α+ dendritic cells are uniquely able to cross-prime CD8+ t cells with antigen acquired from intestinal epithelial cells. Mucosal Immunol 8(1):38–48

5. Bogunovic M, Ginhoux F, Helft J, Shang L, Hashimoto D et al (2009) Origin of the lamina propria dendritic cell network. Immunity 31(3):513–525

6. Lewis KL, Caton ML, Bogunovic M, Greter M, Grajkowska LT et al (2011) Notch2 receptor signaling controls functional differentiation of dendritic cells in the spleen and intestine. Immunity 35(5):780–791

7. Welty NE, Staley C, Ghilardi N, Sadowsky MJ, Igyártó BZ, Kaplan DH (2013) Intestinal lamina propria dendritic cells maintain t cell homeostasis but do not affect commensalism. J Exp Med 210(10):2011–2024

8. Schlitzer A, McGovern N, Teo P, Zelante T, Atarashi K et al (2013) Irf4 transcription factor-dependent CD11b+ dendritic cells in human and mouse control mucosal il-17 cytokine responses. Immunity 38(5):970–983

9. Persson EK, Uronen-Hansson H, Semmrich M, Rivollier A, Hägerbrand K et al (2013) Irf4 transcription-factor-dependent CD103+CD11b+ dendritic cells drive mucosal t helper 17 cell differentiation. Immunity 38(5):958–969

10. Klebanoff CA, Spencer SP, Torabi-Parizi P, Grainger JR, Roychoudhuri R et al (2013) Retinoic acid controls the homeostasis of pre-cDC-derived splenic and intestinal dendritic cells. J Exp Med 210(10):1961–1976

11. Cerovic V, Houston SA, Scott CL, Aumeunier A, Yrlid U et al (2013) Intestinal CD103- dendritic cells migrate in lymph and prime effector t cells. Mucosal Immunol 6(1):104–113

12. Scott CL, Bain CC, Wright PB, Sichien D, Kotarsky K et al (2015) CCR2+CD103- intestinal dendritic cells develop from dc-committed

precursors and induce interleukin-17 production by t cells. Mucosal Immunol 8:327–39

13. Varol C, Vallon-Eberhard A, Elinav E, Aychek T, Shapira Y et al (2009) Intestinal lamina propria dendritic cell subsets have different origin and functions. Immunity 31(3):502–512

14. Guilliams M, Ginhoux F, Jakubzick C, Naik SH, Onai N et al (2014) Dendritic cells, monocytes and macrophages: a unified nomenclature based on ontogeny. Nat Rev Immunol 14(8):571–578

15. Zigmond E, Varol C, Farache J, Elmaliah E, Satpathy AT et al (2012) Ly6chi monocytes in the inflamed colon give rise to proinflammatory effector cells and migratory antigen-presenting cells. Immunity 37(6):1076–1090

16. Mildner A, Jung S (2014) Development and function of dendritic cell subsets. Immunity 40(5):642–656

17. Schreiber HA, Loschko J, Karssemeijer RA, Escolano A, Meredith MM et al (2013) Intestinal monocytes and macrophages are required for t cell polarization in response to citrobacter rodentium. J Exp Med 210(10):2025–2039

18. Niess JH, Brand S, Gu X, Landsman L, Jung S et al (2005) CX3CR1-mediated dendritic cell access to the intestinal lumen and bacterial clearance. Science 307(5707):254–258

19. Tamoutounour S, Henri S, Lelouard H, de Bovis B, de Haar C et al (2012) CD64 distinguishes macrophages from dendritic cells in the gut and reveals the th1-inducing role of mesenteric lymph node macrophages during colitis. Eur J Immunol 42(12):3150–3166

20. Ginhoux F, Jung S (2014) Monocytes and macrophages: developmental pathways and tissue homeostasis. Nat Rev Immunol 14(6):392–404

21. Rivollier A, He J, Kole A, Valatas V, Kelsall BL (2012) Inflammation switches the differentiation program of ly6chi monocytes from antiinflammatory macrophages to inflammatory dendritic cells in the colon. J Exp Med 209(1):139–155

22. Mortha A, Chudnovskiy A, Hashimoto D, Bogunovic M, Spencer SP et al (2014) Microbiota-dependent crosstalk between macrophages and ilc3 promotes intestinal homeostasis. Science 343(6178):1249288

23. Diehl GE, Longman RS, Zhang J-X, Breart B, Galan C et al (2013) Microbiota restricts trafficking of bacteria to mesenteric lymph nodes by CX3CR1hi cells. Nature 494(7435):116–120

24. Zigmond E, Bernshtein B, Friedlander G, Walker CR, Yona S et al (2014) Macrophage-

restricted interleukin-10 receptor deficiency, but not il-10 deficiency, causes severe spontaneous colitis. Immunity 40(5):720–733

25. Pappenheimer AM, Harper AA, Moynihan M, Brockes JP (1982) Diphtheria toxin and related proteins: effect of route of injection on toxicity and the determination of cytotoxicity for various cultured cells. J Infect Dis 145(1):94–102

26. Yamaizumi M, Mekada E, Uchida T, Okada Y (1978) One molecule of diphtheria toxin fragment a introduced into a cell can kill the cell. Cell 15(1):245–250

27. Saito M, Iwawaki T, Taya C, Yonekawa H, Noda M et al (2001) Diphtheria toxin receptor-mediated conditional and targeted cell ablation in transgenic mice. Nat Biotechnol 19(8):746–750

28. Jung S, Unutmaz D, Wong P, Sano G-I, De los Santos K et al (2002) In vivo depletion of CD11c+dendritic cells abrogates priming of CD8+ T cells by exogenous cell-associated antigens. Immunity 17(2):211–220

29. Bar-On L, Jung S (2010) Defining dendritic cells by conditional and constitutive cell ablation. Immunol Rev 234(1):76–89

30. Zheng Y, Valdez PA, Danilenko DM, Hu Y, Sa SM et al (2008) Interleukin-22 mediates early host defense against attaching and effacing bacterial pathogens. Nat Med 14(3):282–289

31. Mangan PR, Harrington LE, O'Quinn DB, Helms WS, Bullard DC et al (2006) Transforming growth factor-beta induces development of the Th17 lineage. Nature 441(7090):231–234

32. Aychek T, Mildner A, Yona S, Ki-Wook Kim, Lampl N, Reich-Zeliger S, Boon L, Yogev N, Waisman A, Cua D. J, Jung S (2015) IL-23-mediated mononuclear phagocyte crosstalk protects mice from Citrobacter rodentium-induced colon immunopathology. Nat Commun 6:6525. doi: 10.1038/ncomms7525PMCID

33. Varol C, Landsman L, Fogg DK, Greenshtein L, Gildor B et al (2007) Monocytes give rise to mucosal, but not splenic, conventional dendritic cells. J Exp Med 204(1):171–180

34. Jung S, Aliberti J, Graemmel P, Sunshine MJ, Kreutzberg GW et al (2000) Analysis of fractalkine receptor CX$_3$CR1 function by targeted deletion and green fluorescent protein reporter gene insertion. Mol Cell Biol 20(11):4106–4114

35. Zaft T, Sapoznikov A, Krauthgamer R, Littman DR, Jung S (2005) CD11c^{high} dendritic cell ablation impairs lymphopenia-driven proliferation of naive and memory CD8$^+$ T cells. J Immunol 175(10):6428–6435

36. Zammit DJ, Cauley LS, Pham Q-M, Lefrançois L (2005) Dendritic cells maximize the memory CD8 T cell response to infection. Immunity 22(5):561–570

37. Van Rijt LS, Jung S, Kleinjan A, Vos N, Willart M et al (2005) In vivo depletion of lung CD11c$^+$ dendritic cells during allergen challenge abrogates the characteristic features of asthma. J Exp Med 201(6):981–991

38. Landsman L, Jung S (2007) Lung macrophages serve as obligatory intermediate between blood monocytes and alveolar macrophages. J Immunol 179(6):3488–3494

39. Sapoznikov A, Fischer JAA, Zaft T, Krauthgamer R, Dzionek A, Jung S (2007) Organ-dependent in vivo priming of naive CD4$^+$, but not CD8$^+$, T cells by plasmacytoid dendritic cells. J Exp Med 204(8):1923–1933

40. Sapoznikov A, Pewzner-Jung Y, Kalchenko V, Krauthgamer R, Shachar I, Jung S (2008) Perivascular clusters of dendritic cells provide critical survival signals to b cells in bone marrow niches. Nat Immunol 9(4):388–395

41. Fogg DK, Sibon C, Miled C, Jung S, Aucouturier P et al (2006) A clonogenic bone marrow progenitor specific for macrophages and dendritic cells. Science 311(5757):83–87

42. Geissmann F, Jung S, Littman DR (2003) Blood monocytes consist of two principal subsets with distinct migratory properties. Immunity 19(1):71–82

43. Landsman L, Varol C, Jung S (2007) Distinct differentiation potential of blood monocyte subsets in the lung. J Immunol 178(4):2000–2007

44. Imai T, Hieshima K, Haskell C, Baba M, Nagira M et al (1997) Identification and molecular characterization of fractalkine receptor CX$_3$CR1, which mediates both leukocyte migration and adhesion. Cell 91(4):521–530

45. Probst HC, Tschannen K, Odermatt B, Schwendener R, Zinkernagel RM, Van Den Broek M (2005) Histological analysis of CD11c-DTR/GFP mice after in vivo depletion of dendritic cells. Clin Exp Immunol 141(3):398–404

46. Hebel K, Griewank K, Inamine A, Chang H-D, Müller-Hilke B et al (2006) Plasma cell differentiation in t-independent type 2 immune responses is independent of cd11c (high) dendritic cells. Eur J Immunol 36(11):2912–2919

47. Lucas M, Schachterle W, Oberle K, Aichele P, Diefenbach A (2007) Dendritic cells prime natural killer cells by trans-presenting interleukin 15. Immunity 26(4):503–517

48. Brocker T, Riedinger M, Karjalainen K (1997) Targeted expression of major histocompatibility complex MHC class II molecules demonstrates that dendritic cells can induce negative but not positive selection of thymocytes in vivo. J Exp Med 185(3):541–550

Chapter 19

In Vivo Imaging of Cutaneous DCs in Mice

Gyohei Egawa and Kenji Kabashima

Abstract

Varieties of cells orchestrate immune responses. To capture such dynamic phenomena, intravital imaging is an important technique, and it may provide substantial information that is not available using conventional histological analyses. Multiphoton microscopy enables the direct, three-dimensional, and minimally invasive imaging of biological samples with high spatiotemporal resolution, and it has now become the leading method for in vivo imaging studies. Here we describe a basic method for in vivo imaging of dendritic cells (DCs) in the mouse ear skin using multiphoton microscopy.

Key words In vivo imaging, Multiphoton microscopy, Dendritic cells

1 Introduction

Intravital imaging has become an important technique to capture dynamic biological events such as immune responses. Conventionally, intravital imaging studies of epidermal Langerhans cells (LCs) were performed using confocal or fluorescence microscopy [1, 2]; however, the observation of dendritic cells (DCs) in the deep dermis is difficult with these microscopies because of their low light-penetration depth.

Multiphoton (MP) microscopy (also referred to as two-photon excitation microscopy) was first demonstrated by Denk [3] and is commonly used in the mid-2000s. Compared with conventional single-photon excitation microscopies, two-photon excitation allows deeper tissue penetration with less photodamage, achieving high spatiotemporal resolution. From these features, MP microscopy has become the gold standard for intravital imaging of biological specimens [4, 5]. It may provide substantial information that is not available using traditional methods such as histological and flow cytometer-based analysis.

Here we describe a basic method for in vivo imaging of dendritic cells (DCs) in the mouse ear skin using MP microscopy.

Elodie Segura and Nobuyuki Onai (eds.), Dendritic Cell Protocols, Methods in Molecular Biology, vol. 1423,
DOI 10.1007/978-1-4939-3606-9_19, © Springer Science+Business Media New York 2016

2 Materials

2.1 Reagents and Equipment

1. Hair removal cream.
2. Phosphate-buffered saline (PBS).
3. Pentobarbital sodium: 10 % (v/v) solution in PBS.
4. Isoflurane.
5. Grease.
6. Immersion oil for objective lens.
7. Cover glass (24×24/ 24×50 mm).
8. Masking tape.
9. Cotton swab.

2.2 Instruments

1. Inverted MP microscope (*see* **Note 1**).
2. Tunable (690–1040 nm) Ti/sapphire laser (Mai Tai DeepSee, Spectra-Physics).
3. Inhalation anesthesia apparatus for small animals.

2.3 Animals

1. CD11c-enhanced yellow fluorescent protein (EYFP) mice (The Jackson Laboratory: stock number 008829) [6].
2. Langerin-enhanced green fluorescent protein (EGFP) mice (The Jackson Laboratory: stock number 016939) [1].
3. C57BL/6-Tg(CAG-EGFP) mice (The Jackson Laboratory: stock number 003291) [7].

3 Methods

3.1 Labeling of Cutaneous DCs

1. If available, use CD11c-EYFP mice (*see* **Note 2**) or Langerin-EGFP mice (*see* **Note 3**).
2. Alternatively, prepare bone marrow chimeric mice with CAG-EGFP mice as a donor (*see* **Note 4**).
3. Alternatively, prepare fluorescent bone marrow-derived DCs [8] and inject them into the dermis (*see* **Note 5**).

3.2 Preimaging Preparations

Place the mouse on a heating pad at 37 °C throughout the imaging procedure.

1. Anesthetize the mouse by intraperitoneal injection of pentobarbital sodium (10 μl per g body weight) (*see* **Note 6**).
2. Apply the hair removal cream to the ventral and dorsal side of the left ear skin. Three minutes later, remove the cream from the ear using wet cotton or running water (*see* **Note 7**).

Fig. 1 Attach the coverslip to the dorsal side of the ear

3. Apply a grease onto the dorsal side of the left ear by cotton swab (*see* **Note 8**).

4. Attach the 24 × 24 mm cover glass to the dorsal side of the left ear (*see* Fig. 1).

3.3 Placing the Mouse onto the Microscope Stage

1. Place a single drop of immersion oil to objective lens.

2. Cover the central hole of microscope stage with 24 × 50 mm cover grass and fix it by masking tape (*see* Fig. 2) (*see* **Note 9**).

3. Place a drop of immersion oil onto the cover glass just above the objective lens.

4. Place the mouse onto the stage. The ear is sandwiched between two cover glasses (*see* Fig. 3) (*see* **Notes 10 and 11**).

5. Connect the mouse to anesthetic apparatus. Flow 1% isoflurane by 1 L/min (*see* **Note 12**).

6. Stabilize the mouse by taping.

3.4 Observation of Cutaneous DCs

1. Set the wavelength of excitation laser on 940 nm (*see* **Note 13**).

2. Commence imaging. We typically take images every 3–5 min for 3–12 h (*see* **Note 14**).

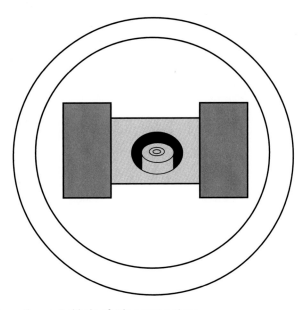

Fig. 2 Cover the central hole of microscope stage

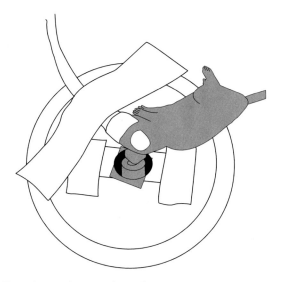

Fig. 3 Stabilize mice on the stage by taping

4 Notes

1. Inverted microscope is more suitable for the observation of the skin rather than upright microscope, because of its ease of fixing the ear.

2. CD11c, also known as integrin αX, is expressed by both epidermal LCs and dermal DCs. Because CD11c expression level is not strong in LCs, this mouse strain is mainly suitable for the observation of dermal DCs.

3. Langerin, a type of C-type lectin family, is specifically and extensively expressed in epidermal LCs and dermal langerin⁺ DCs. This strain is suitable for the observation of LCs.

4. To prepare bone marrow chimeric mice, irradiate 6-week-old C57BL/6 mice (two doses of 550 Rad given 3 h apart) and transfer them with 1×10^7 bone marrow cells from CAG-EGFP mice. Leave untreated for at least 8 weeks before use. Since LCs are radio resistant, only dermal DCs are replaced and labeled with EGFP.

5. Prepare bone marrow-derived DCs from CD11c-EYFP mice or from CAG-EGFP mice, or label DCs with a fluorescent dye such as carboxyfluorescein succinimidyl ester (CFSE) [9].

6. The body temperature of the mouse tends to decrease after the anesthesia. Keep the mouse on the heating pad and warm it sufficiently.

7. Be careful not to allow the hair removal cream to stay on the ear for more than 5 min. The cream is stimulatory and may cause skin inflammation. This is especially important in DC tracking studies.

8. Handle the ear gently to prevent injuries arising from friction.

9. The objective lens and cover glass must be in contact through immersion oil. Do not make bubbles.

10. Do not put strong pressure on the ear, which may disturb blood and lymph circulation. This is important for intravital imaging studies.

11. Do not put masking tape above the ear skin. Masking tape may disturb the fluorescent signals from the specimens.

12. The flow rate of isoflurane must be modulated depending on the body weight of the mouse.

13. The optimum wavelength for two-photon excitation is different from that of single-photon excitation [10].

14. The migratory velocity of dermal DCs is relatively slow (~4 µm/min) when compared to that of skin-infiltrating neutrophils or T cells (6–7 µm/min) [11, 12]. Time interval should be modified depending on the anticipated velocity of target cells.

Acknowledgments

This work was supported by Grants-in-Aid for Scientific Research from the Ministry of Education, Culture, Sports, Science and Technology of Japan.

References

1. Kissenpfennig A, Henri S, Dubois B, Laplace-Builhe C, Perrin P, Romani N, Tripp CH, Douillard P, Leserman L, Kaiserlian D, Saeland S, Davoust J, Malissen B (2005) Dynamics and function of Langerhans cells in vivo: dermal dendritic cells colonize lymph node areas distinct from slower migrating Langerhans cells. Immunity 22:643–654

2. Nishibu A, Ward BR, Boes M, Takashima A (2007) Roles for IL-1 and TNFalpha in dynamic behavioral responses of Langerhans cells to topical hapten application. J Dermatol Sci 45:23–30

3. Denk W, Strickler JH, Webb WW (1990) Two-photon laser scanning fluorescence microscopy. Science 248:73–76

4. Li JL, Goh CC, Keeble JL, Qin JS, Roediger B, Jain R, Wang Y, Chew WK, Weninger W, Ng LG (2012) Intravital multiphoton imaging of immune responses in the mouse ear skin. Nat Protoc 7:221–234

5. Natsuaki Y, Egawa G, Nakamizo S, Ono S, Hanakawa S, Okada T, Kusuba N, Otsuka A, Kitoh A, Honda T, Nakajima S, Tsuchiya S, Sugimoto Y, Ishii KJ, Tsutsui H, Yagita H, Iwakura Y, Kubo M, Lg N, Hashimoto T, Fuentes J, Guttman-Yassky E, Miyachi Y, Kabashima K (2014) Perivascular leukocyte clusters are essential for efficient activation of effector T cells in the skin. Nat Immunol 15:1064–1069

6. Celli S, Albert ML, Bousso P (2011) Visualizing the innate and adaptive immune responses underlying allograft rejection by two-photon microscopy. Nat Med 17:744–749

7. Okabe M, Ikawa M, Kominami K, Nakanishi T, Nishimune Y (1997) Green mice'as a source of ubiquitous green cells. FEBS Lett 407:313–319

8. Inaba K, Inaba M, Romani N, Aya H, Deguchi M, Ikehara S, Muramatsu S, Steinman RM (1992) Generation of large numbers of dendritic cells from mouse bone marrow cultures supplemented with granulocyte/macrophage colony-stimulating factor. J Exp Med 176:1693–1702

9. Matheu MP, Sen D, Cahalan MD, Parker I (2008) Generation of bone marrow derived murine dendritic cells for use in 2-photon imaging. J Vis Exp 9:773

10. Drobizhev M, Makarov NS, Tillo SE, Hughes TE, Rebane A (2011) Two-photon absorption properties of fluorescent proteins. Nat Methods 8:393–399

11. Ng LG, Hsu A, Mandell MA, Roediger B, Hoeller C, Mrass P, Iparraguirre A, Cavanagh LL, Triccas JA, Beverley SM (2008) Migratory dermal dendritic cells act as rapid sensors of protozoan parasites. PLoS Pathog 4, e1000222

12. Biotec M, Gladbach B (2011) In vivo imaging of T-cell motility in the elicitation phase of contact hypersensitivity using two-photon microscopy. J Invest Dermatol 131:977–979

Chapter 20

Analysis of Dendritic Cell Function Using Clec9A-DTR Transgenic Mice

Piotr Tetlak and Christiane Ruedl

Abstract

The Clec9A-diphtheria toxin receptor (DTR) transgenic mouse strain provides a robust animal model to study the function of lymphoid organ-resident CD8$^+$ dendritic cells (DCs) and nonlymphoid organ-specific CD103$^+$ DCs in infectioous diseases and inflammation. Here we describe some basic protocols for CD8$^+$/CD103$^+$ DC isolation, for their in vivo depletion, and for their characterization by multi-color flow cytometry analysis. As an example for in vivo functional characterization of this DC subset, we present here the experimental cerebral malaria model. Furthermore, we illustrate advantages and pitfalls of the Clec9A-DTR system.

Key words Clec9A/DNGR-1, Diphtheria toxin receptor (DTR) transgenic mice, Dendritic cell (DC) subsets, Cross-priming, Interferon-γ experimental cerebral malaria (ECM)

1 Introduction

1.1 Dendritic Cells (DCs), a Heterogeneous Family of Myeloid Cells

Dendritic cells (DCs) are crucial component of the first line of defense against foreign invaders that help to "decide" what kind of an adaptive immune action should be taken in order to fight pathogens and restore normal homeostasis. DCs are heterogeneous group of cells with regard to expression of different cell surface receptors and anatomical localization, morphology, and function [1]. Recently, they have been also classified into different subpopulations based on their ontogenic relationship [2]. For example, in the spleen, at least three different DCs have been identified since they can be divided in CD11b$^+$CD11c^{high}, CD8$^+$CD11c^{high}, and SiglecH$^+$CD11c^{int} plasmacytoid DC (pDCs) subsets [3] (*see* Fig. 1). In other nonlymphoid organs, such as the intestine, respiratory tract, and kidney, the complexity of DC heterogeneity is even higher [4–6]. Conventional DCs can be further divided in migratory DCs (including Langerhans and dermal dendritic cells), lymphoid-tissue resident DCs, and monocyte-derived DCs, which are only recruited into lymphoid organs under inflammatory conditions.

Elodie Segura and Nobuyuki Onai (eds.), *Dendritic Cell Protocols*, Methods in Molecular Biology, vol. 1423,
DOI 10.1007/978-1-4939-3606-9_20, © Springer Science+Business Media New York 2016

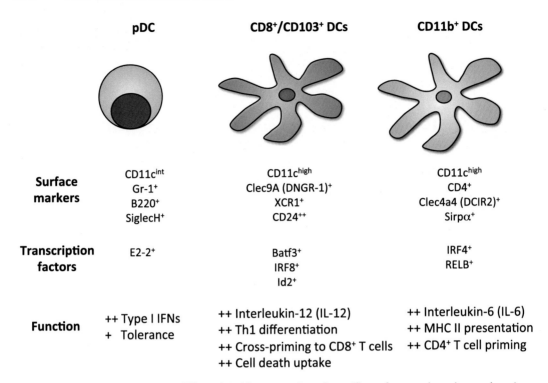

	pDC	CD8+/CD103+ DCs	CD11b+ DCs
Surface markers	CD11c^{int} Gr-1$^+$ B220$^+$ SiglecH$^+$	CD11c^{high} Clec9A (DNGR-1)$^+$ XCR1$^+$ CD24^{++}	CD11c^{high} CD4$^+$ Clec4a4 (DCIR2)$^+$ Sirpα$^+$
Transcription factors	E2-2$^+$	Batf3$^+$ IRF8$^+$ Id2$^+$	IRF4$^+$ RELB$^+$
Function	++ Type I IFNs + Tolerance	++ Interleukin-12 (IL-12) ++ Th1 differentiation ++ Cross-priming to CD8$^+$ T cells ++ Cell death uptake	++ Interleukin-6 (IL-6) ++ MHC II presentation ++ CD4$^+$ T cell priming

Fig. 1 Distinct DC subsets can be differentiated by expression of specific surface markers, by a unique transcriptional control and functional specialization

Recently, extensive gene expression study has identified distinct molecular signatures among different populations of DCs [7]. Besides a clear DC core gene expression signature, new DC subset-restricted markers have been identified. In particular, lymphoid-tissue CD8$^+$ cDCs and nonlymphoid-tissue CD103$^+$ cDCs represent a common lineage since they share a similar gene expression signature independent of their location of residence.

Diversity in DC subsets strongly suggests a functional divergence among them (Fig. 1). Although CD11b$^+$ DCs are functionally specialized in presenting antigens via MHC class II complexes to stimulate helper T cells [8], in vitro experiments have shown that DCs are generally capable of presenting exogenous antigens in association with MHC class I (cross-priming), and the CD8$^+$ DC subset demonstrates this ability in vivo [9]. The same DC subset is the major producer of IL-12 and therefore the dominant DC population involved in priming Th1 responses [10].

Since the cells that form the DC compartment differ in their phenotypic features, surface markers, anatomic locations, and ontogeny, one of perhaps the best ways of categorizing them comes from the function-based studies. This notion prompted generation of different animal models, which allow functional studies of individual DC subsets in vivo.

1.2   Clec9A Expression Profile and Function

Clec9A (C-type lectin domain family 9, member a), also known as DNGR-1, is a C-type lectin-like molecule expressed at high levels on Batf3-dependent murine DCs, such as CD8$^+$ DCs and CD103$^+$ DCs [11], and found at lower levels also on pDCs [12]. In humans, Clec9A is selectively expressed by CD141$^+$ myeloid DCs (reviewed in [13]), which have been characterized as a lineage and functional equivalent of the mouse CD8$^+$ DCs. Similarly to the mouse counterparts, this human DC subset is transcriptionally regulated via Batf3 and IRF8 and resides in several lymphoid organs such as the spleen, lymph nodes, and tonsils as well as in the gut [14]. Recently, Clec9A was described as a key receptor responsible in sensing necrotic debris or damaged cells via their exposed actin filaments [15]. Its function depends on signaling via the tyrosine kinase SYK, which is required for an efficient cross-presentation of dead cell-associated antigens [16]. The fact that Clec9A is expressed on the same murine and human DC subpopulations makes animal models such as Clec9A-diphtheria toxin receptor (DTR) mice useful in vivo research tools for elucidating the exact function of this peculiar DC subset in steady state and in perturbed situations such as infection and inflammation. Results obtained in the Clec9A-DTR mouse model can then hopefully shed more light on function of the human immune system.

1.3   DTR-Mediated DC Ablation in Transgenic Mice

To provide insight into roles of different DC populations, several experimental approaches based on conditional depletion of DCs in vivo have been developed. Initially established conventional techniques, such as injection of depleting antibodies or liposomal clodronate suffered from a lack of efficiency or/and specificity. These limitations were overcome by development of a genetic approach in which high-affinity DTR could be expressed in specific cell types, marking them as targets for diphtheria toxin (DT)-triggered depletion. The DTR-DT system that allows inducible (and temporal) cell depletion offers significant advantages and is especially attractive to immunologists because DT treatment eliminates all DTR-expressing cells, including also the nondividing and terminally differentiated cells such as DCs, and does it by inducing apoptosis. Since subsequent clearance of apoptotic cells doesn't invoke pro-inflammatory immune responses, this method of cell ablation is considered to occur in immunologically silent manner. One of the drawbacks of the DTR-DT system is potential formation of neutralizing antibodies in animal hosts injected repeatedly with DT during a prolong period of time, which could render the cell depletion less effective.

Systemic depletion of DCs by DTR-DT system was achieved for the first time in the transgenic mouse strain, which expresses DTR under control of minimal *Cd11c* promoter [17]. CD11c is expressed by all DCs, and therefore it is not surprising that injection of DT into CD11c-DTR transgenic mice efficiently depletes all conventional DC (cDCs) including pDCs. Unfortunately repeated systemic administration of DT is lethal in these mice, probably due to aberrant expression of DTR on some other, non-related cell types [17]; hence experiments that require prolonged depletion of DCs have to be conducted in the wild-type animals in which, upon lethal irradiation, the whole immune system is reconstituted with the bone marrow (BM) transplanted from CD11c-DTR transgenic donors. In these chimeric mice, only the cells derived from transgenic BM remain DT sensitive and can be later efficiently ablated.

Although the CD11c-DTR animal model has been very useful in studies of DC biology, it has its limitations regarding specificity. The expression of CD11c among immune cells is not strictly restricted to DCs. In reality, CD11c can be also detected on many macrophage subpopulations (e.g., gut and lung tissue macrophages and spleen marginal zone macrophages), plasmablasts, activated T cells, NK cells, and Ly-6C^{low} monocytes, so naturally many of these cell types get depleted in CD11c-DTR mice upon DT treatment. It became very clear that the use of the CD11c-DTR model is limited and new approaches directed toward depletion of specific DC subsets rather than all DCs are needed. Therefore, due to its unique expression pattern on a specific DC subpopulation, Clec9A was selected as a marker for the depletion of CD8$^+$ and CD103$^+$CD11b$^-$ DC subsets and a Clec9A-DTR transgenic mouse was successfully generated [18].

Upon DT treatment, the Clec9A-DTR transgenic mice efficiently ablate the splenic and mesenteric lymph nodes (MLNs) as well as kidney Clec9A$^+$ DCs as shown by flow cytometry analysis (Fig. 2). DT injection eliminates CD11c^{high}CD8$^+$ DCs (but not CD11c^{high}CD8$^-$ DCs) (see Fig. 2, left upper panel), while pDCs (CD11c^{int}Siglec-H$^+$) known to express lower levels of Clec9A were only partially depleted (~50%, data shown in [18]). In MLNs, both classical lymphoid organ-resident CD11c^{high}MHCII$^+$CD8$^+$CD11b$^-$ and lamina propria-derived migratory CD11c^{int}MHCII^{++}CD103$^+$CD11b$^-$ disappear upon DT treatment (Fig. 2 right panel). Classical lymphoid organ-resident CD11c^{high}MHC II$^+$CD8$^-$CD11b$^+$DC fraction is diminished by 90% (Fig. 2, right panel). Also in the case of kidney, as representative of a nonlymphoid tissue, DT treatment efficiently ablates the CD11c^{high}MHCII$^+$CD103$^+$CD11b$^-$ DC subset, but not the CD11c^{high}MHCII$^+$CD103$^-$CD11b$^+$ DC subpopulation (Fig. 2, left lower panel).

1.4 Experimental Cerebral Malaria (ECM) in Clec9A-DTR Mice

Cross-priming, a mechanism by which CD8$^+$ T cells are primed, is mediated by CD8$^+$ DCs [9]; hence the Clec9A-DTR mouse strain and other mouse strains lacking this particular DC subset (e.g., Langerin-DTR and Batf3 $-/-$ mice) provide a valuable models to investigate this type of immune response. DC-mediated

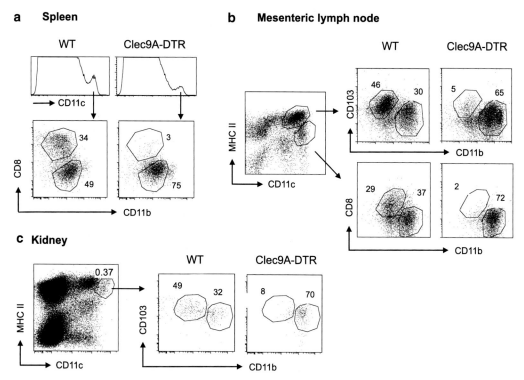

Fig. 2 Specific ablation of CD8[+] and CD103[+] DCs in Clec9A-DTR mice. Spleen, lymph node, and kidney cells were obtained from DT-treated WT controls and Clec9A-DTR mice. (**a**) Spleen cells were analyzed for CD8 and CD11b expression by gating on CD11c[high] cells. (**b**) Lymph node cells were analyzed for CD103 and CD11b expression by gating on CD11c[int]MHCII[++] migratory DCs and classical lymphoid CD11c[high]MHC II[+] DCs. (**c**) Kidney cells were analyzed for CD103 and CD11b expression by gating on CD11c[high]MHCII[++] DCs. Indicated numbers show the percentage of each gated cell subset

cross-priming is not only essential for the initiation of antigen-specific CTL responses to viruses but also to other pathogens such as *Plasmodium*. The innate and adaptive immune responses during malaria infection have to be tightly controlled, since excessive immune activation can lead to severe pathology known as cerebral malaria (CM). The pathogenesis of CM is still not fully understood, but sequestration of parasitized erythrocytes in deep tissue microvasculature is thought to be an essential factor for CM to occur [19]. Although the parasite burden in the brain appears to be the obvious indicator of the disease progression/outcome, it is clear that DCs, the major group of antigen-presenting cells (APCs) initiating malaria immunity, contribute to the ECM pathology [18].

To gain insight into the cellular mechanisms underlying the development of CM, several animal models of experimental CM (ECM) have been established, including infection of mice with *Plasmodium berghei* ANKA (PbA) [18]. In this case sequestration of parasitized erythrocytes in the cerebral microvasculature as well

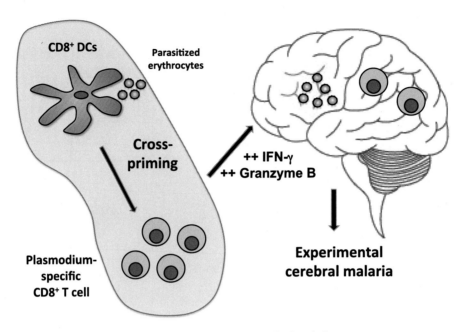

Fig. 3 CD8+ DCs mediate the development of experimental cerebral malaria

as intravascular accumulation of CD8+ T cells in the brain clearly promotes the progression of the disease [20].

We have decided to use our newly generated Clec9A-DTR transgenic mice to analyze the contribution of CD8+ DC in ECM caused by PbA. We were able to demonstrate that CD8+ DCs are the major culprits in activating parasite-specific CD8+ T cells, which subsequently accumulate in the brain promoting the progression of the disease, as illustrated in Fig. 3. In fact, ablation of this specific cross-priming DC subset in PbA infected Clec9A-DTR mice resulted in complete protection from severe cerebral pathology [18]. This resistance observed in Clec9A-DTR mice correlated with diminished IFN-γ and granzyme B response as well as decreased accumulation of parasitized erythrocytes in the brain.

In this chapter, we describe how to successfully deplete CD8+/CD103+ DCs in Clec9A-DTR mice, how to isolate and characterize DCs from lymphoid and nonlymphoid tissues, and how to analyze DC-mediated cross-priming, including T-cell isolation and extracellular and intracellular flow cytometry staining protocols.

2   Materials

2.1 Conditional Depletion of CD8/CD103+ DCs in Clec9A-DTR Mice

1. Clec9A-DTR transgenic mice [18].

2. C57BL/6 mice (control mice).

3. Syringe and needle.

4. PBS.

5. Mouse serum.

6. Diphtheria toxin (DT): prepare stock solution at 1 mg/mL in PBS and store in aliquots at –80 °C. Dilute in PBS 2 % mouse serum before use.

2.2 Isolation of DCs from Lymphoid and Nonlymphoid Tissues

1. Forceps and surgery scissors.

2. PBS.

3. 15 and 50 mL tubes.

4. Petri dishes.

5. Collagenase D (Roche): prepare stock solution at 200 mg/mL in PBS and store in aliquots at –20 °C. Dilute aliquots in PBS before use.

6. Fetal bovine serum (FBS).

7. Iscove's Minimal Essential Medium (IMDM).

8. Digestion medium: IMDM, 7 % FBS, 0.1 mg/mL collagenase D.

9. Tube shaker.

10. 80 µm mesh strainer.

11. Hypotonic buffer: 0.87 % ammonium chloride solution. Dissolve 0.87 g of ammonium chloride in 100 mL distilled water.

12. $10 \times$ PBS.

13. Percoll.

14. 70 % Percoll solution: mix 63 mL Percoll, 7 mL $10 \times$ PBS and 30 mL IMDM.

15. 40 % Percoll solution: mix 36 mL Percoll, 4 mL $10 \times$ PBS and 60 mL IMDM.

16. OptiPrep (Sigma) (density = 1.32 g/mL).

17. Hank's Balanced Salt Solution (HBSS) (without Ca^{2+} and Mg^{2+}).

18. 0.1 M EDTA solution.

19. 0.1 M HEPES solution.

20. Diluent solution: 0.88 % NaCl, 1 mM EDTA, 0.5 % bovine serum albumin (BSA), 10 mM HEPES-NaOH, pH 7.4. Dissolve 0.88 g of NaCl and 0.5 g of BSA in 50 mL of water; add 10 mL of 0.1 M HEPES solution and 1 mL of 0.1 M EDTA solution; adjust to pH 7.4 with NaOH and make up to 100 mL with water.

2.3 Analysis of DC Depletion in Lymphoid and Nonlymphoid Organs

1. D-PBS: Dulbecco's phosphate-buffered saline (PBS without Ca^{++} and Mg^{++}).

2. FBS.

3. FACS buffer: D-PBS 2 % FBS.

4. Phenotyping antibodies: PE-Cy7-labeled rat anti-CD45 (30-F11), PE-labeled hamster anti-CD11c (N418), Pacific Blue-labeled rat anti-MHC class II (M5/114.15.2), PercP-labeled rat anti-CD103 (2E7), APC-labeled anti-CD11b (M1/70), FITC-labeled rat anti-CD8 (53–6.7) antibodies.

5. 96-well v-bottom microtiter plate.

6. 5 mL polystyrene tubes.

7. Flow cytometer with blue (488 nm), red (633 nm), and violet (405 nm) laser.

2.4 In Vivo Analysis of CD8+ DC Function in ECM Model

1. Red blood cells (RBCs) infected with transgenic *Plasmodium berghei* ANKA clone 15Cy1 expressing green fluorescent protein (PbA-GFP) [18, 21].

2. Syringe and needle.

3. Forceps and surgery scissors.

4. PBS.

5. 15 mL and 50 mL tubes.

6. Collagenase D (Roche): prepare stock solution at 200 mg/mL in PBS and store in aliquots at –20 °C. Dilute aliquots in PBS before use.

7. DNase I (10.000 U, Roche): store at –20 °C. Use at final concentration of 2 U/mL.

8. FBS.

9. IMDM.

10. Digestion medium II: IMDM, 7 % FBS, 0.1 mg/mL collagenase D, 2 U/mL DNase I.

11. Tube shaker.

12. 21G needle.

13. 33 % Percoll solution: mix 29.7 mL Percoll, 3.3 mL 10× PBS, and 67 mL IMDM.

14. Hypotonic buffer: 0.87 % ammonium chloride solution. Dissolve 0.87 g of ammonium chloride in 100 mL distilled water.

15. Brefeldin A: prepare stock solution at 5 mg/mL in 100 % ethanol. Store in aliquots at –80 °C. Dilute to a working concentration of 5 μg/mL.

16. D-PBS.

17. FACS buffer: D-PBS 2 % FBS.

18. Antibodies: PE-Cy7-labeled rat anti-mouse CD45 (30-F11), FITC-labeled rat anti-CD4 (GK1.5) and PerCP-labeled rat anti-CD8 (53–6.7), APC-labeled rat anti-IFN-γ (DB-1), and PE-labeled rat anti-granzyme B (NGZB). Dilute in FACS buffer.

19. Fixation and Permeabilization Buffer Set (eBioscience).

20. 96-well v-bottom microtiter plate.

21. 5 mL polystyrene tubes.

22. Flow cytometer with blue (488 nm), red (633 nm), and violet (405 nm) laser.

3 Methods

3.1 Conditional Depletion of CD8/CD103⁺ DCs in Clec9A-DTR Mice

1. Measure the body weight of each Clec9A-DTR transgenic mouse.

2. For systemic depletion of CD8/CD103⁺ DCs, inject Clec9A-DTR transgenic mice with 10 ng DT/g body weight intraperitoneally (i.p.).

3. Repeat injections every 3–4 days to maintain the target DC subset ablated (*see* **Notes 1–5**).

3.2 Isolation of DCs from Lymphoid and Nonlymphoid Tissues

1. Sacrifice mice.

2. Remove the lymphoid organs (spleen and MLNs) and a representative of nonlymphoid organs (e.g., kidney) and place them in a Petri dish containing PBS on ice.

3. Cut the collected organs into small pieces using surgery scissors.

4. Place the fragments of each organ in a tube containing 5 mL of digestion medium.

5. Place the tubes on a shaker at 37 °C for 45 min.

6. Gently pass each cell suspension through an 80 μm mesh strainer and collect the cell suspensions in fresh tubes.

7. Pellet down by centrifugation at $300 \times g$ for 7 min at 4 °C.

8. Wash with 10 mL of cold PBS and centrifuge at $300 \times g$ for 7 min at 4 °C. For spleen cell suspensions, perform **steps 9** and **10**. For kidney cell suspensions, perform **steps 11–15**. For MLN cell suspension, perform **steps 16–21**.

9. For spleen cell suspensions, resuspend in 10 mL of hypotonic buffer.

10. Incubate at room temperature (RT) for 10 min and repeat **steps 7** and **8** (*see* **Note 6**).

11. For kidney cell suspensions, resuspend the cells in 4 mL of 40 % Percoll.

12. Place 4 mL of 70 % Percoll in a fresh tube, and carefully overlay the cell suspension on top of the 70 % Percoll.

13. Centrifuge at RT at $1500 \times g$ for 10 min with acceleration set to "3" and break set to "0."

14. Collect the leukocyte at the gradient interphase.

15. Repeat **steps 7** and **8**.

16. For MLN cell suspension, resuspend the cell pellet in 1 mL of HBSS and mix gently but thoroughly with 3 mL of OptiPrep to give a 15 % iodixanol solution [22–23].

17. Prepare an 11.5 % solution of iodinaxol: you will need 5 mL of solution per sample. Mix 1 volume of OptiPrep with 4.2 volumes of diluent solution.

18. Overlay the 4 mL of the cell suspension with 5 mL of the 11.5 % iodixanol solution and 3 mL of HBSS.

19. Centrifuge at $600 \times g$ for 15 min at RT with the break set to "0."

20. Collect the cells from the top of the 11.5 % iodixanol layer.

21. Repeat steps 7 and 8.

22. Count the cells.

23. Centrifuge at $300 \times g$ for 7 min at 4 °C.

24. Resuspend the cells in FACS buffer to obtain a concentration of 2×10^6 cells/mL.

3.3 Analysis of DC Depletion in Lymphoid and Nonlymphoid Organs of DT-Treated Clec9A DTR Mice

1. Transfer 2×10^5 cells in 100 μL of each cell suspension into the well of a 96-well v-bottom-microtiter plate.

2. Centrifuge the cells for 2 min at $500 \times g$ at 4 °C.

3. Resuspend the cells in 100 μL of FACS buffer containing PE-Cy7-labeled rat anti-CD45 (dilution 1:600), PE-labeled hamster anti-CD11c (dilution 1:800), Pacific Blue-labeled rat anti-MHC II (dilution 1:1000), FITC-labeled rat anti-CD8 (dilution 1:500), PerCP-labeled rat anti-CD103 (dilution 1:600), and APC-labeled rat anti-CD11b (dilution 1:800) (*see* **Note 7**).

4. Incubate on ice for 20 min.

5. Centrifuge the cells for 2 min at $500 \times g$ at 4 °C.

6. Resuspend the cells in 200 μL of FACS buffer and transfer to 5 mL tubes.

7. Acquire on a flow cytometer (*see* **Note 8**).

8. For spleen DCs, gate the cells as follows: gate on CD11c[high] expressing cells and analyze CD11b and CD8 in a two-dimensional dot plot (*see* Fig. 2a).

9. For LN DCs, gate the cells as follows: gate on CD11c[int] MHC II[high] cells (migratory DCs) and analyze CD11b and CD103 in a two-dimensional dot plot; gate on CD11c[high] MHC II[int] cells (resident DCs) and analyze CD11b and CD8 in a two-dimensional dot plot (*see* Fig. 2b).

10. For kidney DCs, gate the cells as follows: gate on CD11c^{high} MHC II^{high} cells and analyze CD11b and CD103 in a two-dimensional dot plot (*see* Fig. 2c)

3.4 In Vivo Analysis of CD8+ DC Function in Experimental Cerebral Malaria Model (ECM)

1. Follow the protocol described in Subheading 3.1 to treat 10 Clec9A-DTR and 10 control WT mice with DT before infection (day -1).

2. Inject the Clec9A-DTR and WT control mice i.p. with 10^6 RBCs infected with PbA-GFP (day 0).

3. Monitor parasitemia (*see* **Note 9**), ECM symptoms (*see* **Note 10**), and survival over a period of 20 days.

4. Inject the mice with DT every 3–4 days during the infection period for constant DC ablation.

5. Monitor parasite-induced T-cell activation at day 7 post infection. Sacrifice three experimental animals.

6. Collect the brains from both experimental groups and place them in a tube containing 3 mL of PBS.

7. Cut the brains into small pieces using surgery scissors.

8. Place the fragments of each brain in a tube containing 3 mL of digestion medium II.

9. Place the tubes on a shaker at 37 °C for 30 min.

10. Homogenize the single-cell suspension obtained by passing it through a 21G needle and collect the cell suspensions in fresh tubes.

11. Pellet down by centrifugation at 300 × g for 7 min at 4 °C.

12. Resuspend the cells in 4 mL of 33 % Percoll (*see* **Note 11**).

13. Centrifuge at RT at 1500 × g for 10 min with acceleration set to "3" and break set to "0."

14. Collect the leukocytes at the gradient interphase.

15. Resuspend the cells in 10 mL of hypotonic buffer.

16. Incubate at room temperature (RT) for 10 min (*see* **Note 6**).

17. Pellet down by centrifugation at 300 × g for 7 min at 4 °C.

18. Wash with 10 mL of cold PBS and centrifuge at 300 × g for 7 min at 4 °C.

19. Count the cells in each tube.

20. Transfer 1 × 10^5 cells of each suspension to the well of a 96-well v-bottom plate.

21. Centrifuge the cells for 2 min at 500 × g at RT.

22. Resuspend the cells in IMDM containing 5 μg/mL Brefeldin A and incubate for 2 h at 37 °C.

23. Centrifuge the cells for 2 min at 500 × g at RT.

24. Resuspend the cells in 100 μL of FACS buffer and add 100 μL of an antibody mix containing PE-Cy7-labeled rat anti-CD45 (dilution 1:600), PerCP-labeled rat anti-CD8 (dilution 1:600), and FITC-labeled rat anti-CD4 (dilution 1:500).

25. Incubate for 20 min on ice.

26. Centrifuge the cells for 2 min at $500 \times g$ at RT.

27. Fix and permeabilize the cells using the Fixation and Permeabilization Buffer Set, by adding 100 μL of Fixation and Permeabilization Buffer, and incubate for 30 min at RT.

28. Spin the cells for 2 min at 500 g at RT.

29. Resuspend in 100 μL of permeabilization buffer containing APC-labeled rat anti-IFN-γ (dilution 1:500) and PE-labeled rat anti-granzyme B (dilution 1:800).

30. Incubate for 20 min at RT.

31. Spin the cells for 2 min at 500 g at RT.

32. Resuspend the cells in 200 μL of FACS buffer and analyze them on a flow cytometer (*see* **Note 8**).

4   Notes

1. Unlike CD11c-DTR mice, the Clec9A-DTR mouse strain does not show any significant inflammation after prolonged DT injections and mice do not succumb the DT treatment. However, repetitive injections of DT induce the production of neutralizing anti-DT antibodies, which impair an efficient ablation after long-term DT regiment. Based on our results, Clec9A-DTR can be injected with 10 ng DT/g body weight for at least 1 month; afterward the depletion efficiency will be decreased. Consequently, DTR transgenic mice, including the Clec9A-DTR mice, are not suitable for chronic infection and disease investigations.

2. Since Clec9A is expressed also on CDPs and pre-DCs in the BM [2], the repetitive DT injections over a long period of time could possibly affect all DC subsets. To exclude this, we analyzed spleen and other tissues 15 days after the first DT injection (followed by further DT injections at days 4 and 8) and could confirm that the spleen CD11b⁺ DC subset was not affected in our Clec9A-DTR mouse. On the contrary CD8⁺ and CD103⁺CD11b⁻ DCs stayed efficiently ablated over the observation period (data not shown).

3. Like other DC-ablating DTR strains, such as the CD11c-DTR [25], DC ablation via DT injections causes a transient neutrophilia and monocytosis in Clec9-DTR mice, and this phenom-

enon has to be taken into account in the interpretation of the obtained results.

4. When using Clec9-DTR mice there are some minor problems, which have to be taken in consideration. First, it is to note that DT treatment of Clec9A-DTR leads to a 50 % reduction of the number of pDCs, since these cells express low amount of this molecule. The "side effect" could be circumvented including an analysis of a pDC-depleting mouse, like the Siglec H-DTR mouse, and verify the contribution of this particular DC subset in the outcome of the experiment.

5. Other transgenic or knock out mouse strains, like Langerin-DTR [26] and CD205-DTR mice [27] and *Batf3* −/− mice [28], deplete or lack the same CD8+/CD103+ DC subpopulation. When Clec9-DTR mice are compared to Langerin-DTR or CD205-DTR mice, the Clec9A-DTR mouse strain offers the advantage of depleting only dermal CD103+ DCs, whereas Langerhans cells remain unaffected. Furthermore, Langerin-DTR mouse does not affect tissue CD103+ DC subpopulation, like in the gut. When compared to *Batf3* −/− mice, plausible explanation for possible differences is inducible versus continuous ablation of this DC subset. The latter has time to develop compensatory adaptations, and therefore, it would be interesting to see the phenotype of mice with inducible knockout of *Batf3*.

6. This step will lyse the erythrocytes.

7. CD45 staining will delineate the leukocyte population (this is especially important for nonlymphoid organs), CD11c and MHC II will mark bona fide DCs and CD8, and CD103 and CD11b will delineate DC subsets.

8. Live cells were gated based on forward and side scatter.

9. Collect blood from the tail vein of each mouse. Dilute 1 μL of blood in 500 μL of PBS and evaluate the percentage of GFP+ RBCs by flow cytometry.

10. ECM development can be monitored every day from days 5 to 12 post infection using the rapid murine coma and behavior scale (RMCBS) score system as previously described [24].

11. This step will enrich brain leukocyte fraction.

Acknowledgments

We thank Klaus Karjalainen for his critical reading of the manuscript. This work was supported by National Medical Research Council grants NMMR/1253/2010, NMRC/1307/2011, and MOE2014-T2-1-011 to C.R. The authors have no conflicting financial interests.

References

1. Hashimoto D, Miller J, Merad M (2011) Dendritic cell and macrophage heterogeneity in vivo. Immunity 35:323–335

2. Schraml BU, van Blijswijk J, Zelenay S, Whitney PG, Filby A et al (2013) Genetic tracing via DNGR-1 expression history defines dendritic cells as a hematopoietic lineage. Cell 154:843–858

3. Hey YY, O'Neill HC (2012) Murine spleen contains a diversity of myeloid and dendritic cells distinct in antigen presenting function. J Cell Mol Med 16:2611–2619

4. Varol C, Vallon-Eberhard A, Elinav E, Aychek T, Shapira Y et al (2009) Intestinal lamina propria dendritic cell subsets have different origin and functions. Immunity 31:502–512

5. Pabst O, Bernhardt G (2010) The puzzle of intestinal lamina propria dendritic cells and macrophages. Eur J Immunol 40:2107–2111

6. Rescigno M (2010) Intestinal dendritic cells. Adv Immunol 107:109–138

7. Miller JC, Brown BD, Shay T, Gautier EL, Jojic V et al (2012) Deciphering the transcriptional network of the dendritic cell lineage. Nat Immunol 13:888–899

8. Vander Lugt B, Khan AA, Hackney JA, Agrawal S, Lesch J et al (2014) Transcriptional programming of dendritic cells for enhanced MHC class II antigen presentation. Nat Immunol 15:161–167

9. den Haan JM, Lehar SM, Bevan MJ (2000) CD8(+) but not CD8(-) dendritic cells cross-prime cytotoxic T cells in vivo. J Exp Med 192:1685–1696

10. Macatonia SE, Hosken NA, Litton M, Vieira P, Hsieh CS et al (1995) Dendritic cells produce IL-12 and direct the development of Th1 cells from naive CD4+ T cells. J Immunol 154:5071–5079

11. Poulin LF, Reyal Y, Uronen-Hansson H, Schraml BU, Sancho D et al (2012) DNGR-1 is a specific and universal marker of mouse and human Batf3-dependent dendritic cells in lymphoid and nonlymphoid tissues. Blood 119:6052–6062

12. Caminschi I, Proietto AI, Ahmet F, Kitsoulis S, Shin Teh J et al (2008) The dendritic cell subtype-restricted C-type lectin Clec9A is a target for vaccine enhancement. Blood 112:3264–3273

13. Villadangos JA, Shortman K (2010) Found in translation: the human equivalent of mouse CD8+ dendritic cells. J Exp Med 207:1131–1134

14. Poulin LF, Salio M, Griessinger E, Anjos-Afonso F, Craciun L et al (2010) Characterization of human DNGR-1+ BDCA3+ leukocytes as putative equivalents of mouse CD8alpha+ dendritic cells. J Exp Med 207:1261–1271

15. Zhang JG, Czabotar PE, Policheni AN, Caminschi I, Wan SS et al (2012) The dendritic cell receptor Clec9A binds damaged cells via exposed actin filaments. Immunity 36:646–657

16. Sancho D, Joffre OP, Keller AM, Rogers NC, Martinez D et al (2009) Identification of a dendritic cell receptor that couples sensing of necrosis to immunity. Nature 458:899–903

17. Jung S, Unutmaz D, Wong P, Sano G, De los Santos K et al (2002) In vivo depletion of CD11c+ dendritic cells abrogates priming of CD8+ T cells by exogenous cell-associated antigens. Immunity 17:211–220

18. Piva L, Tetlak P, Claser C, Karjalainen K, Renia L et al (2012) Cutting edge: Clec9A+ dendritic cells mediate the development of experimental cerebral malaria. J Immunol 189:1128–1132

19. Schofield L, Grau GE (2005) Immunological processes in malaria pathogenesis. Nat Rev Immunol 5:722–735

20. Baptista FG, Pamplona A, Pena AC, Mota MM, Pied S et al (2010) Accumulation of Plasmodium berghei-infected red blood cells in the brain is crucial for the development of cerebral malaria in mice. Infect Immun 78:4033–4039

21. Franke-Fayard B, Trueman H, Ramesar J, Mendoza J, van der Keur M et al (2004) A Plasmodium berghei reference line that constitutively expresses GFP at a high level throughout the complete life cycle. Mol Biochem Parasitol 137:23–33

22. Ruedl C, Rieser C, Bock G, Wick G, Wolf H (1996) Phenotypic and functional characterization of CD11c+ dendritic cell population in mouse Peyer's patches. Eur J Immunol 26:1801–1806

23. Ruedl C, Hubele S (1997) Maturation of Peyer's patch dendritic cells in vitro upon stimulation via cytokines or CD40 triggering. Eur J Immunol 27:1325–1330

24. Carroll RW, Wainwright MS, Kim KY, Kidambi T, Gomez ND et al (2010) A rapid murine coma and behavior scale for quantitative assessment of murine cerebral malaria. PLoS One 5

25. Tittel AP, Heuser C, Ohliger C, Llanto C, Yona S et al (2012) Functionally relevant

neutrophilia in CD11c diphtheria toxin receptor transgenic mice. Nat Methods 9:385–390

26. Bennett CL, van Rijn E, Jung S, Inaba K, Steinman RM et al (2005) Inducible ablation of mouse Langerhans cells diminishes but fails to abrogate contact hypersensitivity. J Cell Biol 169:569–576

27. Fukaya T, Murakami R, Takagi H, Sato K, Sato Y et al (2012) Conditional ablation of CD205+ conventional dendritic cells impacts the regulation of T-cell immunity and homeostasis in vivo. Proc Natl Acad Sci U S A 109:11288–11293

28. Hildner K, Edelson BT, Purtha WE, Diamond M, Matsushita H et al (2008) Batf3 deficiency reveals a critical role for CD8alpha + dendritic cells in cytotoxic T cell immunity. Science 322:1097–1100

Chapter 21

Analysis of DC Functions Using CD205-DTR Knock-In Mice

Tomohiro Fukaya, Hideaki Takagi, Tomofumi Uto, Keiichi Arimura, and Katsuaki Sato

Abstract

Dendritic cells (DCs) are essential antigen-presenting cells (APCs) that consist of heterogeneous subsets, mainly classified as conventional DCs (cDCs) and plasmacytoid DCs (pDCs). CD205, an endocytic type I C-type lectin-like molecule that belongs to the mannose receptor family, is mainly expressed on CD8α^+ cDCs. However, it is unclear how CD205$^+$ cDCs control immune responses in vivo. To evaluate the contribution of CD205$^+$ cDCs to the immune system, we engineered knock-in (KI) mice that express the diphtheria toxin receptor (DTR) under the control of the *Cd205* gene, which allows the selective conditional ablation of CD205$^+$ cDCs in vivo. Conditional ablation of CD205$^+$ cDCs impaired the antigen-specific priming of CD8$^+$ T cells to generate cytotoxic T lymphocytes (CTLs) mediated through cross presentation of soluble antigen. Upon microbial infection, CD205$^+$ cDCs contributed to the cross priming of CD8$^+$ T cells for generating antibacterial CTLs to efficiently eliminate pathogens. Here, we provide a protocol for the generation of bone marrow WT/CD205-DT chimeric mice, depletion of CD205$^+$ DCs and assessment of depletion efficiency, and protocols for in vivo cross presentation assay, CTL generation assay, and antibacterial immunity assay.

Key words Dendritic cells, Cytotoxic T lymphocytes, Cross priming, Microbial infection, Knock-in mice

1 Introduction

Dendritic cells (DCs) are essential antigen-presenting cells (APCs) that consist of heterogeneous subsets, mainly classified as conventional DCs (cDCs) and plasmacytoid DCs (pDCs) [1]. DCs serve as sentinels, recognizing the presence of invading pathogens or virus-infected cells through various pattern-recognition receptors (PRRs), including Toll-like receptor (TLR) [1]. DCs process such exogenous antigens intracellularly and present them to CD4$^+$ T cells via MHC class II (MHC II) for induction of CD4$^+$ effector T (Teff) cells [1–3]. DCs also show an unusual specialization in their MHC class I (MHC I) presentation pathway to prime CD8$^+$ T cells. Although most cells use MHC I molecules to present peptides derived from endogenously synthesized proteins, DCs have

Elodie Segura and Nobuyuki Onai (eds.), *Dendritic Cell Protocols*, Methods in Molecular Biology, vol. 1423,
DOI 10.1007/978-1-4939-3606-9_21, © Springer Science+Business Media New York 2016

the capacity to deliver exogenous antigens to the MHC I pathway, a phenomenon known as cross presentation, that underlies the generation of cytotoxic T lymphocyte (CTL) immunity [1–3]. DCs thereby play a critical role in the link between innate and adaptive immunity.

Mouse cDCs in lymphoid organs are comprised of two major subsets, classified as CD8α$^+$ cDCs and CD8α$^-$ cDCs. CD8α$^+$ cDCs mainly reside in the T-cell zone, while CD8α$^-$ cDCs reside in the red pulp and marginal zone [2, 4]. Series of in vitro and ex vivo studies reported that CD8α$^+$ cDCs are more efficient in the phagocytic uptake of dead cells and in the cross presentation of cell-bound or soluble antigens on MHC I to generate CTLs than other DC subsets [5].

CD205, an endocytic type I C-type lectin-like molecule that belongs to the mannose receptor family, is mainly expressed on CD8α$^+$ cDCs and cortical thymic epithelium as well as interdigitating DCs in cutaneous lymph nodes (LNs) derived from dermal DCs and epidermal Langerhans cells (LCs), usually at a higher level than seen on macrophages and B cells [6–8]. CD205 may function as an endocytic receptor involved in the uptake of extracellular antigens. While an endogenous ligand for CD205 has not been identified, an antigen-conjugated monoclonal antibody (mAb) specific for CD205 was internalized, processed in the endosomal compartment, and presented to both MHC II and MHC I for cross presentation with high efficiency [5, 6]. Although these observations based on analyses in vitro and ex vivo provide the functions of CD205$^+$ cDCs, their role in the immune system under physiological conditions remains unclear due to the lack of a system that selectively eliminates this cell subset in vivo.

To precisely evaluate the contribution of CD205$^+$ cDCs to the immune system, we engineered knock-in mice that express the diphtheria toxin (DT) receptor (DTR) [7, 9, 10] under the control of the *Cd205* gene, which allows the selective conditional ablation of CD205$^+$ cDCs in vivo. Using these mice, we demonstrated the unique role of CD205$^+$ cDCs for the control of antimicrobial immune response vivo [11]. In this chapter, we provide a protocol for the generation of bone marrow WT/CD205-DT chimeric mice, depletion of CD205$^+$ DCs and assessment of depletion efficiency, and protocols for in vivo cross presentation assay, CTL generation assay, and antibacterial immunity assay.

2   Materials

2.1   Mice

All mice are used between 6 and 10 weeks of age and maintained in specific pathogen-free conditions and in accordance with guidelines of the Institutional Animal Care Committee.

1. C57BL/6 (B6) mice used as wild-type (WT) mice.

2. B6.OT-I T-cell receptor (TCR) transgenic mice harboring OVA-specific CD8⁺ T cells.

3. B6.CD45.1⁺ OT-I mice (bred in-house by crossing B6.OT-I mice with CD45.1⁺ B6 mice) [10, 11].

4. B6.Cd205-IRES2DTREGFP knock-in (KI) mice (*Cd205*$^{dtr/dtr}$ mice) (*see* **Note 1**).

2.2 Generation of Bone Marrow Chimeric Mice and DT Treatment

1. Scissors.

2. Pincette.

3. Petri dish.

4. 1-mL tuberculin syringe.

5. 27$^{G3/4}$ needle.

6. 100-μm cell strainer.

7. 50-mL polypropylene conical tube.

8. Dulbecco's phosphate-buffered saline without calcium and magnesium (D-PBS (−)). Store at 4 °C.

9. Solution of 70 % ethanol.

10. 0.04 % trypan blue stain (Gibco/BRL), dilute in D-PBS (−), and store at room temperature.

11. Refrigerated centrifuge.

12. Irradiator with ^{137}Cs source.

13. Weighing scale.

14. 100 μg/mL diphtheria toxin (DT) (Sigma-Aldrich): dissolve in D-PBS (−), and store in single use aliquots at −20 °C.

2.3 Assessment of Depletion Efficiency

1. Culture medium: RPMI 1640 medium containing L-glutamine supplemented with 10 % fetal bovine serum (FBS) and antibiotic-antimycotic mixed stock solution. Store at 4 °C.

2. Dulbecco's phosphate-buffered saline without calcium and magnesium (D-PBS (−)). Store at 4 °C.

3. Red Blood Cell Lysing Buffer Hybri-Max™. Store at room temperature.

4. 1-mL tuberculin syringe.

5. 27$^{G3/4}$ needle.

6. 100-μm cell strainer.

7. 50-mL polypropylene conical tube.

8. 5-mL tuberculin syringe.

9. 35 mm/tissue culture dish.

10. 4000 U/mL collagenase type III (Worthington Biochemical): dissolve in D-PBS (−), and store in single use aliquots at −20 °C.

11. Refrigerated centrifuge.

12. Water jacket-type CO_2 incubator.

13. 0.04 % trypan blue stain (Gibco/BRL), dilute in D-PBS (−), and store at room temperature.

14. AutoMACS Running Buffer (Miltenyi Biotec). Store at 4 °C.

15. Mouse CD11c (clone N418) Microbeads (Miltenyi Biotec). Store at 4 °C.

16. 5-mL polystyrene round-bottom tube with a cell-strainer cap.

17. 1.5-mL snaplock microtube.

18. AutoMACS Separator (Miltenyi Biotec).

19. AutoMACS Separation column (Miltenyi Biotec).

20. 0.5 mg/mL anti-mouse CD16/CD32 mAb (clone 2.4G2), store at 4 °C.

21. 0.5 mg/mL Alexa Fluor@ 488-conjugated anti-mouse CD205 mAb (clone NLDC-145), store at 4 °C.

22. 0.2 mg/mL R-phycoerythrin (R-PE)-conjugated anti-mouse CD8α (clone 53-6.7), store at 4 °C.

23. 0.2 mg/mL APC-conjugated anti-mouse CD11c mAb (clone HL3), store at 4 °C.

24. 25 μg/mL propidium iodide (PI), dissolve in D-PBS (−) and store in single use aliquots at −20 °C.

25. Flow cytometer.

2.4 OVA Cross Presentation Assay

1. Culture medium: RPMI 1640 medium containing L-glutamine supplemented with 10% fetal bovine serum (FBS) and antibiotic-antimycotic mixed stock solution. Store at 4 °C.

2. Dulbecco's phosphate-buffered saline without calcium and magnesium (D-PBS (−)). Store at 4 °C.

3. Red Blood Cell Lysing Buffer Hybri-Max™. Store at room temperature.

4. 1-mL tuberculin syringe.

5. $27^{G3/4}$ needle.

6. 100-μm cell strainer.

7. 50-mL polypropylene conical tube.

8. 5-mL tuberculin syringe.

9. 35 mm/tissue culture dish.

10. 4000 U/mL collagenase type III (Worthington Biochemical): dissolve in D-PBS (−), and store in single use aliquots at −20 °C.

11. Refrigerated centrifuge.

12. Water jacket-type CO_2 incubator

13. 0.04 % trypan blue stain (Gibco/BRL): dilute in D-PBS (−), and store at room temperature.

14. 1× BD IMag buffer. Prepare 1× BD IMag buffer by diluting 10× BD IMag buffer (BD Biosciences) with sterile distilled water (1:10). Store at 4 °C.

15. Mouse CD8 T lymphocyte Enrichment Set-DM (BD Biosciences). Store at 4 °C.

16. BD IMagnet (BD Biosciences).

17. BD IMag Streptavidin Particles Plus—DM (BD Biosciences). Store at 4 °C.

18. 5-mL polystyrene round-bottom tube.

19. 5-mL polystyrene round-bottom tube with a cell-strainer cap.

20. 1.5-mL snaplock microtube

21. 1 mM Carboxyfluorescein diacetate-succinimidyl ester (CFSE) (Molecular Probes): dissolve in DMSO and store in single use aliquots at −20 °C.

22. 1 mg/mL OVA protein (Sigma-Aldrich): dissolve in D-PBS (−), and store in single use aliquots at −20 °C.

23. *Listeria monocytogenes* expressing OVA (LM-OVA) (DMX) [12] (*see* **Note 2**). Store at −80 °C in single use aliquots at 1×10^9 colony-forming units (CFU).

24. 0.5 mg/mL anti-mouse CD16/CD32 mAb (clone 2.4G2), store at 4 °C.

25. 0.2 mg/mL R-phycoerythrin (R-PE)-conjugated anti-mouse CD8α (clone 53-6.7), store at 4 °C.

26. 0.2 mg/mL APC-conjugated anti-mouse CD45.1 (clone A20), store at 4 °C.

27. 25 μg/mL propidium iodide (PI), dissolve in D-PBS (−) and store in single use aliquots at −20 °C.

28. Flow cytometer.

2.5 CTL Generation Assay

1. Culture medium: RPMI 1640 medium containing L-glutamine supplemented with 10 % fetal bovine serum (FBS) and antibiotic-antimycotic mixed stock solution. Store at 4 °C.

2. Dulbecco's phosphate-buffered saline without calcium and magnesium (D-PBS (−)). Store at 4 °C.

3. Red Blood Cell Lysing Buffer Hybri-Max™. Store at room temperature.

4. 1-mL tuberculin syringe.

5. 27$^{G3/4}$ needle.

6. 100-μm cell strainer.

7. 50-mL polypropylene conical tube.

8. 5-mL tuberculin syringe.

9. 35 mm/tissue culture dish.

10. 4000 U/mL collagenase type III (Worthington Biochemical): dissolve in D-PBS (–), and store in single use aliquots at –20 °C.

11. Refrigerated centrifuge.

12. Water jacket-type CO_2 incubator.

13. 0.04 % trypan blue stain (Gibco/BRL), dilute in D-PBS (–), and store at room temperature.

14. 1.5-mL snaplock microtube.

15. 1 mg/mL OVA protein (Sigma-Aldrich): dissolve in D-PBS (–), and store in single use aliquots at –20 °C.

16. 1 mg/mL poly (I:C) (InvivoGen): dissolve in DNase- and RNAse-free distilled water, and store in single use aliquots at –20 °C.

17. 1 mg/mL anti-mouse CD40 mAb (clone 1C10), store at 4 °C.

18. 0.5 mg/mL anti-mouse CD16/CD32 mAb (clone 2.4G2), store at 4 °C.

19. FITC-conjugated anti-mouse CD8α (clone 53-6.7), store at 4 °C.

20. R-PE-conjugated H-2K^b OVA tetramer-SIINFEKL (MBL), store at 4 °C. 0.2 mg/mL.

21. APC-conjugated anti-mouse CD44 mAb (clone IM7), store at 4 °C.

22. 25 μg/mL propidium iodide (PI), dissolve in D-PBS (–) and store in single use aliquots at –20 °C.

23. Flow cytometer.

24. 1 mM synthesized OVA$_{257-264}$ peptide (SIINFEKL; >90 % pure), dissolve in D-PBS (–) and store in single use aliquots at –20 °C.

25. BD Cytofix/Cytoperm™ Fixation/Permeabilization Solution Kit with BD GolgiPlug™ (BD Biosciences), store at 4 °C.

26. APC-conjugated anti-mouse CD8α (clone 53-6.7), store at 4 °C.

27. R-PE-conjugated anti-mouse IFN-γ mAb (clone XMG1.2), store at 4 °C.

2.6 Assay for Antibacterial Immunity

1. Culture medium: RPMI 1640 medium containing L-glutamine supplemented with 10 % fetal bovine serum (FBS) and antibiotic-antimycotic mixed stock solution. Store at 4 °C.

2. Dulbecco's phosphate-buffered saline without calcium and magnesium (D-PBS (–)). Store at 4 °C.

3. Red Blood Cell Lysing Buffer Hybri-Max™. Store at room temperature.

4. 1-mL tuberculin syringe.

5. 27$^{G3/4}$ needle.

6. 100-μm cell strainer.

7. 50-mL polypropylene conical tube.

8. 5-mL tuberculin syringe.

9. 35 mm/tissue culture dish.

10. 4000 U/mL collagenase type III (Worthington Biochemical): dissolve in D-PBS (−), and store in single use aliquots at −20 °C.

11. Refrigerated centrifuge.

12. Water jacket-type CO_2 incubator.

13. 0.04 % trypan blue stain (Gibco/BRL), dilute in D-PBS (−), and store at room temperature.

14. 1.5-mL snaplock microtube.

15. *Listeria monocytogenes* expressing OVA (LM-OVA) (DMX) [12] (*see* **Note 2**). Store at −80 °C in single use aliquots at 1×10^9 colony-forming units (CFU).

16. Petri dish.

17. Oxford-Listeria-Selective-Agar (Merck). Store at 4 °C.

18. Oxford-Listeria-Selective Supplement (Merck). Store at 4 °C.

3 Methods

3.1 Generation of Bone Marrow (BM) Chimeric Mice and DT Treatment

3.1.1 Preparation of Single-Cell Suspensions from the Bone Marrow (BM)

This preparation must be performed in sterile conditions:

1. Sacrifice female WT mice or *Cd205*$^{dtr/dtr}$ mice.

2. Remove tibias and femurs from sacrificed mice and place the bones in a clean petri dish containing 10 mL of ice-cold D-PBS (−) until bone collection is finished.

3. Place the bones in a clean petri dish containing 10 mL of a solution of 70 % ethanol for 2 min.

4. Remove any remaining muscles from the bones using sterile scissors and pincette and rinse the bones with D-PBS (−).

5. Cut both ends of each bone with scissors and transfer them in a clean petri dish containing 10 mL of D-PBS (−).

6. Flush the BM out of the bones using a 1-mL tuberculin syringe with a 27$^{G3/4}$ needle.

7. Force the BM through a 100-μm cell strainer on a 50-mL polypropylene conical tube in order to obtain a single-cell suspension (*see* **Note 3**).

8. Centrifuge the cells ($780 \times g$, 5 min, 4 °C), and carefully discard all the supernatant.

9. Resuspend the cell pellet in 10 mL of D-PBS (−).

10. Count the leukocytes using trypan blue stain, and keep on ice (*see* **Note 4**).

3.1.2 Generation of BM Chimeric Mice

1. Irradiate female recipient WT mice with a total 10 Gy body irradiation, split into two doses separated by 4 h (*see* **Note 10** and **11**) to minimize gastrointestinal toxicity.

2. Inject recipient WT mice intravenously with 5×10^6 BM cells from WT mice or $Cd205^{dtr/dtr}$ mice (WT → WT chimeras or $Cd205^{dtr/dtr}$ → WT chimeras, respectively).

3. Allow the BM chimeric mice to rest for 8 weeks before using for experiments.

3.1.3 DT Treatment

1. Weight each BM chimeric mouse to determine their body weight.

2. Inject BM chimeric mice intraperitoneally with or without DT at a dose of 25 ng/g of body weight using a 1-mL tuberculin syringe with a $27^{G3/4}$ needle.

3. Wait until the next day to use the mice for an experiment.

3.2 Assessment of Depletion Efficiency

Depletion efficiency can be assessed 1 day after DT treatment by staining splenic DC.

3.2.1 Preparation of Single-Cell Suspensions from the Spleen of DT-Treated Mice

1. Sacrifice DT-treated mice.

2. Remove the spleen from sacrificed mice.

3. Keep the spleen in 3 mL of culture medium containing 400 U/mL collagenase type III in a 35 mm/tissue culture dish.

4. Inject the spleen with a total volume of 300 μl of culture medium containing 400 U/mL collagenase type III at several sites using a 1-mL tuberculin syringe with a $27^{G3/4}$ needle.

5. Incubate for 20 min at 37 °C in a Water jacket-type CO_2 incubator.

6. Place the digested spleen on a 100-μm cell strainer and put back in the 35 mm/tissue culture dish in culture medium containing 400 U/mL collagenase type III.

7. Grind the spleen through the cell strainer using the plunger of a 5-mL tuberculin syringe (*see* **Note 5**).

8. Collect the single-cell suspension in 50-mL polypropylene conical tubes, and add 30 mL of D-PBS (−).

9. Centrifuge the cells ($780 \times g$, 5 min, 4 °C), and carefully discard all the supernatant.

10. Resuspend the cell pellet in 10 mL of Red Blood Cell Lysing Buffer Hybri-Max™, vortex, and incubate for 5 min at room temperature.

11. Add 40 mL of D-PBS (−) to the 50-mL polypropylene conical tube containing the hemolyzed cell suspension and vortex.

12. Centrifuge the cells ($780 \times g$, 5 min, 4 °C), and carefully discard all the supernatant.

13. Resuspend the cell pellet in 10 mL of culture medium.

14. Filter the single-cell suspension through a 100-μm cell strainer on 50-mL polypropylene conical tubes (*see* **Note 3**).

15. Count the leukocytes using trypan blue stain, and keep on ice (*see* **Note 4**).

3.2.2 Purification of CD11c⁺ DCs from Single-Cell Suspensions

CD11c⁺ DCs are positively selected with Mouse CD11c (N418) Microbeads and the AutoMACS Separator from splenic single-cell suspensions obtained from BM chimeric mice according to the manufacturer's instructions (*see* **Note 6**) with some modifications.

1. Wash splenic single-cell suspensions with a 10× excess volume of AutoMACS Running Buffer in 50 mL polypropylene conical tubes, centrifuge ($780 \times g$, 5 min, 4 °C), and carefully discard all the supernatant.

2. Resuspend cell pellets in 400 μl of AutoMACS Running Buffer per 10^8 cells.

3. Add 100 μl of Mouse CD11c Microbeads per 10^8 cells.

4. Mix well and incubate for 15 min at 4 °C in the refrigerator.

5. Wash the labeled cells by adding 1–2 mL of AutoMACS Running Buffer per 10^7 total cells and centrifuge ($780 \times g$, 5 min, 4 °C). Carefully discard all the supernatant.

6. Resuspend up to 10^8 cells in 500 μl of AutoMACS Running Buffer, and filter through a 5-mL polystyrene round-bottom tube with a cell-strainer cap.

7. Prepare and prime the AutoMACS Separator (*see* **Note 7**).

8. Apply the tube containing the sample and provide tubes (50-mL polypropylene conical tubes) for collecting the labeled and unlabeled cell fractions.

9. Place sample tube at the uptake port and the fraction collection tubes at port neg1 and port pos2. The positive fraction contains CD11c⁺ DCs with Mouse CD11c Microbeads.

10. Recover CD11c⁺DCs by centrifugation ($780 \times g$, 5 min, 4 °C), and carefully discard all the supernatant.

11. Resuspend cell pellets in 10 mL of D-PBS (−).

12. Count the cells using trypan blue stain, and keep on ice (*see* **Note 8**).

<table>
<tr><td>

3.2.3 Analysis of DC Depletion Efficiency by Flow Cytometry

</td><td>

1. Place 5×10^5 CD11c$^+$ cells from each sample in 1.5-mL snaplock microtubes. Spin down ($151 \times 100 \times g$, 1 min, 4 °C) and carefully discard all the supernatant.

2. Resuspend cell pellets in 1 mL of D-PBS (−) and spin down ($151 \times 100 \times g$, 1 min, 4 °C). Carefully discard all the supernatant.

3. Resuspend cell pellets in 100 µl of D-PBS (−), and keep on ice.

4. Add 1 µg/10^6 cells of anti-mouse CD16/CD32 mAb with pulse vortexing and incubate for 5 min on ice prior to staining.

5. Add 1 µg/10^6 cells of Alexa Fluor@ 488-conjugated anti-mouse CD205 mAb, R-PE-conjugated anti-mouse CD8α, and APC-conjugated anti-mouse CD11c mAb to each tube with pulse vortexing, and incubate for 30 min on ice in the dark.

6. Wash the labeled cells by adding 1 mL of D-PBS (−) and spin down ($151 \times 100 \times g$, 1 min, 4 °C). Carefully discard all the supernatant.

7. Resuspend stained cell pellets in 500 µl of D-PBS (−) containing PI (*see* **Note 9**) with pulse vortexing and analyze samples on a flow cytometer. Example results are shown in Fig. 1a.

</td></tr>
</table>

3.3 OVA Cross Presentation Assay In Vivo

In this assay, the BM chimeric mice are injected with DT on day 1. On day 2, the mice are transferred with CFSE-labeled OT-I T cells. On day 3, mice are immunized with OVA or OVA-expressing *Listeria monocytogenes* bacteria (LM-OVA). On day 6, the mice are sacrificed and CD8 T cells are isolated from the spleens. CD8 T-cell activation is assessed by flow cytometry.

3.3.1 Preparation of Single-Cell Suspensions from the Spleen of OT-I Mice

1. Sacrifice B6.CD45.1$^+$ OT-I mice.

2. Follow **steps 2–15** from Subheading 3.2.1.

3.3.2 Purification of CD8$^+$ T Cells from Single-Cell Suspensions

CD8$^+$ T cells are negatively selected with mouse CD8 T lymphocyte Enrichment Set-DM and BD IMagnet according to the manufacturer's instructions (*see* **Note 10**) with some modifications.

1. Wash the splenic single-cell suspension with a 10× excess volume of 1× BD IMag buffer in a 50-mL polypropylene conical tube, centrifuge ($780 \times g$, 5 min, 4 °C), and carefully discard all the supernatant.

2. Resuspend the cell pellet in 1× BD IMag buffer at a concentration of 20×10^6 cells/mL.

3. Add the Biotinylated Mouse CD8 T Lymphocyte Enrichment Cocktail at 5 µl per 1×10^6 cells, vortex and incubate on ice for 15 min.

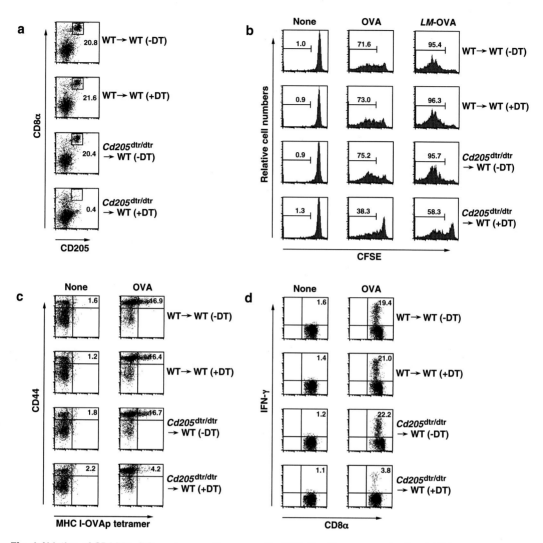

Fig. 1 Ablation of CD205+ cDCs reduces antigen-specific CD8+ T-cell responses in vivo. (**a**) WT → WT chimeras and *Cd205*^dtr/dtr → WT chimeras were injected with DT, and the spleen was obtained 2 days after the injection. The frequency of CD11c+CD8α+CD205+ cDCs was analyzed by flow cytometry. Data are represented by a dot plot, and numbers represent the proportion of CD8α+CD205+ cDCs among CD11c+ cells in each quadrant (*see* **Note 13**). (**b**) CFSE-labeled CD45.1+OT-I CD8+ T cells were transferred into WT → WT chimeras and *Cd205*^dtr/dtr → WT chimeras that had been treated with DT, and then the mice were immunized with or without OVA protein. Alternatively, the chimeric mice were uninfected or infected with LM-OVA. Antigen-specific division of CD45.1+OT-I CD8+ T cells was analyzed at 3 days after the immunization or infection by flow cytometry. Data are represented by a histogram, and numbers represent the proportion of CFSE dilution among gated CD45.1+OT-I CD8+ T cells in each quadrant (*see* **Note 13**). (**c** and **d**) WT → WT chimeras and *Cd205*^dtr/dtr → WT chimeras that had been treated with DT were immunized with or without OVA protein in combination with poly (I:C) plus anti-CD40 mAb. At 6 days after the immunization, splenocytes were analyzed for the generation of MHC I-OVA tetramer+CD44^high CD8+ T cells (**c**) and for intracellular IFN-γ-producing CD8+ T cells (**d**) by flow cytometry. Data are represented by a dot plot, and numbers represent the proportion of MHC I-OVA tetramer+CD44^high cells (**c**; *see* **Note 13**) and IFN-γ+ cells (**d**; *see* **Note 14**) among gated CD8+ T cells in each quadrant

4. Wash the labeled cells with a 10× excess volume of 1× BD IMag buffer, centrifuge ($780 \times g$, 5 min, 4 °C), and carefully discard all the supernatant.

5. Vortex the BD™ IMag Streptavidin Particles Plus—DM thoroughly—and add to the cells 5 µl of particles for every 1×10^6 cells.

6. Mix thoroughly and refrigerate for 30 min at 4 °C.

7. Bring the labeling volume up to 20×10^6 cells/mL with 1× BD IMag buffer (maximum volume added not to exceed 3 mL).

8. Transfer the labeled cells to a 5-mL polystyrene round-bottom tube.

9. Place this "positive fraction" tube on the BD IMagnet (horizontal position) for 6 min.

10. With the tube on the BD IMagnet and using a 1 mL tip, carefully aspirate the supernatant ("enriched fraction") and place in a new sterile tube.

11. Remove the "positive fraction" tube from the BD IMagnet, and add 1× BD IMag buffer to the same volume as in **step 7**. Resuspend the positive fraction well by pipetting up and down 15 times, and place the tube back on the BD IMagnet for 6 min.

12. Using a 1 mL tip, carefully aspirate the supernatant and combine with the enriched fraction from **step 10**.

13. Repeat **steps 11** and **12**. The combined enriched fraction contains CD8⁺ T cells with no bound antibodies or magnetic particles.

14. Centrifuge the CD8⁺ T cells ($780 \times g$, 5 min, 4 °C), and carefully discard all the supernatant.

15. Resuspend the cell pellet in 10 mL of D-PBS (−), and filter using a 5-mL polystyrene round-bottom tube with a cell-strainer cap.

16. Count the cells using trypan blue stain, and keep on ice (*see* **Note 11**).

3.3.3 CFSE Labeling of OT-I T Cells

1. Prepare 5×10^6 CD45.1⁺ OT-I CD8⁺ T cells in 1 mL of D-PBS (−) per mouse to be injected, and place the cells in a 1.5-mL snaplock microtube.

2. Add 2.5 µl of 1 mM CFSE per 5×10^6 (final concentration of 2.5 µM), vortex and incubate at 37 °C for 10 min in a water jacket-type CO_2 incubator.

3. Spin down the cells ($151 \times 100 \times g$, 1 min, 4 °C), and carefully discarded all the supernatant.

4. Resuspend the cell pellets in 1 mL of D-PBS (−).

5. Spin down the cells ($151 \times 100 \times g$, 1 min, 4 °C), and carefully discard all the supernatant.

6. Resuspend the cell pellet in 250 μl of D-PBS (−), and transfer the labeled cells into a clean 1.5-mL snaplock microtube. Keep on ice.

3.3.4 Adoptive Transfer and Immunization

1. Inject DT-treated BM chimeric mice intravenously with CFSE-labeled CD45.1+ OT-I CD8+ T cells using a 1-mL tuberculin syringe with a 27$^{G3/4}$ needle.

2. The next day, prepare 50 μg of OVA protein in 250 μl of D-PBS (−) per mouse or 5×10^5 CFU LM-OVA in 250 μl of D-PBS (−) per mouse into a 1.5-mL snaplock microtube.

3. Inject mice intraperitoneally with 50 μg of OVA protein or 5×10^5 CFU LM-OVA using a 1-mL tuberculin syringe with a 27$^{G3/4}$ needle.

4. 3 days after immunization or bacterial infection, sacrifice the mice.

5. Remove the spleens and prepare splenic single-cell suspensions followed by CD8+ T-cell isolation as described above (Subheadings 3.3.1 and 3.3.2).

6. Count the cells using trypan blue stain, and keep on ice (*see* **Note 11**).

3.3.5 Analysis of CD8+ T-Cell Proliferation by Flow Cytometry

1. Place 5×10^5 cells from each CD8+ T-cell preparation in 1.5-mL snaplock microtube. Spin down ($151 \times 100 \times g$, 1 min, 4 °C) and carefully discard all the supernatant.

2. Resuspend the cell pellets in 1 mL of D-PBS (−) and spin down ($151 \times 100 \times g$, 1 min, 4 °C). Carefully discard all the supernatant.

3. Resuspend the cell pellets in 100 μl of D-PBS (−), and keep on ice.

4. Add 1 μg/10^6 cells of anti-mouse CD16/CD32 mAb, pulse vortex, and incubate for 5 min on ice prior to staining.

5. Add 1 μg/10^6 cells of R-PE-conjugated anti-mouse CD8α and APC-conjugated anti-mouse CD45.1 to each tube with pulse vortexing, and incubate for 30 min on ice in the dark.

6. Wash the labeled cells by adding 1 mL of D-PBS (−) and spin down ($151 \times 100 \times g$, 1 min, 4 °C). Carefully discard all the supernatant.

7. Resuspend stained cell pellets in 500 μl of D-PBS (−) containing PI (*see* **Note 9**) with pulse vortexing and analyze samples on a flow cytometer. Example results are shown in Fig. 1b.

3.4 CTL Generation Assay

In this assay, the BM chimeric mice are injected with DT on day 1. On day 2, the mice are immunized with a mixture of OVA, poly (I:C), and anti-CD40 mAb. On day 8, the mice are sacrificed and CTL generation is assessed by flow cytometry using tetramer staining and intracellular IFN-γ^+ staining.

3.4.1 Immunization

1. Prepare the mixture 100 µg of OVA protein, 50 µg poly (I:C), and 10 µg anti-CD40 mAb in 250 µl D-PBS (–)/mouse in a 1.5-mL snaplock microtube.

2. Inject DT-treated BM chimeric mice intraperitoneally with the mixture of OVA protein, poly (I:C), and anti-CD40 mAb using a 1-mL tuberculin syringe with a 27$^{G3/4}$ needle.

3. 6 days after immunization, remove spleens from treated mice, and prepare hemolyzed single-cell suspensions as described above (Subheading 3.2.1).

4. Count leukocytes using trypan blue stain, and keep on ice (*see* **Note 4**).

3.4.2 Tetramer Staining of CD8+ T Cells

1. Place 5×10^5 cells from each splenic preparation in 1.5-mL snaplock microtube. Spin down ($151 \times 100 \times g$, 1 min, 4 °C) and carefully discard all the supernatant.

2. Resuspend the cell pellets in 1 mL of D-PBS (–) and spin down ($151 \times 100 \times g$, 1 min, 4 °C). Carefully discard all the supernatant.

3. Resuspend the cell pellets in 100 µl of D-PBS (–), and keep on ice.

4. Add 1 µg/10^6 cells of anti-mouse CD16/CD32 mAb, pulse vortex, and incubate for 5 min on ice prior to staining.

5. Add 1 µg/10^6 cells of FITC-conjugated anti-mouse CD8α mAb, 1 test/10^6 cells R-PE-conjugated H-2K^b OVA tetramer-SIINFEKL, and 1 µg/10^6 cells of APC-conjugated anti-mouse CD44 mAb to each tube with pulse vortexing, and incubate for 30 min on ice in the dark.

6. Wash the labeled cells by adding 1 mL of D-PBS (–) and spin down ($151 \times 100 \times g$, 1 min, 4 °C). Carefully discard all the supernatant.

7. Resuspend stained cell pellets in 500 µl of D-PBS (–) containing PI (*see* **Note 9**) with pulse vortexing and analyze samples on a flow cytometer. Example results are shown in Fig. 1c.

3.4.3 Intracellular Staining for CD8+IFN-γ^+ T Cells

The following protocol for intracellular staining of IFN-γ is performed using R-PE-conjugated anti-mouse IFN-γ mAb and BD Cytofix/Cytoperm™ Fixation/Permeabilization Solution Kit with BD GolgiPlug™ according to the manufacturer's instructions (*see* **Note 12**) with some modifications.

1. Place single-cell suspensions of 5×10^6 hemolyzed splenocytes in 1.5-mL snaplock microtubes. Spin down ($151 \times 100 \times g$, 1 min, 4 °C) and carefully discard all the supernatant.

2. Resuspend cell pellets in 1 mL of culture medium in a 1.5-mL snaplock microtube, and keep on ice.

3. Add 10 μl of 1 mM OVA$_{257-264}$ peptide per 5×10^6 cells (final concentration of 10 μM), vortex, and incubate at 37 °C for 2 h in water jacket-type CO$_2$ incubator.

4. Add 1 μl of GolgiPlug™ to each tube, vortex, and incubate at 37 °C during the final 2 h in water jacket-type CO$_2$ incubator.

5. Spin down ($151 \times 100 \times g$, 1 min, 4 °C) and carefully discard all the supernatant.

6. Resuspend cell pellets in 1 mL of D-PBS (−) and spin down ($151 \times 100 \times g$, 1 min, 4 °C). Carefully discard all the supernatant.

7. Resuspend cell pellets in 100 μl of D-PBS (−), and keep on ice.

8. Add 1 μg/10^6 cells of anti-mouse CD16/CD32 mAb with pulse vortexing and incubate for 5 min on ice prior to staining.

9. Add 1 μg/10^6 cells of APC-conjugated anti-mouse CD8α mAb to each tube with pulse vortexing, and incubate for 30 min on ice in the dark.

10. Wash the labeled cells by adding 1 mL of D-PBS (−) and spin down ($151 \times 100 \times g$, 1 min, 4 °C). Carefully discard all the supernatant.

11. Resuspend cell pellets in 250 μl of fixation/permeabilization solution in a 1.5-mL snaplock microtube, and incubate for 20 min at 4 °C in the dark.

12. Spin down ($151 \times 100 \times g$, 1 min, 4 °C) and carefully discard all the supernatant.

13. Wash the labeled cells by adding 1 mL of BD Perm/Wash™ buffer and spin down ($151 \times 100 \times g$, 1 min, 4 °C). Carefully discard all the supernatant.

14. Repeat **step 9**.

15. Resuspend cell pellets in 50 μl of BD Perm/Wash™ buffer.

16. Add 1 μg/10^6 cells of R-PE-conjugated anti-mouse IFN-γ mAb to each tube with pulse vortexing, and incubate for 30 min on ice in the dark.

17. Wash the labeled cells by adding 1 mL of BD Perm/Wash™ buffer and spin down ($151 \times 100 \times g$, 1 min, 4 °C). Carefully discard all the supernatant.

18. Repeat **step 13**.

19. Resuspend stained cell pellets in 500 µl of D-PBS (–) with pulse vortexing and analyze samples on a flow cytometer. Example results are shown in Fig. 1d.

3.5 Assay for Antibacterial Immunity

In this assay, the BM chimeric mice are injected with DT on day 1. On day 2, the mice are immunized with OVA-expressing *Listeria monocytogenes* bacteria (LM-OVA). Mouse survival is monitored for 14 days. On day 8, a cohort of mice are sacrificed, and bacterial burden is assessed by culturing single-cell splenic suspensions from infected mice.

3.5.1 Bacterial Infection

1. Prepare LM-OVA (5×10^5 CFU/mouse) in 250 µl of D-PBS (–) per mouse in a 1.5-mL snaplock microtube.

2. Inject DT-treated BM chimeric mice intraperitoneally with LM-OVA using a 1-mL tuberculin syringe with a $27^{G3/4}$ needle.

3. Monitor survival for 14 days. Example results are shown in Fig. 2a.

3.5.2 Determination of Bacterial Burden

1. Sacrifice BM chimeric mice 6 days after bacterial infection and remove their spleens.

2. Prepare splenic single-cell suspensions as described above (Subheading 3.2.1).

3. Place 10^7 cells of each splenic single-cell suspension in 1 mL D-PBS (–) in a 1.5-mL snaplock microtube.

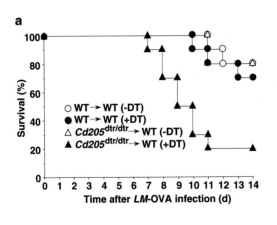
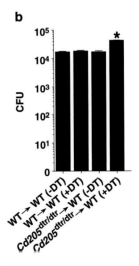

Fig. 2 Ablation of CD205⁺ cDCs controls host defenses against bacterial infections in vivo. WT → WT chimeras and *Cd205*^dtr/dtr → WT chimeras that had been treated with DT were uninfected or infected with LM-OVA. (**a**) Survival was monitored for 14 days. (**b**) Bacterial burden in the spleen was determined as CFU 6 days after infection. *$P < 0.01$ compared with WT → WT chimeras. Data are the mean ± s.d

4. Prepare twofold serial dilutions of splenic single-cell suspensions in 500 μl of D-PBS (−) in a 1.5-mL snaplock microtube.

5. Transfer the diluted splenic single-cell suspensions to a clean petri dish containing Oxford-Listeria-Selective-Agar supplemented with Oxford-Listeria-Selective Supplement prepared according to the manufacturer's instructions.

6. Count bacterial colonies after an overnight incubation at 37 °C. Example results are shown in Fig. 2b.

4   Notes

1. B6.Cd205-IRES2DTREGFP KI mice were created by the introduction of the knock-in cassette containing cDNA encoding the human DTR fused to enhanced GFP (EGFP) and an internal ribosome entry site (IRES) [7, 10, 11] into the 3′ untranslated region of the *Cd205* gene . B6.Cd205-IRES2DTREGFP KI mice can be obtained from RIKEN BioResource Center (http://en.brc.riken.jp/index.shtmL; RBRC05659) through a material transfer agreement (MTA).

2. *Listeria monocytogenes* expressing OVA (LM-OVA) are grown in brain-heart infusion broth (BHI) (Sigma-Aldrich) medium.

3. Cell debris are removed by filtration through a 100-μm cell strainer.

4. The number of leukocytes in BM and spleen obtained from B6 mice is approximately 3×10^7/mouse and 1.5×10^8/mouse, respectively.

5. Undigested fibrous material is removed by filtration through a 100-μm cell strainer.

6. Technical data sheet is available at https://www.miltenyibiotec.com/~/media/Images/Products/Import/0001200/IM0001247.ashx.

7. AutoMACS Separator program is "Posseld2".

8. The number of CD11c⁺DCs obtained from total splenocytes (1.5×10^8/mouse) of B6 mice is approximately 1.5×10^6/mouse.

9. PI-stained cells are excluded from the analysis.

10. Technical data sheet is available at

    https://www.bdbiosciences.com/external_files/pm/doc/tds/cell_sep/live/web_enabled/558471.pdf#search='mouse+CD8+T+lymphocyte+Enrichment+SetDM'.

11. The number of CD8⁺ T cells obtained from total splenocytes (1.5×10^8/mouse) of B6 mice is approximately 5×10^6/mouse.

12. Technical data sheet is available at http://www.bdbiosciences.com/external_files/pm/doc/manuals/live/web_enabled/00-81014-4-C.pdf.

13. Data are the results of cell surface staining.

14. Data are the results of intracellular staining.

Acknowledgments

The author would like to thank Yumiko Sato for excellent technical assistance. This work was supported by Grants-in-Aid for Scientific Research from the Ministry of Education, Science and Culture of Japan (B) 25293117 (K.S.) and Japan Science and Technology Agency, Precursory Research for Embryonic Science and Technology (PRESTO) (K.S).

References

1. Villadangos JA, Schnorrer P (2007) Intrinsic and cooperative antigen-presenting functions of dendritic-cell subsets in vivo. Nat Rev Immunol 27:543–555

2. Dudziak D, Kamphorst AO, Heidkamp GF et al (2007) Differential antigen processing by dendritic cell subsets in vivo. Science 315:107–111

3. Hildner K, Edelson BT, Purtha WE et al (2008) Batf3 deficiency reveals a critical role for CD8α+ dendritic cells in cytotoxic T cell immunity. Science 322:1097–1100. doi:10.1126/science.1164206

4. Yamazaki S, Dudziak D, Heidkamp GF et al (2008) CD8+ CD205+ splenic dendritic cells are specialized to induce Foxp3+ regulatory T cells. J Immunol 181:6923–6933. doi:10.4049/jimmunol.1300975

5. Shortman K, Heath WR (2010) The CD8+ dendritic cell subset. Immunol Rev 234:18–31. doi:10.1111/j.0105-2896.2009.00870.x

6. Jiang W, Swiggard WJ, Heufler C et al (1995) The receptor DEC-205 expressed by dendritic cells and thymic epithelial cells is involved in antigen processing. Nature 375:151–155

7. Kissenpfennig A, Henri S, Dubois B et al (2005) Dynamics and function of Langerhans cells in vivo: dermal dendritic cells colonize lymph node areas distinct from slower migrating Langerhans cells. Immunity 22:643–654

8. Kamphorst AO, Guermonprez P, Dudziak D et al (2010) Route of antigen uptake differentially impacts presentation by dendritic cells and activated monocytes. J Immunol 185:3426–3435. doi:10.4049/jimmunol.1001205

9. Jung S, Unutmaz D, Wong P et al (2002) In vivo depletion of CD11c+ dendritic cells abrogates priming of CD8+ T cells by exogenous cell-associated antigens. Immunity 17:211–220

10. Takagi H, Fukaya T, Eizumi K et al (2011) Crucial role of plasmacytoid dendritic cells in the initiation of inflammation and T cell immunity in vivo. Immunity 35:958–971. doi:10.1016/j.immuni.2011.10.014

11. Fukaya T, Murakami R, Takagi H et al (2012) Conditional ablation of CD205+ conventional dendritic cells impacts the regulation of T-cell immunity and homeostasis in vivo. Proc Natl Acad Sci U S A 109:11288–11293. doi:10.1073/pnas.1202208109

12. Foulds KE, Zenewicz LA, Shedlock DJ et al (2002) Cutting edge: CD4 and CD8 T cells are intrinsically different in their proliferative responses. J Immunol 168:1528–1532

Chapter 22

Generation of Humanized Mice for Analysis of Human Dendritic Cells

Yasuyuki Saito, Jana M. Ellegast, and Markus G. Manz

Abstract

Transplantation of human CD34+ hematopoietic stem and progenitor cells into severe immunocompromised newborn mice allows the development of a human hemato-lymphoid system (HHLS) including dendritic cells (DCs) in vivo. Therefore, it can be a powerful tool to study human DC subsets, residing in different lymphoid and nonlymphoid organs. We have recently generated novel mouse strains called human cytokine knock-in mice in which human versions of several cytokines are knocked into Rag2$^{-/-}$γC$^{-/-}$ strains. In addition, human SIRPα, which is a critical factor to prevent donor cell to be eliminated by host macrophages, is expressed as transgene. These mice efficiently support human myeloid cell development and, indeed, allow the analysis of three major subsets of human DC lineages, plasmacytoid DCs and CD1c+ and CD141+ classical DCs. Moreover, these strains also support cytokine-mobilized peripheral blood CD34+ cell engraftment and subsequent DC development. Here we describe our standard methods to characterize DCs developed in human cytokine knock-in mice.

Key words Humanized mouse, CD34+ hematopoietic stem and progenitor cell, Stem cell transplantation, Human dendritic cells

1 Introduction

Human dendritic cells (DCs) are heterogeneous and exist in different tissues [1–3]. Although several DC subsets have been identified and each DC subset has been functionally characterized in mice, human counterparts of DC subsets have been less well characterized. Human DCs are not only circulating in blood but can be also found in a variety of tissues including lymphoid and nonlymphoid organs. These tissue DCs appear to be different in terms of origin and function from monocyte-derived DCs that are commonly used in in vitro experiments. In addition, some markers to distinguish DC subsets are differently expressed in humans and mice. Thus, it is difficult to characterize the function of human DC subsets obtained from peripheral blood of patients or healthy donors.

Upon transplantation of human CD34+ hematopoietic stem cells (HSC) and progenitor cells into severe immunocompromised

Elodie Segura and Nobuyuki Onai (eds.), *Dendritic Cell Protocols*, Methods in Molecular Biology, vol. 1423,
DOI 10.1007/978-1-4939-3606-9_22, © Springer Science+Business Media New York 2016

newborn mice, all subpopulations of classical DCs (cDC) and plasmacytoid DCs (pDC) develop in lymphoid organs and peripheral blood; strikingly percentages of subsets of DCs are comparable to those found in humans under physiological conditions [4, 5]. In addition to the lymphoid tissue DCs in humans, some DC subsets in nonlymphoid tissues can be found in HHLS mouse models [6].

We recently generated mouse strains called human cytokine knock-in mice in which human versions of cytokines (M-CSF, TPO, and IL-3/GM-CSF) are knocked into the respective locus of Rag2$^{-/-}$γC$^{-/-}$ strains [7]. In addition, human SIRPα, which is a critical factor to prevent donor cells to be eliminated by host macrophages [8, 9], is expressed as a transgene. Compared with second generation such as NOD.SCID.γC$^{-/-}$ (NSG) strains, these humanized mice show superior engraftment and development of human cells, especially of the myeloid lineage [7]. Human hematopoietic stem and progenitor cells (HSPCs) can be purified from different sources such as bone marrow (BM), cytokine (mostly G-CSF)-mobilized peripheral blood, umbilical cord blood, or fetal liver. The amount of CD34$^+$ cells needed for HHLS models are tenfold higher using adult-derived samples in comparison to neonatal and prenatal samples. Many protocols have been published regarding the isolation of CD34$^+$ cells from cord blood or fetal liver [10, 11], while there is few literature on CD34$^+$ cell isolation from cytokine-mobilized peripheral blood. We have recently shown substantial engraftment of cytokine-mobilized CD34$^+$ cells as well as DC development in human cytokine knock-in mice ([7] and unpublished data (YS, JME, and MGM, manuscript in preparation)). Here we provide our standard protocol to isolate CD34$^+$ cells from peripheral blood after mobilization with G-CSF as well as our protocol to characterize human dendritic cell lineages in HHLS models.

2   Materials

2.1   Isolation of CD34$^+$ Hematopoietic Stem and Progenitor Cells

1. Human cytokine-mobilized peripheral blood, within 24 h after mobilization. Store at 4 °C.

2. Ficoll-Hypaque Plus, 1.078 g/mL (GE healthcare). Store in the dark.

3. Sterile phosphate-buffered saline (PBS).

4. Fetal calf serum (FCS).

5. MACS buffer: PBS, 2 % FCS, 2 mM EDTA.

6. CD34 Microbeads kit (Miltenyi Biotec).

7. Cryopreservation solution: FCS, 10 % DMSO.

2.2 Intrahepatic Injection of Human CD34⁺ Cells

1. Newborn humanized mice (within 3 days).
2. Purified CD34⁺ cells.
3. Defreeze solution: IMDM, 10% FCS, 50 µg/mL DNaseI.
4. Sterile phosphate-buffered saline (PBS).
5. 70% ethanol.
6. Irradiator.
7. Auto-craved animal cage.
8. 15-mL polypropylene conical tube.
9. 1.5-mL polypropylene microcentrifuge tube.
10. Sterile paper.
11. 25-µL syringe.
12. 30G needle.
13. Scissors.

2.3 FACS Analysis of Human Dendritic Cells

1. Transplanted mice.
2. 70% ethanol.
3. Heparin-Natrium 25.000IE/5 mL.
4. Phosphate-buffered saline (PBS).
5. Mortar and pestle.
6. 70-µm nylon mesh.
7. 12-well plate.
8. Staining buffer: PBS, 2 mM EDTA, 2% human AB serum.
9. 4% paraformaldehyde (PFA) in PBS.
10. ACK lysis buffer: 150 mM NH_4Cl, 10 mM $KHCO_3$, 0.1 mM EDTA.
11. Digestion medium: RPMI 10% FCS, 2 mg/mL collagenase D (Roche), 50 µg/mL DNase I bovine pancreas grade II (Sigma).
12. Ficoll-Hypaque Plus, 1.078 g/mL (GE healthcare). Store in the dark.
13. 96-well plate.
14. Mouse FcγR-blocking reagent (clone 2.4G2).
15. Tricolor/PE-Cy5-conjugated antibodies against lineage antigens (hCD3, AKA7D6; hCD19, SJ25-C1; hCD20, HI47; hCD56, MEM-188; CD235a (glycophorin A), GAR-2 (HIR2); mouse CD45, 30-F11).
16. Primary antibodies (for 4 laser instruments): eFlour450-conjugated anti-hCD45 (HI30), FITC-conjugated anti-hCD1c (BDCA-1) (AD5-8E7), PE-Cy5.5-conjugated anti-hCD14 (M5E2), PE-conjugated anti-hHLA-DR (TU36), PE-Cy7-conjugated anti-hCD11c (3.9), APC-conjugated

anti-hCD141 (BDCA-3) (AD5-14H12), APC-Cy7-conjugated anti-hCD16 (3G8), biotin-conjugated anti-hCD304 (BDCA-4) (AD5-17F6).

17. Secondary antibody (for 4 laser instruments): Qdot 605 streptavidin.

18. Primary antibodies (for 3 laser instruments): e450-conjugated anti-hCD123 (6H6), V500-conjugated anti-hCD45 (HI30), FITC-conjugated anti-hCD1c (BDCA-1) (AD5-8E7), PE-conjugated anti-hHLA-DR (TU36), PE-Cy7-conjugated anti-hCD14 (M5E2), APC-conjugated anti-hCD141 (BDCA-3) (AD5-14H12), APC-Cy7-conjugated anti-hCD16 (3G8).

19. Live/dead marker: Aqua live/dead (for 4 laser) or propidium iodide (PI, for 3 laser).

20. 4-laser FACS instruments (e.g., BD LSRII Fortessa, BD FACS Aria) or 3-laser FACS instruments (e.g., BD FACS CantoII).

3   Methods

3.1   Isolation of CD34+ Hematopoietic Stem and Progenitor Cells

3.1.1   Isolation of Mononuclear Cells from Cytokine-Mobilized Peripheral Blood

1. Open PBMC collection bag in a biosafety hood (*see* Fig. 1).

2. Flush PBMC collection bag with 50 mL of sterile PBS to collect remaining cells.

3. Centrifuge at $400 \times g$ for 5 min at 18–20 °C, aspirate supernatant, and add 50 mL of sterile PBS.

4. Prepare two fresh 50-mL polypropylene tubes containing 15-mL Ficoll-Hypaque Plus isotonic solution.

5. Load the PBMC sample onto the Ficoll solution, being careful to minimize mixing of PBMC with the Ficoll solution (*see* **Note 1**).

6. Centrifuge for 30 min at $400 \times g$ at 18–20 °C, without acceleration and brake.

7. Carefully remove the interface containing mononuclear cells (MNCs) using pasteur pipette and transfer into a new 50-mL conical tube.

8. Add FACS buffer up to 50 mL and centrifuge for 5 min at $400 \times g$ at 4 °C.

9. Aspirate supernatant and resuspend the cells in 10 mL of FACS buffer and transfer them to a 15-mL conical tube. Count cell numbers.

3.1.2   Magnetic Labeling and Separation

1. Centrifuge for 5 min at $400 \times g$ at 4 °C, aspirate supernatant, and resuspend the cells in 300 μL of MACS buffer per 2×10^8 cells.

2. Add 100 μL of FcR-blocking reagent per 2×10^8 MNCs, mix gently, and incubate for 10 min on ice.

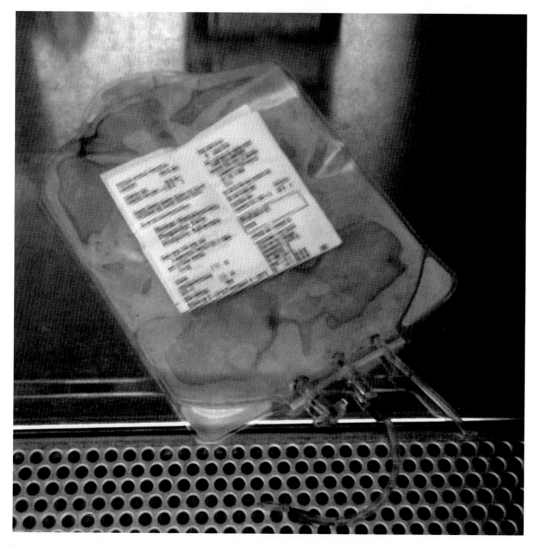

Fig. 1 PBMC collection bag after transplantation. A blood bag after allogenic stem cell transplantation is shown. We usually get $3-5 \times 10^8$ mononuclear cells and $3-5 \times 10^6$ CD34$^+$ cells from one bag

3. Add 100 μL of CD34 microbeads per 2×10^8 MNCs, mix gently, and incubate for 30 min at 4 °C to avoid unspecific binding.

4. Add 10 mL of MACS buffer and centrifuge for 5 min at $400 \times g$ at 4 °C.

5. Resuspend the cells in 500 μL of MACS buffer per 2×10^8 cells, and keep the cells on ice.

6. Prepare LS column(s) (up to 1×10^9 cells per column): place new 15-mL tubes under the columns and wash the columns with 3 mL of MACS buffer.

7. Apply the cell suspension onto the first column and collect the flow-through fraction.

8. Wash the column twice with 3 mL of MACS buffer, remove the LS column from the magnetic field and put it on a new 15-mL conical tube.

9. Load 3 mL of MACS buffer onto the LS column and flush the CD34+ fraction by gently pushing the plunger into the column.

10. To increase the purity of CD34+ cells, repeat **steps 6–9**.

11. Centrifuge collected effluent for 5 min at $400 \times g$ at 4 °C.

12. Count the cells, both from the flow-through fraction and from the CD34+ enriched cells.

13. Freeze aliquots of flow-through fraction and of CD34+ cells in cryopreservation solution at –80 °C. Keep a small fraction (e.g., 1×10^4 cells) of CD34+ cells for FACS staining to check for the purity of the isolated cells (*see* **Note 2**).

3.2 Intrahepatic Injection of Human CD34+ Cells

In the authors' laboratory, we normally inject cells into the liver of newborn mice within 24–48 h after birth. As the liver contributes to perinatal hematopoiesis, and the hemato-lymphoid system expands significantly during the first weeks of life, HSPCs should find optimal conditions to engraft, expand, and reconstitute a hemato-lymphoid system. This method allows us to inject without using special restraining devices or anesthesia. Animals should be bread and maintained under specific pathogen-free conditions in accordance with institutional guidelines.

3.2.1 Finger Cutting and Genotyping

In the authors' laboratory, we use genetically modified immunocompromised mice carrying a human *SIRPA* transgene [7, 9], and we maintain them as heterozygous breeding. We thus need to check whether newborns carry the transgene or not. We usually extract DNA from the toe(s) after birth in order to identify the transgene from individuals as well as to mark the individual animals.

1. Check breeding pairs for newborns after day 18 of gestation (*see* **Note 3**).

2. Place the cage with newborns in a sterile hood and lift off the mouse cover.

3. Hold the mouse and cut toe(s) with scissors.

4. Extract DNA from toe(s) and check genotype by PCR analysis (*see* **Note 4**).

3.2.2 Irradiation of Newborn Mice

To enhance engraftment of the human CD34+ cells into newborn mice, sublethal irradiation is believed to create "space" in the stem cell niche. Besides different sublethal dose between strains, the irradiation dose may vary between irradiation sources; therefore, it should be titrated carefully before starting experiments (*see* **Note 5**). In the author's laboratory (using a Rad Source RS-2000 instrument), a suitable dose is 1.5Gy for NSG, MISTRG, and MITRG mice, whereas we irradiate with 3.75Gy split into two doses separated by 4 h for Rag2$^{-/-}$γC$^{-/-}$ (RG) and RG. h*SIRPA*-transgenic newborns [7, 9].

1. Irradiate newborns in a sterile container with a uniform irradiation dose.

2. Carefully put irradiated mice back into their breeding cage.

3.2.3 Preparation of CD34$^+$ Cells

1. Thaw isolated human CD34$^+$ cells in a 37 °C water bath.

2. Place thawed cells in a 15-mL polypropylene tube and add a tenfold volume of defreeze solution.

3. Centrifuge for 5 min at 400×*g* at 4 °C.

4. Remove the supernatant, resuspend the cells in 1 mL of fresh PBS and place into a 1.5-mL polyethylene microcentrifuge tube.

5. Centrifuge for 5 min at 400×*g* at 4 °C.

6. Aspirate the supernatant, resuspend the cells with 200–500 µL of PBS and count the cells.

7. Centrifuge for 5 min at 400×*g* at 4 °C.

8. Calculate the volume needed (depending on the number of newborns), based on 1–2×10^5 cells in 25 µL of injection volume per mouse.

9. Remove PBS and resuspend cells with the appropriate volume of fresh sterile PBS.

3.2.4 Cell Injection for Transplantation

1. Place the cage with newborns in a sterile hood and lift off the mouse cage cover. Pick up all the newborns and place them on sterile tissue beside the cage.

2. Load the solution to be injected into a 25-µL Hamilton syringe with a 30G needle.

3. Carefully grasp one newborn by the shoulders and back of the neck with the thumb and forefinger. Grab enough of the skin to hold the mouse tightly. Without releasing the skin, lift the mouse with one hand. Turn it over (belly side up).

4. Pick up the syringe with the free hand. Hold the syringe in a way that the index finger is on the end of the syringe plunger.

5. Insert the needle subcutaneously (s.c.) at a 10° angle above the sternum.

6. Without changing the position or angles of the syringe/needle, carefully insert the needle into the skin and move subcutaneously through the costal margin toward the abdomen. The liver is visible under the right costal margin.

7. As soon as the needle has reached the site of the liver, dip the bevel into the liver and move forward a few millimeters.

Hold the animal and the syringe very still and carefully inject the cells.

8. Remove the needle. Turn the animal over and release it into the cage (*see* **Note 6**). If necessary, cut toe(s) with sterile scissors to mark the newborns.

3.3 FACS Analysis of Human Dendritic Cells from Humanized Mice

It is necessary to check transplanted animals to verify engraftment and differentiation of the human HSPCs as early as 4 weeks after transplantation by taking small amounts of blood (50–100 µL) (*see* **Note 7**). This protocol describes removal of lymphoid organs and nonlymphoid organs such as the liver and lung and preparation of single-cell suspensions.

3.3.1 Preparation of Cell Suspension from Peripheral Blood

1. Sacrifice transplanted mice according to institutional guidelines.

2. Fix the mice on a flat surface with needles and spray the mice with 70 % ethanol.

3. Bleed mice from the heart immediately after sacrifice and transfer the blood into 1.5-mL tubes containing 10 µL of heparin.

4. Centrifuge for 5 min at $400 \times g$ at 4 °C.

5. Discard the supernatants and add 1 mL of ACK buffer. Incubate for 2 min at room temperature to remove red blood cells.

6. Repeat **steps 4–5**.

7. Add 500 µL of ice-cold PBS and centrifuge for 5 min at $400 \times g$ at 4 °C.

3.3.2 Preparation of Cell Suspension from the Bone Marrow

1. Remove femurs, tibias, pelvis bones, and spine, and place them into ice-cold PBS. Remove the muscles from the bones.

2. Transfer the bones into a mortar and crash/grind the bones using a pestle, or add 10 mL of ice-cold PBS and flush out the marrow using a syringe and needle to obtain a bone marrow cell suspension from bone shaft. Pass the cell suspension through a 70-µm nylon mesh to remove debris.

3. Centrifuge for 5 min at $400 \times g$ at 4 °C.

4. Aspirate the supernatants, add 1 mL of ACK buffer and incubate for 2 min at room temperature to remove red blood cells.

5. Wash cells twice in PBS (5 min centrifugation at $400 \times g$ at 4 °C).

6. Resuspend the cells in 1 mL of staining buffer.

7. Count cell numbers and store about half of the cells in freezing solution at –80 °C.

3.3.3 Preparation of Cell Suspension from the Spleen and Thymus

1. Collect organs and place them in ice-cold PBS.

2. Mesh the organs in the well of a 12-well plate containing 1 mL of ice-cold PBS.

3. Add 4 mL of ice-cold PBS, transfer into a 15-mL tube through a 70-μm nylon mesh to remove debris.

4. Centrifuge for 5 min at $400 \times g$ at 4 °C.

5. Add 1 mL of ACK lysis buffer and incubate for less than 2 min at room temperature.

6. Wash the cells twice in 5 mL of ice-cold PBS (5 min centrifugation at $400 \times g$ at 4 °C).

7. Resuspend the cells in 1 mL of staining buffer.

8. Count cell numbers.

3.3.4 Preparation of Cell Suspension from the Liver and Lung

1. Collect organs and place them in ice-cold PBS.

2. Cut the organs into small fragments and treat for 30 min at 37 °C with digestion medium.

3. Wash the cells twice in PBS (5 min centrifugation at $400 \times g$ at 4 °C) and resuspend the cells in 10 mL of PBS.

4. Prepare fresh 15-mL polypropylene tubes containing 4 mL of Ficoll-Hypaque Plus isotonic solution.

5. Load the samples onto the Ficoll-Hypaque solution, being careful to minimized mixing with Ficoll-Hypaque solution.

6. Centrifuge for 30 min at $400 \times g$ at 18–20 °C, without acceleration and brake.

7. Carefully collect the interface containing mononuclear cells (MNCs) using pasteur pipette and transfer into new 15-mL conical tubes.

8. Wash the cells twice in 5 mL of PBS (5 min centrifugation at $400 \times g$ at 4 °C).

9. Resuspend the cells in 1 mL of staining buffer.

10. Count cell numbers.

3.4 Antibody Staining

1. Place 2.5 million cells of each sample in a well of a 96-well plate.

2. Centrifuge the cell suspension at $400 \times g$ for 5 min at 4 °C and discard the supernatant.

3. Add 50 μL of staining buffer containing mouse FcγR blocker (1:200), mix well, and incubate for 15 min on ice.

4. Wash the cells with ice-cold staining buffer, centrifuge for 5 min at $400 \times g$ at 4 °C, and discard the supernatant.

5. Add 50 μL of staining buffer containing primary antibodies, mix well, and incubate for 30 min on ice in the dark.

6. Wash the cells with ice-cold staining buffer, centrifuge for 5 min at $400 \times g$ at 4 °C, and discard the supernatant.

7. Add 50 μL of staining buffer containing secondary antibody to the cells, mix well, and incubate for 30 min on ice in the dark.

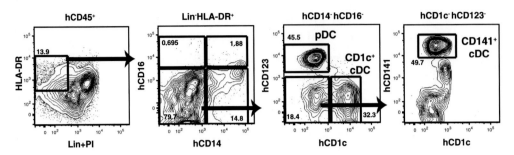

Fig. 2 FACS analysis of human DC subsets in the bone marrow of engrafted animal. Gating strategy of human DC subsets by 3-laser instrument. Viable human CD45$^+$ cells (gated in the first panel) were subdivided on the basis of lineage (Lin) and HLA-DR expression (second panel). Lin$^-$HLA-DR$^+$ cells were then subdivided as CD14$^-$CD16$^-$ cells to gate out monocyte lineages (third panel). Consequently, plasmacytoid DCs (pDC) were defined as hCD123$^+$ cells, CD1c$^+$ classical DCs (cDCs) as hCD123$^-$ hCD1c$^+$ cells, and CD141$^+$ cDCs as hCD123$^-$ hCD1c$^-$ hCD141^{hi} cells (fourth and fifth panels)

8. Wash the cells with ice-cold staining buffer in excess, centrifuge for 5 min at $400 \times g$ at 4 °C, and discard the supernatant.

9. Resuspend the cells in PBS containing Aqua (for 4 laser) or PI (for 3 laser) for 15 min to exclude dead cells by FACS.

10. Wash the cells with ice-cold staining buffer in excess, centrifuge for 5 min at $400 \times g$ at 4 °C, and discard the supernatant.

11. Fix the cells with 4 % PFA/PBS for 10 min at room temperature (*see* **Note 8**).

12. Analyze the cell fraction by FACS instrument (*see* Fig. 2).

4   Notes

1. We usually get a high purity of CD34$^+$ cells after magnetic cell separation when we isolate mononuclear cells by density gradient centrifugation on Ficoll-Hypaque.

2. We check the cells for the expression of human CD3, human CD19, human CD34, and HLA-A2. We use the cells with a CD34$^+$ purity ≥95 % and T cell contamination <0.1 % to avoid xenogeneic graft-versus-host disease. If this grade of purity is not achieved after two rounds of purification, the sample of hematopoietic stem cells should be discarded. HLA-A2 positivity will be needed when we evaluate HLA-restricted immune responses.

3. We need to select the pups having milk in their stomach, because it can be the typical sign the mother recognize for feeding the pups. It is also important to transplant human CD34$^+$ cells as soon as possible after birth (within 3 days) because there is a strict inverse correlation between the success of engraftment and time of injection after birth [12]. In order

to make sure of the age of the pups, the poster from Jackson laboratory will help confirming the age (http://jaxmice.jax.org/images/literature/pupsposter-large.jpg).

4. We have optimized the protocol to check the transgene quickly by Q-PCR-based method. It takes about 3 h to check the positivity of transgene. Briefly, toes (1–2 mm) are digested with 180 μL of 50 mM NaOH for 1 h at 95 °C, after that 1 M Tris (pH 8.0) is added and then diluted in H_2O (1:20). 2xFast SYBR Green Master Mix is used for Q-PCR reaction, and Ct value is measured by 7500 Fast Real-Time PCR System (Applied Biosystem).

5. The irradiation dose for newborn mice should be adapted to each institute and rechecked frequently because both irradiator strength and radiation sensitivity of the colony might vary slightly over time.

6. It is important to adapt the odor of our gloves by touching urine and cage bedding during the procedure. Otherwise, the mother will not recognize the pups after injection.

7. We usually check by flow cytometry the expression of human CD45 together with that of human CD3, CD19, CD33, and mouse CD45 in the peripheral blood to evaluate human cell engraftment and lineage cell development. We normally evaluate human cell engraftment (human CD45+ cells out of total CD45+ cells), as well as myeloid cell (human CD33+), B cell (human CD19+), and T cell (human CD3+) development.

8. In the author's facility, it is mandatory to fix human-derived cells before FACS analysis.

References

1. Shortman K, Liu Y-J (2002) Mouse and human dendritic cell subtypes. Nat Rev Immunol 2:151–161

2. Palucka K, Banchereau J (2013) Human dendritic cell subsets in vaccination. Curr Opin Immunol 25:396–402

3. Merad M, Sathe P, Helft J et al (2013) The dendritic cell lineage: ontogeny and function of dendritic cells and their subsets in the steady state and the inflamed setting. Ann Rev Immunol 31:563–604

4. Ishikawa F, Niiro H, Iino T et al (2007) The developmental program of human dendritic cells is operated independently of conventional myeloid and lymphoid pathways. Blood 110:3591–3660

5. Tanaka S, Saito Y, Kunisawa J et al (2012) Development of mature and functional human myeloid subsets in hematopoietic stem cell-engrafted NOD/SCID/IL2rγKO Mice. J Immunol 188:6145–6155

6. Poulin LF, Reyal Y, Uronen-Hansson H et al (2012) DNGR-1 is a specific and universal marker of mouse and human Batf3-dependent dendritic cells in lymphoid and nonlymphoid tissues. Blood 119:6052–6062

7. Rongvaux A, Willinger T, Martinek J et al (2014) Development and function of human innate immune cells in a humanized mouse model. Nat Biotech 32:364–372

8. Takenaka K, Prasolava TK, Wang JCY et al (2007) Polymorphism in *Sirpa* modulates engraftment of human hematopoietic stem cells. Nat Immunol 8:1313–1323

9. Strowig T, Rongvaux A, Rathinam C et al (2011) Transgenic expression of human signal regulatory protein alpha in Rag2-/- gamma c-/- mice improves engraftment of human

hematopoietic cells in humanized mice. Proc Natl Acad Sci U S A 108:13218–13223

10. Ziegler P, Manz MG (2007) Mouse models for human hemato-lymphopoiesis. Curr Protocol Toxicol. Chapter 2, Unit2.13

11. van Lent AU, Centlivre M, Nagasawa M et al (2010) In vivo modulation of gene expression by lentiviral transduction in "human immune system" Rag2-/- gamma c -/- mice. Methods Mol Biol 595:87–115

12. Gimeno R, Weijer K, Voordouw A et al (2004) Monitoring the effect of gene silencing by RNA interference in human CD34+ cells injected into newborn RAG2-/- gammac-/- mice: functional inactivation of p53 in developing T cells. Blood 104:3886–3893

INDEX

Elodie Segura and Nobuyuki Onai (eds.), *Dendritic Cell Protocols*, Methods in Molecular Biology, vol. 1423,
DOI 10.1007/978-1-4939-3606-9, © Springer Science+Business Media New York 2016